# Molecular Toxicology and Cancer Prevention

# Molecular Toxicology and Cancer Prevention

Editors

**Guohui Sun**
**Chongwen Wang**

MDPI

Basel • Beijing • Wuhan • Barcelona • Belgrade • Novi Sad • Cluj • Manchester

*Editors*
Guohui Sun
Beijing University of Technology
Beijing, China

Chongwen Wang
Guangdong Provincial People's Hospital
Guangzhou, China

*Editorial Office*
MDPI
St. Alban-Anlage 66
4052 Basel, Switzerland

This is a reprint of articles from the Special Issue published online in the open access journal *Molecules* (ISSN 1420-3049) (available at: https://www.mdpi.com/journal/molecules/special_issues/Molecules_Toxicology_Cancer).

For citation purposes, cite each article independently as indicated on the article page online and as indicated below:

| |
|---|
| Lastname, A.A.; Lastname, B.B. Article Title. *Journal Name* **Year**, *Volume Number*, Page Range. |

**ISBN 978-3-0365-9668-6 (Hbk)**
**ISBN 978-3-0365-9669-3 (PDF)**
**doi.org/10.3390/books978-3-0365-9669-3**

# Contents

# About the Editors

**Guohui Sun**

Dr. Guohui Sun is an Associate Professor at the Beijing Key Laboratory of Environmental and Viral Oncology, Faculty of Environment and Life, Beijing University of Technology. His research interests are tumor resistance, pharmacology, molecular toxicology, environmental pollutants, and computational toxicology. He is the Editorial Broad Member of Frontiers in Bioscience-Landmark (IF:3.1) and World Journal of Clinical Oncology (IF:2.8); Young Editorial Board Member: Chemical Reagents (Chinese Core Journal); Guest Editor: Molecules (SCI, IF:4.6), Cells (SCI, IF:6.0) and Frontiers in Molecular Biosciences (SCI, IF:5.0)

**Chongwen Wang**

Dr. Chongwen Wang is currently an Associate Professor at Guangdong Provincial People's Hospital. His work focuses on the preparation and application of novel nanomaterial-based biosensors. He is a member of the Biomarker Committee of the Chinese Society of Research Hospitals, the Laboratory Medicine Branch of the Chinese Medical Equipment Association and the Field Rapid Detection (POCT) branch of the China Medical Device Industry Association.

# Preface

Molecular toxicology is a field that investigates the interactions between chemical or biological molecules and organisms at the molecular level. In this Special Issue, we focus on the toxic effects and mechanisms of action of chemical and biological molecules, will be of great interest of readers in molecular toxicology and applied pharmacology. We sincerely hope this Special Issue will prompt advances in molecular toxicology, as well its application in cancer prevention. Science is an evolving process, with progress made through affirmation and negation. Therefore, the conclusions and viewpoints of all articles purely reflect the outcomes under specific conditions. We look forward to seeing more valuable research in the field of molecular toxicology with the aid of this Special Issue, contributing to the advancement of science and the enhancement of human health. As Guest Editors of this Special Issue, we thank all of the authors for their contributions, and hope that the contents of this publication will help readers to further develop their research.

**Guohui Sun and Chongwen Wang**
*Editors*

*molecules*

MDPI

*Editorial*

# Molecular Toxicology and Cancer Prevention

**Guohui Sun [1,*] and Chongwen Wang [2,*]**

1 Beijing Key Laboratory of Environmental and Viral Oncology, Faculty of Environment and Life, Beijing University of Technology, Beijing 100124, China
2 Laboratory Medicine, Guangdong Provincial People's Hospital, Guangdong Academy of Medical Sciences, Guangzhou 510000, China
* Correspondence: sunguohui@bjut.edu.cn (G.S.); wangchongwen1987@126.com (C.W.); Tel.: +86-10-67391917 (G.S.)

Molecular toxicology is a field that investigates the interactions between chemical or biological molecules and organisms at the molecular level. Chemical molecules, such as environmental organic contaminants and therapeutic drugs, are closely associated with the development of diseases such as cancer, and therapeutic effects [1–8]. Biomolecules, such as lncRNA, microRNA, and pathogenic proteins have important regulating effects on disease development [9–12]. In this Special Issue, we focus on the toxic effects and mechanisms of action of chemical and biological molecules. We collected 11 high-quality papers that will be of interest for researchers in molecular toxicology and applied pharmacology. Of the eleven papers, four papers are related to the environmental contaminants, four papers are related to biomolecules, two papers are related to natural organic small molecule products, and one paper is related to the application of biosensors in the detection of pathogenic microorganisms. We sincerely hope this Special Issue will prompt advances in molecular toxicology, as well its application in cancer prevention.

**Environmental contaminants.** The opening paper reported the effects of smoking on inflammatory-related cytokine levels in human serum [contribution 1]. In this study, Wang et al. [contribution 1] delved into the underlying mechanisms of cytokine production associated with tobacco smoking. A cytokine is a critical factor involved in the development of chronic systemic inflammation, which is the initial hallmark of cardiovascular and respiratory diseases as well as certain cancers. To explore this, blood samples from 78 male volunteers were categorized into three groups: non-smokers (30 individuals), current smokers (30 individuals), and ex-smokers (18 individuals). They used an advanced technique known as the liquid suspension chip method to analyze and compare the levels of 17 different cytokines and chemokines in the participants' serum. The results unveiled distinct patterns in cytokine expression in response to smoking and smoking cessation. This study sheds light on the production of cytokines and chemokines under various smoking conditions, providing valuable insights into the early stages of smoking-related chronic diseases and cancers.

In the next paper, Sun et al. [contribution 2] focused on the impact of nicotine and a tobacco-specific carcinogen called 4-(methylnitrosamino)-1-(3-pyridyl)-1-butanone (NNK) on lung cells, particularly in the context of the cytochrome enzyme P450 2A13 (CYP2A13), which plays a role in metabolizing these substances. Past research has hinted at nicotine's potential to reduce the harmful effects of NNK, but no comprehensive in vitro investigations have been conducted in lung cells, which are a key target organ for tobacco-related damage. To address this gap, they engineered a special cell line, BEAS-2B cells with a stable expression of CYP2A13 (referred to as B-2A13 cells), to explore how nicotine influences the cytotoxic and genotoxic effects of NNK. The results revealed that B-2A13 cells were more susceptible to NNK-induced cytotoxicity compared to regular BEAS-2B cells and control cells lacking CYP2A13 (B-vector cells). Notably, NNK induced significant DNA damage, caused cell cycle disruptions, and led to chromosomal damage in B-2A13 cells, whereas it

**Citation:** Sun, G.; Wang, C. Molecular Toxicology and Cancer Prevention. *Molecules* **2023**, *28*, 7730. https://doi.org/10.3390/molecules28237730

Received: 3 November 2023
Revised: 18 November 2023
Accepted: 21 November 2023
Published: 23 November 2023

had no significant impact on BEAS-2B cells or the control cells. This study demonstrates that CYP2A13 enhances the cytotoxic and genotoxic effects of NNK in BEAS-2B cells, and the presence of nicotine has the potential to mitigate the toxicity of NNK, providing valuable insights into the interplay between these substances in lung cells.

The presence of *Aspergillus flavus* and aflatoxins in grain crops poses a significant threat to food security and results in substantial economic losses. Aflatoxin B1, in particular, has been classified as a Group 1 carcinogen to humans by IARC. Traditional detection methods are often time-consuming and may not be as sensitive as required. However, Surface-Enhanced Raman Scattering (SERS) has emerged as a promising approach for the rapid and nondestructive detection of *Aspergillus flavus* and aflatoxins in grain crops due to its speed and sensitivity. Wang et al. [contribution 3] presented a rapid method for detecting *Aspergillus flavus* and quantitatively determining aflatoxin B1 in grain crops. This approach utilizes a portable Raman spectrometer in conjunction with colloidal gold nanoparticles (AuNPs). As the concentration of *Aspergillus flavus* spore suspension increased within the range of $10^2$–$10^8$ colony-forming units per milliliter (CFU/mL), a more effective combination of *Aspergillus flavus* spores was observed and AuNPs resulted in an improved enhancement of the AuNPs solution's effect on *Aspergillus flavus*. The authors successfully determined various concentrations of aflatoxin B1 in methanol solution by utilizing SERS, and the obtained spectra closely resembled those of solid powder samples. This rapid detection method has the potential to significantly reduce the detection time, from several hours or even tens of hours to just a few minutes. Such efficiency enables swift and effective measures to be taken to prevent substantial economic losses in the agricultural sector.

Polybrominated diphenyl ethers (PBDEs) are a class of both classic and emerging pollutants with the potential to harm the human immune system. Among PBDE congeners, 2,2′,4,4′-tetrabrominated biphenyl ether (BDE-47) stands out as the most biotoxic. In this study, Gao et al. [contribution 4] assessed the toxicity of BDE-47 on RAW264.7 cells, a type of mouse macrophage. The results reveal that exposure to BDE-47 resulted in a substantial reduction in cell viability and a notable increase in apoptosis. Furthermore, BDE-47 hindered the phagocytic activity of RAW264.7 cells, altered immune-related factors, and compromised immune function. Furthermore, a substantial increase in cellular reactive oxygen species (ROS) levels was noted. Transcriptome sequencing revealed the regulation of genes associated with oxidative stress. These findings strongly indicate that oxidative damage triggered by BDE-47 plays a pivotal role in mitochondrial apoptosis in RAW264.7 macrophages, ultimately resulting in the suppression of immune function. This study underscores the adverse effects of BDE-47 on immune cells and highlights the pivotal role of oxidative damage in producing these effects, with potential implications for understanding the immunotoxicity of PBDEs.

**Natural products.** Euphorbia factors are lathyrane-type diterpenoids found in the medicinal herb *Euphorbia lathyris L.* from the Euphorbiaceae family. These compounds have been associated with intestinal irritation toxicity, but the precise mechanisms behind this toxicity have remained a mystery. In the study by Zou et al. [contribution 5], they aimed to uncover the mechanisms by evaluating the transcriptome and microRNA (miRNA) profiles in human Caco-2 colon cancer cells after exposure to Euphorbia factors L1 (EFL1) and EFL2. The research protocol entailed subjecting Caco-2 cells to a 200 μM EFL treatment for a duration of 72 h. In the EFL1 group compared to the control group, there were alterations in the expression of 16 miRNAs and 154 mRNAs. Similarly, in the EFL2 group compared to the control group, 47 miRNAs and 1101 mRNAs exhibited differential expression. An analysis of these sequenced mRNAs highlighted their association with various critical processes, including transcription, post-translational modification, chaperones, protein turnover, secretion, vesicular transport, signal transduction mechanisms, intracellular trafficking, and cytoskeletal functions. The functions and pathways associated with differentially expressed mRNAs were enriched in areas such as transmembrane transport, T cell extravasation, the IL-17 signaling pathway, apoptosis, and the cell cycle. Additionally, the study identified that EFLs induced alterations in gene structures, encompassing processes like alternative splicing, insertion and deletion events, and single nucleotide polymorphisms. These findings offer valuable insights into the mechanisms that underlie the intestinal toxicity induced by EFLs in intestinal cells.

Conditionally replicating adenoviruses (CRAds) are a type of oncolytic adenovirus designed for tumor-targeted therapy. They can selectively enter cancer cells through coxsackievirus-Ad receptors (CARs), allowing them to replicate and destroy cancer cells while sparing normal cells. However, not all tumor cells express high levels of CARs, which can limit the effectiveness of CRAd therapy. Lu et al. [contribution 6] explored the use of 6-cyclohexyl methyl-β-D-maltoside (6-β-D) as a maltoside transfection agent. They found several advantages to using 6-β-D, including high transfection efficiency, low toxicity, and ease of use. By pretreating cancer cells with a low concentration of 6-β-D ($\leq 5$ μg/mL), they achieved an 18-fold improvement in the transduction efficiency of a "model" adenovirus (eGFP-Ad) compared to using eGFP-Ad alone. The addition of 6-β-D not only enhanced the transduction efficiency of CRAds but also improved their anti-tumor effects. Importantly, this combination did not harm normal cells. With 6-β-D treatment, CRAd at a lower multiplicity-of-infection ratio of 10 (MOI 10) achieved oncolytic outcomes similar to using a higher MOI of 50. This suggests that by combining CRAd with 6-β-D, the amount of CRAd required in clinical practice could be significantly reduced without compromising its therapeutic efficacy or exposing patients to potential side effects associated with high CRAd titers. This study provides a promising approach to improve adenovirus-mediated cancer gene therapy in clinical practice.

**Biomolecules.** The role of the neuroblast differentiation-associated protein AHNAK in biological processes has remained somewhat enigmatic. AHNAK is known to exhibit both suppressive and progressive functions in different types of cancers. In this study, Li et al. [contribution 7] sought to investigate the specific role of AHNAK in hepatocellular carcinoma (HCC). To understand its role, they conducted cell viability assays to assess the impact of AHNAK knockdown on cell proliferation in a stable HepG2 cell line. Additionally, co-immunoprecipitation (Co-IP) and LC-MS/MS were employed to analyze proteins in both HCC and matched paracancerous (MPC) tissues. The study revealed that silencing AHNAK led to a reduction in the viability of HepG2 cells. The analysis of protein interactions in HCC and MPC tissues identified 204 enriched pathways and processes. Furthermore, the study showed that AHNAK can co-localize and interact with the insulin-like growth factor 1 receptor (IGF-1R). This study sheds light on the involvement of AHNAK in HCC, emphasizing its potential contribution to HCC growth by interacting with IGF-1R.

Gene silencing is a crucial strategy in biology for understanding gene functions, investigating disease mechanisms, and developing potential therapeutics. The 8–17 DNAzyme is a promising tool for gene silencing due to its strong RNA-cleaving activity. However, its practical use has been limited by its dependence on divalent cations and a lack of comprehensive understanding of its cellular mechanisms. Zhou et al. [contribution 8] have investigated the activity of the 8–17 DNAzyme both in vitro and within cells. They discovered that this DNAzyme can effectively cleave RNA substrates under conditions that simulate physiological environments. Furthermore, its gene-silencing activity is enhanced by its compatibility with RNase H, allowing it to provide both cleavage and antisense activities within cells. They also found that chemical modifications can improve the stability, substrate binding affinity, and gene-silencing activity of the 8–17 DNAzyme. These findings suggest that this DNAzyme exhibits significant activity in cells, making it a valuable tool for exploring various biomedical applications.

Liu et al. (contribution 9) have introduced a novel prognostic signature based on aging-related long non-coding RNAs (lncRNAs) that is associated with immune cell infiltration and responses to breast cancer (BC) immunotherapy. They obtained BC samples from the breast-invasive carcinoma cohort within The Cancer Genome Atlas (TCGA) database and identified differentially expressed aging-related lncRNAs (DEarlncRNAs) using Pearson correlation analysis. Their analysis of the TCGA cohort led to the identification of a six aging-related lncRNA signature, comprising MCF2L-AS1, USP30-AS1, OTUD6B-AS1, MAPT-AS1, PRR34-AS1, and DLGAP1-AS1. The time-dependent ROC curve demonstrated the model's strong prognostic predictability for BC patients. Patients in the low-risk group exhibited better overall survival and lower total tumor mutational burden. Additionally, the high-risk group had a smaller proportion of tumor-killing immune cells. It was also noted that the low-risk group derived more significant benefits from immunotherapy and specific chemotherapeutics compared to the high-risk group. This aging-related lncRNA signature offers new insights and strategies for early BC diagnosis and potential therapeutic targets, particularly in the context of tumor immunotherapy. We also published a review paper by Sumneang et al. [contribution 10], related to the toll-like receptor 4 (TLR4) inflammatory perspective on doxorubicin (Dox)-induced cardiotoxicity. In this review, the authors provide an overview of the existing evidence that supports the involvement of the TLR4 signaling pathway in various models of Dox-induced cardiotoxicity. They also discuss the impact of the TLR4 signaling pathway on Dox-induced cardiotoxicity. Gaining a better understanding of the role of the TLR4 signaling pathway in Dox-induced cardiac inflammation can aid in the development of therapeutic strategies to mitigate Dox-induced cardiotoxicity. This knowledge could be instrumental in the development of potential therapeutic approaches to address Dox-induced cardiotoxicity and improve the overall safety and efficacy of cancer treatment with Dox.

**Biosensors.** In this Special Issue, we published an interesting paper related to the graphene oxide (GO)-sensitized surface plasmon resonance (SPR) biosensor of porcine reproductive and respiratory syndrome virus. In this study, Liu et al. [contribution 11] investigated the impact of GO modification on the sensitivity of an SPR biosensor and utilized this GO-modified sensor for the detection of the porcine reproductive and respiratory syndrome virus (PRRSV) in cell cultures. The results revealed that GO modification significantly enhanced the sensitivity of the Fourier-transform SPR sensor. The GO-modified sensor, therefore, presents a promising alternative for virus detection. This study demonstrates the potential of GO-modified SPR biosensors for highly sensitive and label-free virus detection, paving the way for applications in virus diagnostics and research in biological samples.

To date, the toxicity of many chemical molecules and hypothetic molecules to special targets can be predicted by structure-based approaches like quantitative structure/activity/toxicity relationship (QSAR/QSTR), machine learning and artificial intelligence [1,2,4,5,8]. In addition, their toxic mechanisms can be further understood following advancements in bioinformatics, genomics, and proteomics. This Special Issue reported the toxic effects and mechanisms of four environmental contaminants and two natural products, as well as the role of three biomolecules in cancer development, gene silencing, and cancer prognosis. Additionally, the analytic determination of GO-based biosensor in pathogenic microorganisms was also reported. It should be noted that these advances represent a mere glimpse into the realm of molecular toxicology research. We sincerely hope that the published articles can inspire researchers in the field and stimulate further valuable investigations. Science is an evolving process, with progress made through affirmation and negation. Therefore, the conclusions and viewpoints of all articles purely reflect the outcomes under specific conditions. We look forward to seeing more valuable research in the field of molecular toxicology with the aid of this Special Issue, contributing to the advancement of science and the enhancement of human health.

**Funding:** This Special Issue was inspired by the National Natural Science Foundation of China (No. 82003599, 32200076) and The Project of Cultivation for Young Top-Motch Talents of Beijing Municipal Institutions (No. BPHR202203016).

**Acknowledgments:** As Guest Editors of this Special Issue, we thank all of the authors for their contributions, and hope that the contents of this publication will help readers to further develop their research.

**Conflicts of Interest:** The authors declare no conflict of interest.

**List of Contributions**

1. Wang, H.; Chen, H.; Fu, Y.; Liu, M.; Zhang, J.; Han, S.; Tian, Y.; Hou, H.; Hu, Q. Effects of Smoking on Inflammatory-Related Cytokine Levels in Human Serum. *Molecules* **2022**, *27*, 3715. https://doi.org/10.3390/molecules27123715.
2. Sun, Y.; Wang, H.; Chen, H.; Zhang, S.; Li, J.; Zhang, J.; Tian, J.; Zhang, Y.; Hou, H.; Hu, Q. Nicotine Inhibits the Cytotoxicity and Genotoxicity of NNK Mediated by CYP2A13 in BEAS-2B Cells. *Molecules* **2022**, *27*, 4851. https://doi.org/10.3390/molecules27154851.
3. Wang, H.; Liu, M.; Zhang, Y.; Zhao, H.; Lu, W.; Lin, T.; Zhang, P.; Zheng, D. Rapid Detection of *Aspergillus flavus* and Quantitative Determination of Aflatoxin $B_1$ in Grain Crops Using a Portable Raman Spectrometer Combined with Colloidal Au Nanoparticles. *Molecules* **2022**, *27*, 5280. https://doi.org/10.3390/molecules27165280.
4. Gao, Q.; Zhou, Z.-Y.; He, Y.-N.; Dong, M.-H.; Wang, Z.-N.; Chen, H.-M. BDE-47 Induces Immunotoxicity in RAW264.7 Macrophages through the Reactive Oxygen Species-Mediated Mitochondrial Apoptotic Pathway. *Molecules* **2023**, *28*, 2036. https://doi.org/10.3390/molecules28052036.
5. Zou, L.; Bao, W.; Gao, Y.; Chen, M.; Wu, Y.; Wang, S.; Li, C.; Zhang, J.; Zhang, D.; Wang, Q.; et al. Integrated Analysis of Tran-scriptome and microRNA Profile Reveals the Toxicity of Euphorbia Factors toward Human Colon Adenocarcinoma Cell Line Caco-2. *Molecules* **2022**, *27*, 6931.
6. Lu, W.; Fang, Y.; Meng, X.; Wang, X.; Liu, W.; Liu, M.; Zhang, P. Improving the Transduction Efficiency and Antitumor Effect of Conditionally Replicative Adenovirus by Application of 6-cyclohexyl Methyl-β-D-maltoside. *Molecules* **2023**, *28*, 528. https://doi.org/10.3390/molecules28020528.
7. Li, K.; Song, K.; Hou, Y.; Tian, Y.; Wang, H.; Sun, L.; Li, A.; Zhang, Y. AHNAK Contributes to Hepatocellular Carcinoma Growth by Interacting with IGF-1R. *Molecules* **2022**, *27*, 8680. https://doi.org/10.3390/molecules27248680.
8. Zhou, Z.; Sun, W.; Huang, Z. 8–17 DNAzyme Silencing Gene Expression in Cells via Cleavage and Antisense. *Molecules* **2023**, *28*, 286. https://doi.org/10.3390/molecules28010286.
9. Liu, Z.; Ren, C.; Cai, J.; Yin, B.; Yuan, J.; Ding, R.; Ming, W.; Sun, Y.; Li, Y. A Novel Aging-Related Prognostic lncRNA Signature Correlated with Immune Cell Infiltration and Response to Immunotherapy in Breast Cancer. *Molecules* **2023**, *28*, 3283. https://doi.org/10.3390/molecules28083283.
10. Sumneang, N.; Tanajak, P.; Oo, T.T. Toll-like Receptor 4 Inflammatory Perspective on Doxorubicin-Induced Cardiotoxicity. *Molecules* **2023**, *28*, 4294.
11. Liu, X.; Xu, C.; Fu, C.; Xia, D.; Wang, F.; Yin, H.; Peng, J. Graphene Oxide-Sensitized Surface Plasmon Resonance Biosensor of Porcine Reproductive and Respiratory Syndrome Virus. *Molecules* **2022**, *27*, 3942.

## References


1. Chen, S.; Sun, G.; Fan, T.; Li, F.; Xu, Y.; Zhang, N.; Zhao, L.; Zhong, R. Ecotoxicological QSAR study of fused/non-fused polycyclic aromatic hydrocarbons (FNFPAHs): Assessment and priority ranking of the acute toxicity to *Pimephales promelas* by QSAR and consensus modeling methods. *Sci. Total Environ.* **2023**, *876*, 162736. [CrossRef] [PubMed]
2. Sosnowska, A.; Bulawska, N.; Kowalska, D.; Puzyn, T. Towards higher scientific validity and regulatory acceptance of predictive models for PFAS. *Green Chem.* **2023**, *25*, 1261–1275. [CrossRef]
3. Keith, L.H.; Telliard, W.A. Priority Pollutants I-A Perspective View. *Environ. Sci. Technol.* **1979**, *13*, 416–423. [CrossRef]
4. Gramatica, P.; Papa, E.; Sangion, A. QSAR modeling of cumulative environmental end-points for the prioritization of hazardous chemicals. *Environ. Sci. Process. Impacts* **2018**, *20*, 38–47. [CrossRef]
5. Hassaan, M.A.; El Nemr, A. Pesticides pollution: Classifications, human health impact, extraction and treatment techniques. *Egypt. J. Aquat. Res.* **2020**, *46*, 207–220. [CrossRef]
6. Kaina, B.; Christmann, M. DNA repair in personalized brain cancer therapy with temozolomide and nitrosoureas. *DNA Repair* **2019**, *78*, 128–141. [CrossRef] [PubMed]
7. Wu, Q.; Qian, W.; Sun, X.; Jiang, S. Small-molecule inhibitors, immune checkpoint inhibitors, and more: FDA-approved novel therapeutic drugs for solid tumors from 1991 to 2021. *J. Hematol. Oncol.* **2022**, *15*, 143. [CrossRef] [PubMed]

8. Yang, L.; Wang, Y.; Hao, W.; Chang, J.; Pan, Y.; Li, J.; Wang, H. Modeling pesticides toxicity to Sheepshead minnow using QSAR. *Ecotoxicol. Environ. Saf.* **2020**, *193*, 110352. [CrossRef] [PubMed]
9. Bai, P.; Fan, T.; Wang, X.; Zhao, L.; Zhong, R.; Sun, G. Modulating MGMT expression through interfering with cell signaling pathways. *Biochem. Pharmacol.* **2023**, *215*, 115726. [CrossRef] [PubMed]
10. Cheng, M.; Wang, Q.; Chen, L.; Zhao, D.; Tang, J.; Xu, J.; He, Z. LncRNA UCA1/miR-182-5p/MGMT axis modulates glioma cell sensitivity to temozolomide through MGMT-related DNA damage pathways. *Hum. Pathol.* **2022**, *123*, 59–73. [CrossRef] [PubMed]
11. Li, B.Y.; Shen, S.S.; Zhang, W.M.; Qi, T.J.; Hu, Q.; Cheng, Y. Long non-coding RNA TUG1 as a potential novel biomarker for predicting the clinical outcome of cancer patients: A meta-analysis. *Clin. Lab.* **2018**, *64*, 1799–1808. [CrossRef]
12. Ghaffar, M.; Khodahemmati, S.; Li, J.T.; Shahzad, M.; Wang, M.; Wang, Y.J.Q.; Li, C.S.; Chen, S.; Zeng, Y. Long non-coding RNA LINC01234 regulates proliferation, invasion and apoptosis in esophageal cancer cells. *J. Cancer* **2018**, *9*, 4242–4249. [CrossRef]

*molecules*

MDPI

*Article*

# Effects of Smoking on Inflammatory-Related Cytokine Levels in Human Serum

**Hongjuan Wang, Huan Chen, Yaning Fu, Min Liu, Jingni Zhang, Shulei Han, Yushan Tian, Hongwei Hou * and Qingyuan Hu ***

China National Tobacco Quality Supervision and Test Center, Key Laboratory of Tobacco Biological Effects, Zhengzhou, No. 6, Cuizhu Street, Gaoxin District, Zhengzhou 450001, China; redbri2013@163.com (H.W.); hunny_ch@163.com (H.C.); fynmail@126.com (Y.F.); ggayt0616@163.com (M.L.); zhangjnlife@163.com (J.Z.); hsl1983@163.com (S.H.); yushantian@126.com (Y.T.)

* Correspondence: qsfctc@163.com (H.H.); huqy@ztri.com.cn (Q.H.); Tel.: +86-135-9809-8330 (H.H.); +86-139-0384-3190 (Q.H.); Fax: +86-0371-67672625 (H.H. & Q.H.)

**Abstract:** Cardiovascular and respiratory diseases, and several cancers resulting from tobacco smoking, are initially characterized by chronic systemic inflammation. Cytokine imbalances can result in inflammation, making it important to understand the pathological mechanisms behind cytokine production. In this study, we collected blood samples from 78 healthy male volunteers, including non-smokers ($n$ = 30), current smokers ($n$ = 30), and ex-smokers ($n$ = 18), and utilized the liquid suspension chip technique to investigate and compare the expression levels of 17 cytokines and chemokines in the human serum of these volunteers. The results demonstrated that the expression levels of CXCL9/MIG and sIL-6R significantly increased after smoking, and continued to increase after quitting smoking. The expression levels of TARC, ITAC, and sVEGFR-3 increased after smoking but decreased after quitting smoking; the expression level of SAA significantly decreased after smoking and showed an upward trend after quitting smoking. Seven cytokines (IL-1$\beta$, BCA-1, TNF-$\alpha$, CRP, ENA-78, MDC, and TNFRII) did not vary between the three groups, while four cytokines (IL-1$\alpha$, IL-6, IL-8, and SCF) were not detected in any serum sample. In conclusion, this study assessed the physiological production of cytokines and chemokines, highlighting the differences in each due to smoking status. Our results could help evaluate the early development of smoking-related chronic diseases and cancers.

**Keywords:** smoking; inflammatory; cytokine; human serum

**Citation:** Wang, H.; Chen, H.; Fu, Y.; Liu, M.; Zhang, J.; Han, S.; Tian, Y.; Hou, H.; Hu, Q. Effects of Smoking on Inflammatory-Related Cytokine Levels in Human Serum. *Molecules* **2022**, *27*, 3715. https://doi.org/10.3390/molecules27123715

Academic Editors: Guohui Sun and Chongwen Wang

Received: 9 April 2022
Accepted: 7 June 2022
Published: 9 June 2022



## 1. Introduction

Inflammatory responses are driven by a complex network of mediators and signaling pathways. For example, cytokines regulating inflammatory responses include interleukins, whose core function is the orchestration of different cells of the immune system during host defense against pathogens, chemokines that promote chemotaxis, and interferons. Moreover, these molecules are involved in innate and adaptive immunity, and play a significant physiological role in lymphoid tissue ontogenesis, organogenesis, vasculogenesis, and tissue repair [1]. Diseases often occur when the expression of these molecules is chronically altered. Indeed, chronic inflammatory diseases are recognized as the most significant cause of death globally, with more than 50% of all deaths attributable to inflammation-related diseases [2]. Treating inflammatory diseases poses a significant challenge to medical science. Environmental factors play important roles in the development and progression of inflammatory-related diseases, including tobacco smoking, which can upset the homeostasis of the immune system by modulating immune-regulatory activities, leading to inflammation [3–5].

Despite widespread knowledge of the health risks associated with smoking, tobacco use remains high in developing countries [6]. Worldwide, tobacco smoking is reported to have killed almost 6 million people annually, including approximately 600,000 non-smokers who died from second-hand smoke exposure [7–9]. Therefore, tobacco smoking is a threat to public health, and a worldwide epidemic [10].

There are several ways that tobacco smoking can lead to inflammatory and autoimmune diseases, including genetic/epigenetic modifications, increased oxidative stress, and free radical production [11–14]. These effects can increase the proliferation of B and T cells; reduce the generation, activity, and autoantibody of immuno-suppressive T regulatory (Treg) cells; enhance the expression of pro-inflammatory mediators, such as Interleukin-1 beta (IL-1β), Interleukin-6 (IL-6), Interleukin-8 (IL-8), and tumor necrosis factors (TNFs); and enhance the expression of chemotactic cytokines, such as recombinant human C-X-C Motif Chemokine Ligand 9 (CXCL9/MIG), thymus and activation-regulated chemokine (TARC), and IFN-inducible T cell α chemoattractant (ITAC) [15–23]. Chemokines are small-molecule secretory proteins that can mediate cell directional migration, activate cellular immune activity, and participate in immune regulation [24]. CXCL9 is a member of the CXC chemokine family, and plays an important role in the chemotaxis of immune cells. T lymphocytes can be recruited to inflammatory sites through CXCL9 chemotaxis, enhancing the proliferation of T lymphocytes and the production of cytokines in allogeneic reaction [25,26]. Studies have found that CXCL9 plays an important role in many diseases, including external infection, autoimmune diseases, tumor treatment, lymphoma [27], and fatty livers [28]. Changes in these inflammatory markers and cytokines can lead to cancers in 18 different tumor sites and a range of other chronic diseases, including coronary heart disease, stroke, and chronic obstructive pulmonary disease [29,30]. Exposure to tobacco smoke can also regulate the expression of cytokine receptors [31]. IL-6 is involved in many biological processes, such as inflammation and immune regulation, while the imbalance of immune regulation of the IL-6/ IL-6 receptor axis can lead to various inflammatory diseases, such as rheumatoid arthritis and chronic hepatitis [32]. After specifically binding with sIL-6R, it can activate the downstream classical signal pathway, such as the JAK/STAT3 and PI3K/Akt signaling pathways; it can then activate T cells to mediate the secretion of inflammatory factors by immune cells such as neutrophils, fibroblasts, and macrophages, and promote the occurrence of inflammation [33,34]. These interleukins, chemokines, and other inflammatory-related cytokines play an important role in transmitting information, activating and regulating immune cells, and mediating inflammatory responses, all of which are closely related to tobacco smoke intake. Most previous studies have described the association between tobacco smoking and diseases. For instance, smoking is known to affect the concentration and activation of some white blood cells, including leukocytes, which are associated with increased concentrations of the inflammatory markers C-reactive protein (CRP) and IL-6 [35]. However, the changes to these cytokines in the serum after smoking cessation requires further study. ITAC is distributed in a small amount in the lung, pancreas, spleen, and other organs or tissues of normal individuals, mainly expressed by bronchial epithelial cells and vascular endothelial cells, which mostly induce the directional migration of T cells and trigger an inflammatory response; inhibit the growth of vascular endothelial cells; and regulate angiogenesis, among other biological functions [36]. Soluble vascular endothelial growth factor receptor 3 (sVEGFR-3) is a novel receptor that regulates lymphangiogenesis by inhibiting vascular endothelial growth factor (VEGF)-C and making it unable to activate cognate receptors; it is a potential biomarker for the antiangiogenic activity of tyrosine kinase inhibitors [37–39]. Serum amyloid A (SAA)is a pro-inflammatory factor, because it is an effective chemokine that mediates leukocyte migration and stimulates the expression of pro-inflammatory mediators in vitro and in vivo. Since SAA is common in chronic obstructive pulmonary disease (COPD) and lung cancer, it can predict the severity of these two diseases, and has been identified as a biomarker of lung cancer severity and a potential target for treatment [40–42].

Smoking causes 140,000 premature deaths from cardiovascular disease (CVD) annually in the United States, representing approximately 30% of all smoking-related deaths [43]. Inflammation is one mechanism by which cigarette smoking could affect CVD [44,45].

A growing body of evidence outlines the relationship between individual inflammatory markers and smoking status. While smoking can cause immune-related diseases, the mechanism behind how it affects inflammation is poorly understood. To understand the changes to typical inflammatory cytokines with smoking and cessation of smoking, and to explore smoking-related inflammation, we measured 17 inflammatory cytokines in the serum of non-smokers, ex-smokers, and current smokers. Bio-plex suspension chip technology was used to investigate the changes in the 17 inflammatory cytokines in the serum of these subjects. This study was designed to provide insights into potential health risks related to the expression of cytokines and chemokines in human serum, based on smoking status.

## 2. Materials and Methods

### 2.1. Study Design and Participants

This study was approved by the Life Science Ethics Committee of Zhengzhou University. Healthy male non-smokers, and smokers between 18 and 60 years of age, were recruited for this study, while any subject with the following conditions was excluded: an acute or chronic infectious disease; a clinically significant disorder; or medicating with drugs known to influence immunological factors (e.g., corticosteroids). All subjects lived in a typical rural area in central China and had similar occupations, lifestyles, and environmental exposure factors. There was no significant difference between or within these groups. Once informed consent was obtained, each participant completed a self-evaluation questionnaire assessing their history of cigarette smoking. According to relevant WHO standards, participants were categorized by smoking status as follows: "non-smoker" (total lifetime smoking <100 cigarettes, 30 males); "current smoker" (≥5 cigarettes/day; smoking time ≥6 months; 30 males); or "ex-smoker" (smoking cessation ≥2 years; 18 males).

### 2.2. Serum Isolation

Blood samples were collected without anticoagulants in a vacuum-sealed blood collection tube, and left undisturbed at room temperature for 20 min to allow them to clot (the participant had fasted after 5 p.m. the previous day). The clots were then removed by centrifuging the samples at 1500 g for 10 min in a refrigerated centrifuge. The resulting supernatant was considered serum. Following centrifugation, the serum was immediately transferred to a clean polypropylene tube using a Pasteur pipette. The samples were maintained at 2–8 °C during handling and were immediately analyzed, avoiding the freeze-thaw cycle that could harm certain serum components.

### 2.3. Assessment of Inflammatory Markers

Seventeen cytokines and chemokines (IL-α, IL-1β, IL-6, IL-8, CXCL9/MIG, sIL-6R, SAA, BCA-1, CRP, TARC, MDC, TNFRII, ITAC, SCF, ENA-78, sVEGFR3, and TNF-α) from the serum samples were analyzed using the Bio-Plex 200 system (Bio-Rad, Hercules, CA, USA) according to the manufacturer's instructions, and the lower limits of detection (LOD) for these cytokines and chemokines were 81.8, 0.4, 2.3, 1.9, 1.8, 257.6, 1100, 0.7, 4, 1.7, 0.9, 30.3, 0.1, 1.5, 7.3, 18.01, and 0.9 pg/mL, respectively. The magnetic bead-based multiplex immunoassay principle is based on the sandwich enzyme-linked immunosorbent assay (ELISA) model, where magnetic beads are covalently bound to antibodies that react to the targeted biomarker. Measurements were performed twice, in accordance with the manufacturer's instructions. The human serum samples were stored at 2–8 °C, and determined within 4 h. The cytokine assay plates were wet with an assay buffer and washed twice with a wash buffer. Two magnetic beads were added to the 96-well plate and serial dilutions of the reconstituted standard and samples. Second detection antibodies with streptavidin–phycoerythrin conjugate completed the sandwich complex. Data from the reactions were acquired using the Bio-Plex200 reader, a digital processor managed data output, and the Bio-Plex Manager software generated data as Median Fluorescence Intensity (MFI) and concentration (pg/mL).

### 2.4. Statistical Analysis

Data were analyzed for each set of experiments, by calculating medians and interquartile ranges (IQRs). Scatter plots were used to show the values of the median, IQR, and Tukey whiskers. The nonparametric Kruskal–Wallis test, and Dunn's multiple comparison test, were used when appropriate. Probability ($p$) values were calculated, based on two-tailed tests. Data analysis was performed using GraphPad Prism software version 5.0 (GraphPad Software, Inc., La Jolla, CA, USA). A $p$-value of less than 0.05 was considered statistically significant.

## 3. Results

The study included 78 subjects with a mean age of 37.5 years. These included non-smokers (30 males), current smokers (30 males), and ex-smokers (18 males). The characteristics of gender and smoking status of the study population are shown in Table 1:

**Table 1.** Characteristics of the subject sample ($n$ = 78).

| Group | Frequency | Age (Median; Min/Max) | Sex | Race |
|---|---|---|---|---|
| non-smokers | 30 | 36.6 yr; 19/57 yr | Male | Han |
| current-smokers | 30 | 35.5 yr; 24/49 yr | Male | Han |
| ex-smokers | 18 | 42.2 yr; 26/56 yr | Male | Han |
| Total | 78 | 37.5 yr; 19/57 yr | Male | Han |

A total of 17 cytokines and chemokines were tested in 78 samples, and were roughly divided into four groups based on their expression across the three groups. CXCL9/MIG and sIL-6R were continuously up-regulated in the three groups, and TARC, ITAC, sVEGFR3, and serum amyloid A protein (SAA) returned to normal levels after smoking cessation. IL-1β, BCA-1, tumor necrosis factor-α (TNF-α), CRP, ENA-78, MDC, and TNFRII did not significantly differ across the groups. The production of these molecules was constant. Several cytokines (IL-1α, IL-6, IL-8, and SCF) were below the lower LOD in all subjects, either because levels were very low or because the molecules were not produced by healthy subjects. The expression levels of various cytokines and chemokines in different groups were shown in Table 2:

**Table 2.** Expression of cytokines and chemokines in three groups (pg/mL).

| Expression Trends | Cytokine | Non-Smokers | Current-Smokers | Ex-Smokers | $p$-Value |
|---|---|---|---|---|---|
| Continuous increase | CXCL9/MIG | 1320.65 | 1493.70 | 1928.06 | 0.03/0.03 |
| | sIL-6R | 57,248.93 | 70,286.72 | 76,155.75 | $2.96 \times 10^{-3}/6.84 \times 10^{-3}$ |
| Recovery after smoking cessation | TARC | 19.47 | 25.47 | 22.39 | $1.28 \times 10^{-4}/1.61 \times 10^{-2}$ |
| | ITAC | 48.09 | 105.34 | 84.09 | $0.07 \times 10^{-4}/3.89 \times 10^{-2}$ |
| | sVEGFR3 | 20,058.64 | 27,610.30 | 26,732.49 | $0.06 \times 10^{-4}/2.53 \times 10^{-2}$ |
| | SAA | 8833.60 | 4771.36 | 7609.59 | $0.03 \times 10^{-4}/9.37 \times 10^{-3}$ |
| No significant difference | IL-1β | 12.35 | 22.21 | 38.24 | 0.35/0.43 |
| | BCA-1 | 24.92 | 39.37 | 27.55 | 0.84/0.35 |
| | TNF-α | 11.21 | 13.95 | 10.63 | 0.99/0.191 |
| | CRP | 9055.50 | 7058.28 | 11,859.6 | 0.918/0.30 |
| | ENA-78 | 547.60 | 457.60 | 649.28 | 0.51/0.34 |
| | MDC | 408.08 | 430.63 | 396.81 | 0.47/0.34 |
| | TNFRII | 14,775.96 | 15,543.39 | 13,188.03 | 0.94/0.17 |
| Not detected | IL-1α | — | — | — | / |
| | IL-6 | — | — | — | / |
| | IL-8 | — | — | — | / |
| | SCF | — | — | — | / |

Values are expressed as pg/mL, median (IQR); differences between groups were not statistically significant ($p > 0.05$) after performing a Kruskal–Wallis test and Dunn's multiple comparison test. The left side of the $p$-value column indicates the result of the differential analysis between non-smokers and current-smokers, and the right side is the result of the differential analysis between current-smokers and ex-smokers. Cytokines and chemokines were not detected in any group, either because they were under the lower limit of detection (LOD, pg/mL) or because they were not produced.

### 3.1. Cytokine Levels Significantly Increased in Non-Smokers, Current Smokers, and Ex-Smokers

Levels of CXCL9/MIG and sIL-6R in the non-smoker group were significantly lower when compared to the current smoker and ex-smoker groups. Additionally, the levels of both CXCL9/MIG and interleukin 6 soluble receptor (sIL-6R) gradually increased in the non-smokers ($n = 30$), current smokers ($n = 30$), and ex-smokers ($n = 18$) (Figure 1).

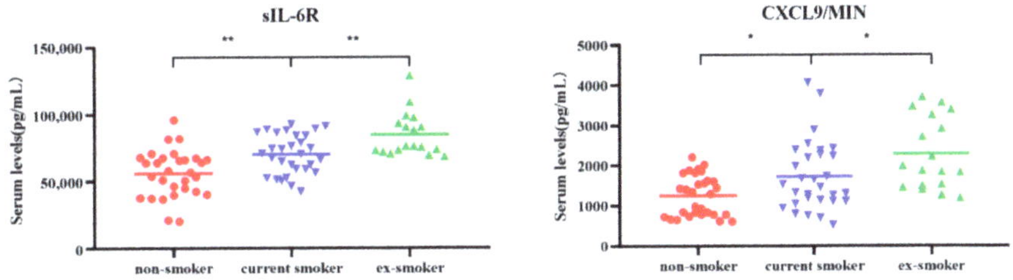


**Figure 1.** Scatter plots of serum levels (pg/mL) of cytokines CXCL9/MIG and sIL-6R for non-smokers ($n = 30$), current smokers ($n = 30$), and ex-smokers ($n = 18$). Whiskers were calculated using the Tukey method. Outliers not shown. * $p < 0.05$, ** $p < 0.01$: Kruskal–Wallis test and Dunn's multiple comparison test.

### 3.2. Changes in Four Types of Cytokines

The levels of human TARC, ITAC, and sVEGFR3, were all significantly higher in current smokers than in non-smokers and ex-smokers. However, SAA content was lowest in current smokers and highest in ex-smokers (Figure 2).

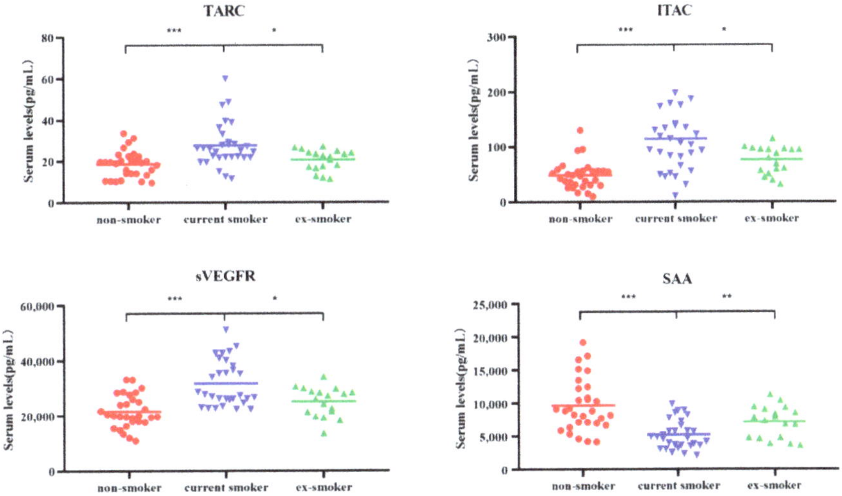


**Figure 2.** Scatter plots of serum levels (pg/mL) of cytokines TARC, ITAC, sVEGFR3, and SAA for non-smokers ($n = 30$), current smokers ($n = 30$), and ex-smokers ($n = 18$). Whiskers calculated using the Tukey method. Outliers not shown. * $p < 0.05$, ** $p < 0.01$, *** $p < 0.001$: Kruskal–Wallis test and Dunn's multiple comparison test.

## 4. Discussion

This study assessed the effects of smoking on the inflammatory cytokine levels of 78 adult male subjects, identifying substantial differences in several immune cytokines between non-smokers ($n$ = 30), current smokers ($n$ = 30) and ex-smokers ($n$ = 18). These cytokines mediate various mechanisms of the immune and inflammation responses, such as the chemotaxis of T cells and eosinophils, inflammation, and anti-inflammation processes. In addition, cytokines are also involved in cell development/differentiation, cell growth and activation, angiogenesis, and hematopoiesis. These findings provide strong evidence that smoking could affect systemic immunity and inflammation.

Toxicants in cigarette smoke can alter several pulmonary and systemic immune characteristics related to several immune cells, including increases in macrophages, neutrophils, eosinophils, and mast cells, and the functionality of various immune cells [46]. Compared to the non-smokers, the expression of CXCL9/MIG and sIL-6R in the serum of the current smokers and ex-smokers continued to increase. CXCL9/MIG serves as an important antiviral defense, and plays an important role in the development or prevention of certain lung diseases [47–49]. CXCL9/MIG is one of three chemokines, along with CXCL10 and CXCL11, that are highly induced by IFN-$\gamma$, as well as by type I and III IFNs [48,50,51]. Our results demonstrate that smoking increases cytokine CXCL9/MIG expression, which can partially explain the pro-inflammatory and pro-atherosclerotic properties of smoking, and could be one of the adverse side effects of long-term nicotine use. In the blood of healthy human body, sIL-6R can be detected [52], and sIL-6R expression could increase in some immune diseases, such as peritonitis [53] and rheumatoid arthritis [54]. This suggests that sIL-6R play an important role in immune diseases. Studies have reported the immune effects of IL-6 and sIL-6R in HBV-infected humans. Increased serum sIL-6R concentrations have been observed in patients with hepatitis B treated with interferon [55]. The concentration of serum sIL-6R did not increase in patients who experienced no effects after interferon treatment. Studies have demonstrated that smoking status is closely related to increases in sIL-6R receptor levels in human serum [56], which is consistent with the results of this study. In the former smoker group, the expression of sIL-6R continued to increase, which could be due to overcompensation caused by metabolic adaptation after smoking cessation [56]. In this study, we found that the levels of CXCL9/MIG and sIL-6R in serum increased after smoking, while the levels of these two cytokines did not decrease after smoking cessation. These results indicate that smoking could exacerbate inflammation, and that inflammation did not decrease or increase continuously after quitting smoking. This suggests that continuous smoking could increase some inflammatory factors. At the same time, the body adapted, and quitting smoking disrupted this stability, further increasing relevant inflammatory factors. It remains unclear whether inflammatory responses decrease after smoking cessation or how long the inflammatory response would decrease after smoking cessation for the body to re-adapt.

Our results demonstrated that there were significant differences in inflammatory cell molecules after quitting smoking. Among ex-smokers, the levels of chemokines (TARC, ITAC) and growth factor receptors (sVEGFR3) approached levels observed in non-smokers, as the time since smoking cessation increased. Therefore, smoking could change those cytokines that gradually revert to normal levels once an individual stops smoking. It has been reported that concentrations of TARC in the serum decrease once an individual quits smoking, and are associated with a decrease in pulmonary inflammation [57]. In addition, the levels of ITAC and sVEGFR3 in ex-smokers were significantly lower than the levels in the smoker group, which was consistent with the results of Sami, et al., [58,59]. ITAC is related to triggering an inflammatory response and inhibiting the growth of vascular endothelial cells [36]. sVEGFR3 regulates lymphangiogenesis by inhibiting VEGF-C, and preventing it from activating its homologous receptor. Overexpression of sVEGFR3 in lung cancer cells can effectively inhibit the density of lymphatic vessels in the tumor, and reduce the incidence of lymph node metastasis [60]. Smoking can damage microvascular function and vascular endothelial function, and reduce the synthesis of oxyhemoglobin [61].

The decreased expression of ITAC after smoking cessation could be related to reduced inflammation and/or the normal growth of vascular endothelial cells [36]. sVEGFR-3 is a novel soluble vascular endothelial growth factor receptor [37], and it is a potential biomarker for the antiangiogenic activity of tyrosine kinase inhibitors [38]. Gene therapy for endometrial cancer lymph node and lung metastases is made possible using muscle-mediated sVEGFR-3 expression [39]. This study demonstrates that the expression levels of ITAC and sVEGFR3 increased in current smokers, and that these expression levels significantly decreased after smoking cessation, which could be because smoking triggers an inflammatory response that leads to vascular endothelial dysfunction and a certain degree of recovery after smoking cessation.

Our findings suggest that current smokers may have suppressed levels of systemic immune markers, with reduced systemic levels of certain immune/inflammatory cytokines (SAAs), compared to non-smokers, and increased levels after smoking cessation, and that the SAA level of current smokers significantly differed from that of non-smokers and ex-smokers. This was different from the finding that there was no difference in SAA levels before and after smoking [62]. However, this could be due to the small sample size (10 normal smokers and 10 moderate smokers with persistent asthma) and individual differences. Our findings also differed from the findings that SAA low-density lipoprotein levels showed a significant upward trend from non-smokers to current smokers [63]. SAA is pro-inflammatory because it is a potent chemokine that mediates leukocyte migration and can also stimulate the expression of pro-inflammatory mediators under in vitro and in vivo conditions [40–42]. Systemic inflammation is common in COPD and lung cancer, and SAA is predictive of severity in both diseases, but the subject group was 578 obese Japanese outpatients [63]. Therefore, the different results obtained by different studies might be related to the sample size, race, health status, age, and other factors of the test group, which requires further discussion and research.

Four kinds of cytokines were not detected in the serum of all healthy subjects in this study. Some of these molecules are unique to certain pathological situations or are related to an inflammatory response. For example, IL-1α, IL-6, and IL-8 are all molecules with a strong inflammatory role [21,64–66]. Previous studies have found that levels of IL-6, IL-1 α, and IL-8 in smokers were higher than those in ex-smokers because exposure to cigarette smoke increases oxidative stress [18,23]. We did not detect IL-6, IL-1 α, or IL-8 in our study, which could be due to the sample size. In addition, previous studies have found SCF to be elevated in pancreatic ductal adenocarcinoma patients, and SCF plays an important role in the pathophysiology of mast cells [67,68]. Other cytokines (IL-1β, BCA-1, TNF-α, CRP, ENA-78, MDC, and TNFRII) showed no obvious differences between the non-smoker ($n = 30$), current smoker ($n = 30$), and former smoker groups ($n = 18$). This suggests that smoking has no obvious effect on the expression pathways of MDC and TNFRII.

We analyzed the serum collected from the three types of subjects, so the levels we detected were those of mature and circulating cytokines and chemokines. We did not analyze the relationship between the number of cigarettes smoked and the inflammatory factor levels. However, this study could be useful as a reference to help evaluate the serum levels of cytokines and chemokines in healthy subjects of different smoking statuses. Our results suggest that inflammatory cytokines such as CXCL9/MIG, sIL-6R, and TARC could add moderate predictive value when evaluating smoking-related lung diseases. Additional studies, ideally with larger sample sizes and more diverse subject groups, are needed to better understand how these cytokines and chemokines are produced and modified under different environmental and physiological conditions.

## 5. Conclusions

In this study, serum samples from 78 healthy male volunteers (aged 18–60) were analyzed using the suspension chip technique to measure 17 different inflammatory and immune cytokines related to three smoking statuses. Long-term smoking can result in the up-regulation of CXCL9/MIG and sIL-6R, a trend that continues to increase after quitting. Smoking caused the levels of TARC, ITAC, and sVEGFR3 to be up-regulated. After quitting, these levels gradually decreased, and almost declined to levels observed in the non-smokers. This study found that smoking had a decreasing effect on SAA, but levels of SAA expression increased to that of non-smokers after smoking cessation. Smoking does not affect the expression of MDC, CRP, IL-1$\beta$, BCA-1, TGF-a, ENA-78, or TNFRII, which could require a larger sample size for a more refined analysis. Assessing the levels of IL-1$\alpha$, IL-6, IL-8, and SCF will also require a larger sample size along with a more sensitive detection method. Therefore, smoking or smoking cessation will cause significant changes in cytokine levels. Changes in related cytokines after smoking cessation are mainly restorative, while some cytokines further strengthen the trend of smoking-related changes.

**Author Contributions:** H.W.: methodology, data collection, and writing; H.C.: study design, data interpretation, and manuscript edits; Y.F.: resources, data analysis, and figures; M.L.: data interpretation and figures; S.H.: literature search and data analysis; J.Z.: data collection; Y.T.: literature search; H.H.: funding acquisition, study design, and supervision; Q.H.: funding acquisition, supervision, and manuscript edits. All authors have read and agreed to the published version of the manuscript.

**Funding:** This work was supported by the Provincial and Ministerial Major Project of China (Grant numbers 110202001031(JY-14), 552018CA0080).

**Institutional Review Board Statement:** The study was conducted in accordance with the Declaration of Helsinki, and approved by the Life Science Ethics Committee of Zhengzhou University (approved on 21 July 2016).

**Informed Consent Statement:** Informed consent was obtained from all subjects involved in the study.

**Data Availability Statement:** Not applicable.

**Conflicts of Interest:** All authors declare no conflict of interest.

**Sample Availability:** Samples of some compounds are available from the authors.

## References

1. Wong, M.M.; Fish, E.N. Chemokines: Attractive mediators of the immune response. *Semin. Immunol.* **2003**, *15*, 5–14. [CrossRef]
2. Furman, D.; Campisi, J.; Verdin, E.; Carrera-Bastos, P.; Targ, S.; Franceschi, C.; Ferrucci, L.; Gilroy, D.W.; Fasano, A.; Miller, G.W.; et al. Chronic inflammation in the etiology of disease across the life span. *Nat. Med.* **2019**, *25*, 1822–1832. [CrossRef] [PubMed]
3. Cooper, G.S.; Bynum, M.L.; Somers, E.C. Recent insights in the epidemiology of autoimmune diseases: Improved prevalence estimates and understanding of clustering of diseases. *J. Autoimmun.* **2009**, *33*, 197–207. [CrossRef] [PubMed]
4. Hartel, C.; Adam, N.; Strunk, T.; Temming, P.; Muller-Steinhardt, M.; Schultz, C. Cytokine responses correlate differentially with age in infancy and early childhood. *Clin. Exp. Immunol.* **2005**, *142*, 446–453. [CrossRef]
5. Benahmed, M.; Meresse, B.; Arnulf, B.; Barbe, U.; Mention, J.J.; Verkarre, V.; Allez, M.; Cellier, C.; Hermine, O.; Cerf-Bensussan, N. Inhibition of TGF-beta signaling by IL-15: A new role for IL-15 in the loss of immune homeostasis in celiac disease. *Gastroenterology* **2007**, *132*, 994–1008. [CrossRef]
6. Lee, J.; Taneja, V.; Vassallo, R. Cigarette smoking and inflammation: Cellular and molecular mechanisms. *J. Dent. Res.* **2012**, *91*, 142–149. [CrossRef]
7. Oberg, M.; Jaakkola, M.S.; Woodward, A.; Peruga, A.; Pruss-Ustun, A. Worldwide burden of disease from exposure to second-hand smoke: A retrospective analysis of data from 192 countries. *Lancet* **2011**, *377*, 139–146. [CrossRef]
8. Mathers, C.D.; Loncar, D. Projections of global mortality and burden of disease from 2002 to 2030. *PLoS Med.* **2006**, *3*, e442. [CrossRef]
9. Thun, M.J.; DeLancey, J.O.; Center, M.M.; Jemal, A.; Ward, E.M. The global burden of cancer: Priorities for prevention. *Carcinogenesis* **2010**, *31*, 100–110. [CrossRef]
10. Ezzati, M.; Lopez, A.D. Estimates of global mortality attributable to smoking in 2000. *Lancet* **2003**, *362*, 847–852. [CrossRef]
11. Sokolowska, M.; Quesniaux, V.F.J.; Akdis, C.A.; Chung, K.F.; Ryffel, B.; Togbe, D. Acute Respiratory Barrier Disruption by Ozone Exposure in Mice. *Front. Immunol.* **2019**, *10*, 2169. [CrossRef] [PubMed]
12. Durazzo, T.C.; Mattsson, N.; Weiner, M.W.; Alzheimer's Disease Neuroimaging Initiative. Smoking and increased Alzheimer's disease risk: A review of potential mechanisms. *Alzheimers Dement.* **2014**, *10* (Suppl. 3), S122–S245. [CrossRef] [PubMed]
13. Shein, M.; Jeschke, G. Comparison of Free Radical Levels in the Aerosol from Conventional Cigarettes, Electronic Ciga-rettes, and Heat-Not-Burn Tobacco Products. *Chem. Res. Toxicol.* **2019**, *32*, 1289–1298. [CrossRef] [PubMed]
14. Bergstrom, J. Tobacco smoking and chronic destructive periodontal disease. *Odontology* **2004**, *92*, 1–8. [CrossRef] [PubMed]

15. Qiu, F.; Liang, C.L.; Liu, H.; Zeng, Y.Q.; Hou, S.; Huang, S.; Lai, X.; Dai, Z. Impacts of cigarette smoking on immune responsiveness: Up and down or upside down? *Oncotarget* **2017**, *8*, 268–284. [CrossRef] [PubMed]
16. Heyn, J.; Luchting, B.; Azad, S.C. Smoking Associated T-Cell Imbalance in Patients with Chronic Pain. *Nicotine Tob. Res.* **2020**, *22*, 111–117. [CrossRef]
17. Andersen, A.M.; Lei, M.K.; Beach, S.R.H.; Philibert, R.A.; Sinha, S.; Colgan, J.D. Cigarette and Cannabis Smoking Effects on GPR15+ Helper T Cell Levels in Peripheral Blood: Relationships with Epigenetic Biomarkers. *Genes* **2020**, *11*, 149. [CrossRef]
18. Hartmann, T.; Marino, F.; Duffield, R. Tobacco smoking and acute exercise on immune-inflammatory responses among relative short and longer smoking histories. *Cytokine* **2019**, *123*, 154754. [CrossRef]
19. Jamil, A.; Rashid, A.; Majeed, A. Correlation between Genotoxicity and Interleukin-6 in Smokers: A Rodent Model. *J. Coll. Physicians Surg. Pak.* **2018**, *28*, 821–823. [CrossRef]
20. Gonzalez-Quintela, A.; Alende, R.; Gude, F.; Campos, J.; Rey, J.; Meijide, L.M.; Fernandez-Merino, C.; Vidal, C. Serum levels of immunoglobulins (IgG, IgA, IgM) in a general adult population and their relationship with alcohol consumption, smoking and common metabolic abnormalities. *Clin. Exp. Immunol.* **2008**, *151*, 42–50. [CrossRef]
21. Yao, L.K.; Liu, G.N.; Huang, S.M.; Li, W.T.; Li, Y. Relationship between expression of HDAC2, IL-8, TNF-alpha in lung adenocarcinoma tissues and smoking. *Natl. Med. J. China* **2016**, *96*, 1410–1413.
22. Pace, E.; Di Vincenzo, S.; Di Salvo, E.; Genovese, S.; Dino, P.; Sangiorgi, C.; Ferraro, M.; Gangemi, S. MiR-21 upregulation increases IL-8 expression and tumorigenesis program in airway epithelial cells exposed to cigarette smoke. *J. Cell. Physiol.* **2019**, *234*, 22183–22194. [CrossRef] [PubMed]
23. Li, D.; Li, J.; Wang, L.; Zhang, Q. Association between IL-1beta, IL-8, and IL-10 polymorphisms and risk of acute pancreatitis. *Genet. Mol. Res.* **2015**, *14*, 6635–6641. [CrossRef] [PubMed]
24. Griffith, J.W.; Sokol, C.L.; Luster, A.D. Chemokines and chemokine receptors: Positioning cells for host defense and immunity. *Annu. Rev. Immunol.* **2014**, *32*, 659–702. [CrossRef] [PubMed]
25. Romagnani, P.; Rotondi, M.; Lazzeri, E.; Lasagni, L.; Francalanci, M.; Buonamano, A.; Milani, S.; Vitti, P.; Chiovato, L.; Tonacchera, M.; et al. Expression of IP-10/CXCL10 and MIG/CXCL9 in the thyroid and increased levels of IP-10/CXCL10 in the serum of patients with recent-onset Graves' disease. *Am. J. Pathol.* **2002**, *161*, 195–206. [CrossRef]
26. Yun, J.J.; Fischbein, M.P.; Whiting, D.; Irie, Y.; Fishbein, M.C.; Burdick, M.D.; Belperio, J.; Strieter, R.M.; Laks, H.; Berliner, J.A.; et al. The role of MIG/CXCL9 in cardiac allograft vasculopathy. *Am. J. Pathol.* **2002**, *161*, 1307–1313. [CrossRef]
27. Maruoka, H.; Inoue, D.; Takiuchi, Y.; Nagano, S.; Arima, H.; Tabata, S.; Matsushita, A.; Ishikawa, T.; Oita, T.; Takahashi, T. IP-10/CXCL10 and MIG/CXCL9 as novel markers for the diagnosis of lymphoma-associated hemophagocytic syndrome. *Ann. Hematol.* **2014**, *93*, 393–401. [CrossRef]
28. Semba, T.; Nishimura, M.; Nishimura, S.; Ohara, O.; Ishige, T.; Ohno, S.; Nonaka, K.; Sogawa, K.; Satoh, M.; Sawai, S.; et al. The FLS (fatty liver Shionogi) mouse reveals local expressions of lipocalin-2, CXCL1 and CXCL9 in the liver with non-alcoholic steatohepatitis. *BMC Gastroenterol.* **2013**, *13*, 120. [CrossRef]
29. Onor, I.O.; Stirling, D.L.; Williams, S.R.; Bediako, D.; Borghol, A.; Harris, M.B.; Darensburg, T.B.; Clay, S.D.; Okpechi, S.C.; Sarpong, D.F. Clinical Effects of Cigarette Smoking: Epidemiologic Impact and Review of Pharmacotherapy Options. *Int. J. Environ. Res. Public Health* **2017**, *14*, 1147. [CrossRef]
30. Shiels, M.S.; Katki, H.A.; Freedman, N.D.; Purdue, M.P.; Wentzensen, N.; Trabert, B.; Kitahara, C.M.; Furr, M.; Li, Y.; Kemp, T.J.; et al. Cigarette smoking and variations in systemic immune and inflammation markers. *J. Natl. Cancer. Inst.* **2014**, *106*, dju294. [CrossRef]
31. Chahal, N.; McLain, A.C.; Ghassabian, A.; Michels, K.A.; Bell, E.M.; Lawrence, D.A.; Yeung, E.H. Maternal Smoking and Newborn Cytokine and Immunoglobulin Levels. *Nicotine Tob. Res.* **2017**, *19*, 789–796. [CrossRef] [PubMed]
32. Schett, G. Physiological effects of modulating the interleukin-6 axis. *Rheumatology* **2018**, *57* (Suppl. 2), ii43–ii50. [CrossRef] [PubMed]
33. Rose-John, S. The Soluble Interleukin 6 Receptor: Advanced Therapeutic Options in Inflammation. *Clin. Pharmacol. Ther.* **2017**, *102*, 591–598. [CrossRef] [PubMed]
34. Karatayli, E.; Hall, R.A.; Weber, S.N.; Dooley, S.; Lammert, F. Effect of alcohol on the interleukin 6-mediated inflammatory response in a new mouse model of acute-on-chronic liver injury. *Biochim. Biophys. Acta. Mol. Basis. Dis.* **2019**, *1865*, 298–307. [CrossRef] [PubMed]
35. Speyer, C.B.; Costenbader, K.H. Cigarette smoking and the pathogenesis of systemic lupus erythematosus. *Expert Rev. Clin. Immunol.* **2018**, *14*, 481–487. [CrossRef]
36. Xian-jin, Z.; Dong-hui, S. I-TAC: A Novel Chemokine. *Int. J. Med. Microbiol.* **2006**, *29*, 330–334.
37. Albuquerque, R.J. The newest member of the VEGF family. *Blood* **2013**, *121*, 4015–4016. [CrossRef]
38. Kanefendt, F.; Lindauer, A.; Mross, K.; Fuhr, U.; Jaehde, U. Determination of soluble vascular endothelial growth factor receptor 3 (sVEGFR-3) in plasma as pharmacodynamic biomarker. *J. Pharm. Biomed. Anal.* **2012**, *70*, 485–491. [CrossRef]
39. Takahashi, K.; Mizukami, H.; Saga, Y.; Takei, Y.; Urabe, M.; Kume, A.; Machida, S.; Fujiwara, H.; Suzuki, M.; Ozawa, K. Suppression of lymph node and lung metastases of endometrial cancer by muscle-mediated expression of soluble vascular endothelial growth factor receptor-3. *Cancer Sci.* **2013**, *104*, 1107–1111. [CrossRef]

40. Su, S.B.; Gong, W.; Gao, J.L.; Shen, W.; Murphy, P.M.; Oppenheim, J.J.; Wang, J.M. A seven-transmembrane, G pro-tein-coupled receptor, FPRL1, mediates the chemotactic activity of serum amyloid A for human phagocytic cells. *J. Exp. Med.* **1999**, *189*, 395–402. [CrossRef]
41. He, R.; Sang, H.; Ye, R.D. Serum amyloid A induces IL-8 secretion through a G protein-coupled receptor, FPRL1/LXA4R. *Blood* **2003**, *101*, 1572–1581. [CrossRef] [PubMed]
42. Bozinovski, S.; Uddin, M.; Vlahos, R.; Thompson, M.; McQualter, J.L.; Merritt, A.S.; Wark, P.A.; Hutchinson, A.; Irving, L.B.; Levy, B.D.; et al. Serum amyloid A opposes lipoxin $A_4$ to mediate glucocorticoid refractory lung inflammation in chronic obstructive pulmonary disease. *Proc. Natl. Acad. Sci. USA* **2012**, *109*, 935–940. [CrossRef] [PubMed]
43. Paffenbarger, R.S., Jr.; Hyde, R.T.; Wing, A.L.; Hsieh, C.C. Physical activity, all-cause mortality, and longevity of college alumni. *N. Engl. J. Med.* **1986**, *314*, 605–613. [CrossRef]
44. Bendlin, B.B.; Carlsson, C.M.; Gleason, C.E.; Johnson, S.C.; Sodhi, A.; Gallagher, C.L.; Puglielli, L.; Engelman, C.D.; Ries, M.L.; Xu, G.; et al. Midlife predictors of Alzheimer's disease. *Maturitas* **2010**, *65*, 131–137. [CrossRef] [PubMed]
45. Krintus, M.; Kozinski, M.; Kubica, J.; Sypniewska, G. Critical appraisal of inflammatory markers in cardiovascular risk stratification. *Crit. Rev. Clin. Lab. Sci.* **2014**, *51*, 263–279. [CrossRef]
46. Arnson, Y.; Shoenfeld, Y.; Amital, H. Effects of tobacco smoke on immunity, inflammation and autoimmunity. *J. Autoimmun.* **2010**, *34*, J258–J265. [CrossRef]
47. House, I.G.; Savas, P.; Lai, J.; Chen, A.X.Y.; Oliver, A.J.; Teo, Z.L.; Todd, K.L.; Henderson, M.A.; Giuffrida, L.; Petley, E.V.; et al. Macrophage-Derived CXCL9 and CXCL10 Are Required for Antitumor Immune Responses Following Immune Checkpoint Blockade. *Clin. Cancer Res.* **2020**, *26*, 487–504. [CrossRef]
48. Koper, O.M.; Kaminska, J.; Sawicki, K.; Kemona, H. CXCL9, CXCL10, CXCL11, and their receptor (CXCR3) in neuroinflammation and neurodegeneration. *Adv. Clin. Exp. Med.* **2018**, *27*, 849–856. [CrossRef]
49. Bonecchi, R.; Galliera, E.; Borroni, E.M.; Corsi, M.M.; Locati, M.; Mantovani, A. Chemokines and chemokine receptors: An overview. *Front. Biosci.* **2009**, *14*, 540–551. [CrossRef]
50. Pekarek, V.; Srinivas, S.; Eskdale, J.; Gallagher, G. Interferon lambda-1 (IFN-lambda1/IL-29) induces ELR(-) CXC chem-okine mRNA in human peripheral blood mononuclear cells, in an IFN-gamma-independent manner. *Genes Immun.* **2007**, *8*, 177–180. [CrossRef]
51. Sheppard, P.; Kindsvogel, W.; Xu, W.; Henderson, K.; Schlutsmeyer, S.; Whitmore, T.E.; Kuestner, R.; Garrigues, U.; Birks, C.; Roraback, J.; et al. IL-28, IL-29 and their class II cytokine receptor IL-28R. *Nat. Immunol.* **2003**, *4*, 63–68. [CrossRef] [PubMed]
52. Honda, M.; Yamamoto, S.; Cheng, M.; Yasukawa, K.; Suzuki, H.; Saito, T.; Osugi, Y.; Tokunaga, T.; Kishimoto, T. Human soluble IL-6 receptor: Its detection and enhanced release by HIV infection. *J. Immunol.* **1992**, *148*, 2175–2180. [PubMed]
53. Hurst, S.M.; Wilkinson, T.S.; McLoughlin, R.M.; Jones, S.; Horiuchi, S.; Yamamoto, N.; Rose-John, S.; Fuller, G.M.; Topley, N.; Jones, S.A. Il-6 and its soluble receptor orchestrate a temporal switch in the pattern of leukocyte recruitment seen during acute inflammation. *Immunity* **2001**, *14*, 705–714. [CrossRef]
54. Richards, P.J.; Nowell, M.A.; Horiuchi, S.; McLoughlin, R.M.; Fielding, C.A.; Grau, S.; Yamamoto, N.; Ehrmann, M.; Rose-John, S.; Williams, A.S.; et al. Functional characterization of a soluble gp130 isoform and its therapeutic capacity in an experimental model of inflammatory arthritis. *Arthritis Rheum.* **2006**, *54*, 1662–1672. [CrossRef]
55. Heinz, D.; Peters, M.; Prange, R.; Gerken, G.; Rose-John, S. Possible role of human interleukin-6 and soluble interleukin-6 receptor in hepatitis B virus infection. *J. Viral. Hepat.* **2001**, *8*, 186–193. [CrossRef]
56. Salvi, M.; Pedrazzoni, M.; Girasole, G.; Giuliani, N.; Minelli, R.; Wall, J.R.; Roti, E. Serum concentrations of proin-flammatory cytokines in Graves' disease: Effect of treatment, thyroid function, ophthalmopathy and cigarette smoking. *Eur. J. Endocrinol.* **2000**, *143*, 197–202. [CrossRef]
57. Ritter, M.; Göggel, R.; Chaudhary, N.; Wiedenmann, A.; Jung, B.; Weith, A.; Seither, P. Elevated expression of TARC (CCL17) and MDC (CCL22) in models of cigarette smoke-induced pulmonary inflammation. *Biochem. Biophys. Res. Commun.* **2005**, *334*, 254–262. [CrossRef]
58. Aldaham, S.; Foote, J.A.; Chow, H.H.; Hakim, I.A. Smoking Status Effect on Inflammatory Markers in a Randomized Trial of Current and Former Heavy Smokers. *Int. J. Inflam.* **2015**, *2015*, 439396. [CrossRef]
59. Shiels, M.S.; Shu, X.O.; Chaturvedi, A.K.; Gao, Y.T.; Xiang, Y.B.; Cai, Q.; Hu, W.; Shelton, G.; Ji, B.T.; Pinto, L.A.; et al. A prospective study of immune and inflammation markers and risk of lung cancer among female never smokers in Shanghai. *Carcinogenesis* **2017**, *38*, 1004–1010. [CrossRef] [PubMed]
60. He, Y.; Kozaki, K.; Karpanen, T.; Koshikawa, K.; Yla-Herttuala, S.; Takahashi, T.; Alitalo, K. Suppression of tumor lymphangiogenesis and lymph node metastasis by blocking vascular endothelial growth factor receptor 3 signaling. *J. Natl. Cancer Inst* **2002**, *94*, 819–825. [CrossRef]
61. Ting-ting, L.; Zhong-lan, Y.; Ze-wei, Y. Carotid ultrasound screening and related behavioral factors in high risk population of cerebral hemorrhage. *China J. Mod. Med.* **2015**, *25*, 64–68.
62. Papaioannou, A.I.; Koutsokera, A.; Tanou, K.; Kiropoulos, T.S.; Tsilioni, I.; Oikonomidi, S.; Liadaki, K.; Pournaras, S.; Gourgoulianis, K.I.; Kostikas, K. The acute effect of smoking in healthy and asthmatic smokers. *Eur. J. Clin. Investig.* **2010**, *40*, 103–109. [CrossRef] [PubMed]

63. Kotani, K.; Satoh-Asahara, N.; Kato, Y.; Araki, R.; Himeno, A.; Yamakage, H.; Koyama, K.; Tanabe, M.; Oishi, M.; Okajima, T.; et al. Serum amyloid A low-density lipoprotein levels and smoking status in obese Japanese patients. *J. Int. Med. Res.* **2011**, *39*, 1917–1922. [CrossRef]
64. Higham, A.; Bostock, D.; Booth, G.; Dungwa, J.V.; Singh, D. The effect of electronic cigarette and tobacco smoke exposure on COPD bronchial epithelial cell inflammatory responses. *Int. J. Chron. Obstruct. Pulmon. Dis.* **2018**, *13*, 989–1000. [CrossRef] [PubMed]
65. Oh, J.Y.; Lee, Y.S.; Min, K.H.; Hur, G.Y.; Lee, S.Y.; Kang, K.H.; Rhee, C.K.; Park, S.J.; Shim, J.J. Difference in systemic inflammation and predictors of acute exacerbation between smoking-associated COPD and tuberculosis-associated COPD. *Int. J. Chron. Obstruct. Pulmon. Dis.* **2018**, *13*, 3381–3387. [CrossRef]
66. Goldbach-Mansky, R. Immunology in clinic review series; focus on autoinflammatory diseases: Update on monogenic autoinflammatory diseases: The role of interleukin (IL)-1 and an emerging role for cytokines beyond IL-1. *Clin. Exp. Immunol.* **2012**, *167*, 391–404. [CrossRef]
67. Gao, C.; Li, S.; Zhao, T.; Chen, J.; Ren, H.; Zhang, H.; Wang, X.; Lang, M.; Liu, J.; Gao, S.; et al. SCF, regulated by HIF-1alpha, promotes pancreatic ductal adenocarcinoma cell progression. *PLoS ONE* **2015**, *10*, e0121338.
68. Wang, Y.; Ma, H.; Tao, X.; Luo, Y.; Wang, H.; He, J.; Fang, Q.; Guo, S.; Song, C. SCF promotes the production of IL-13 via the MEK-ERK-CREB signaling pathway in mast cells. *Exp. Ther. Med.* **2019**, *18*, 2491–2496. [CrossRef]

molecules

MDPI

Article

# Nicotine Inhibits the Cytotoxicity and Genotoxicity of NNK Mediated by CYP2A13 in BEAS-2B Cells

Yulin Sun [1,2,3], Hongjuan Wang [1,2], Huan Chen [1,2], Sen Zhang [4], Jun Li [1,2], Jingni Zhang [1,2], Jianlu Tian [1,2], Youyu Zhang [3,*], Hongwei Hou [1,2,*] and Qingyuan Hu [1,2,*]

1 China National Tobacco Quality Supervision & Test Center, Zhengzhou 450001, China; sunyl@tbescience.com (Y.S.); redbri2013@163.com (H.W.); hunny_ch@163.com (H.C.); lijun@tbescience.com (J.L.); zhangjn@tbescience.com (J.Z.); tianjl@tbescience.com (J.T.)
2 Key Laboratory of Tobacco Biological Effects, Zhengzhou 450001, China
3 Key Laboratory of Chemical Biology and Traditional Chinese Medicine Research (Ministry of Education of China), College of Chemistry and Chemical Engineering, Hunan Normal University, Changsha 410081, China
4 Shaanxi Key Laboratory of Degradable Biomedical Materials, School of Chemical Engineering, Northwest University, Xi'an 710069, China; liqvanzs@163.com
* Correspondence: zhangyy@hunnu.edu.cn (Y.Z.); houhw@tbescience.com (H.H.); huqy@tbescience.com (Q.H.)

**Abstract:** Both tobacco-specific carcinogen 4-(methylnitrosamino)-1-(3-pyridyl)-1-butanone (NNK) and nicotine can be metabolized by cytochrome P450 2A13 (CYP2A13). Previous studies have shown that nicotine has a potential inhibitory effect on the toxicity of NNK. However, due to the lack of CYP2A13 activity in conventional lung cell lines, there had been no systematic in vitro investigation for the key target organ, the lung. Here, BEAS-2B cells stably expressing CYP2A13 (B-2A13 cells) were constructed to investigate the effects of nicotine on the cytotoxicity and genotoxicity of NNK. The results showed more sensitivity for NNK-induced cytotoxicity in B-2A13 cells than in BEAS-2B and B-vector cells. NNK significantly induced DNA damage, cell cycle arrest, and chromosomal damage in B-2A13 cells, but had no significant effect on BEAS-2B cells and the vector control cells. The combination of different concentration gradient of nicotine without cytotoxic effects and a single concentration of NNK reduced or even counteracted the cytotoxicity and multi-dimensional genotoxicity in a dose-dependent manner. In conclusion, CYP2A13 caused the cytotoxicity and genotoxicity of NNK in BEAS-2B cells, and the addition of nicotine could inhibit the toxicity of NNK.

**Keywords:** NNK; nicotine; cytotoxicity; genotoxicity; cytochrome P450 2A13

**Citation:** Sun, Y.; Wang, H.; Chen, H.; Zhang, S.; Li, J.; Zhang, J.; Tian, J.; Zhang, Y.; Hou, H.; Hu, Q. Nicotine Inhibits the Cytotoxicity and Genotoxicity of NNK Mediated by CYP2A13 in BEAS-2B Cells. *Molecules* **2022**, *27*, 4851. https://doi.org/10.3390/molecules27154851

Academic Editors: Guohui Sun and Chongwen Wang

Received: 12 June 2022
Accepted: 25 July 2022
Published: 29 July 2022



## 1. Introduction

4-(Methylnitrosamino)-1-(3-pyridyl)-1-butanone (NNK), the strongest carcinogen of tobacco-specific N-nitrosamines (TSNAs) in cigarette smoke, is listed as a Group 1 human carcinogen by the International Agency for Research of Cancer. NNK is also a key component of monitored tobacco carcinogens and is thought to play an important role in smoking-related cancers in humans [1–4]. NNK must be metabolically activated to exert its carcinogenic and toxicological properties [2,5]. It is difficult to study the toxic effects of NNK due to the complexity of NNK metabolism, including the enzyme system and activation process of the catalysis [1]. NNK leads to the α-hydroxylation of methyl or methylene carbon via bioactivation of cytochrome P450 enzymes (CYP), and CYP2As play pivotal roles in the metabolic activation of NNK in the human pulmonary system, such as CYP2A6, CYP2A7, and CYP2A13. Among these, CYP2A7 is defective in incorporating heme molecules and therefore is an inactive enzyme in vitro. The tissue distribution of these enzymes revealed that CYP2A6 is mainly expressed in the liver, and a minimally in the lungs, and CYP2A13 is predominantly expressed in the respiratory tract, including lung, trachea, and nasal mucosa [6,7]. The metabolism of NNK form unstable intermediates that are then broken down into electrophilic methyl-, pyridine-, oxa-nbutyl-, and pyridylhydroxybutyl-diazonium ions. These ions may attack DNA molecules, and lead to the formation of methyl and butylpyridoxine DNA adducts, other DNA damage and chromosome damage, which could be involved in the toxicological properties of NNK [7,8].

Of all the human CYP enzymes detected thus far, CYP2A13 is the most active and efficient enzyme for NNK metabolic activation in vitro and in vivo [2,9]. CYP2A13 is also a catalyst for nicotine, which is the main alkaloid in tobacco and the primary addictive substance in cigarette smoke, maintaining the long-term use of tobacco products [10,11]. Previous studies showed that irreversible inactivation of CYP2A13 enzyme activity occurs in a time and dose-dependent manner after nicotine metabolism [12]. This affects the metabolic activation of NNK, and then affects the toxic effect of NNK, so it is of great significance to explore the effect of nicotine on the toxicity of NNK. Nicotine has been reported to prevent NNK-induced DNA damage in human hepatic-derived cells, possibly by inhibiting CYP enzymes [13]. However, that study only used a single genotoxicity index to investigate, and did not conclusively demonstrate that nicotine inhibited NNK-induced DNA damage mediated by CYP enzyme. NNK is a more potent lung carcinogen than that of the liver [14], and the main target organ for cancer caused by NNK is the lungs [15]. Therefore, it is necessary to study the effect of nicotine on NNK-induced toxicity in the lung, especially genotoxicity.

Typically, the ideal system to study lung carcinogenicity would be airway cells with a metabolic capacity [13]. Immortalized human bronchial epithelial (BEAS-2B) cells can be used as a substitute in toxicological and pharmacological studies due to their relatively high homology to human lung tissues and primary cells in gene expression patterns [15], but studies have demonstrated low activity and low expression of CYP2A13 in BEAS-2B cells [16]. Moreover, the CYP enzymes required for NNK metabolism are lacking in conventional test lung cell lines [16,17]. Thus, this study established genetic engineered BEAS-2B cells that could stably express CYP2A13 (B-2A13) using the lentiviral system, and BEAS-2B cells and the vector control cells were used as the control. Based on these cell lines, this study explored the role of CYP2A13 in cytotoxicity and genotoxicity of NNK and the effect of nicotine on the toxic effects induced by NNK in lung cells.

## 2. Results and Discussion

Tobacco is a potent multisite carcinogen with a substantial worldwide impact [18]. Both nicotine and NNK are important components in cigarette smoke that contribute to cancer development [19,20], and nicotine has been reported to inhibit the metabolism of NNK by inhibiting CYP2A13 [21,22], so it is of great significance to investigate the effect of CYP2A13-mediated nicotine on the cytotoxicity and genotoxicity of NNK on lung cells. Therefore, based on the constructed B-2A13 cells, the toxic effects of NNK alone and NNK combined with nicotine were studied in this study.

### *2.1. Identification of BEAS-2B Cells That Stably Express CYP2A13*

BEAS-2B cells stably expressing CYP2A13 (B-2A13) or vector (B-Vector) were constructed using the lentiviral system. The expression abundance of mRNA of the CYP2A13 gene in BEAS-2B cells, B-Vector cells, and B-2A13 cells was determined by quantitative Real-Time PCR (qRT-PCR), and the results showed that the expression abundance of the CYP2A13 in B-2A13 cells was significantly higher than that in B-Vector and BEAS-2B cells ($p < 0.001$) (Figure 1A). For example, the expression abundance of the CYP2A13 gene in B-2A13 cells was 438,836.7 times higher than that in BEAS-2B cells (Figure 1A). Western blotting showed that the single CYP2A13 protein band was detected in B-2A13 cells, while there were no obvious bands observed in the B-Vector and BEAS-2B cells (Figure 1B). This indicates that CYP2A13 was successfully expressed in BEAS-2B cells, which was consistent with the results obtained by qRT-PCR. It indicated that BEAS-2B cells stably expressing CYP2A13 were successfully established.

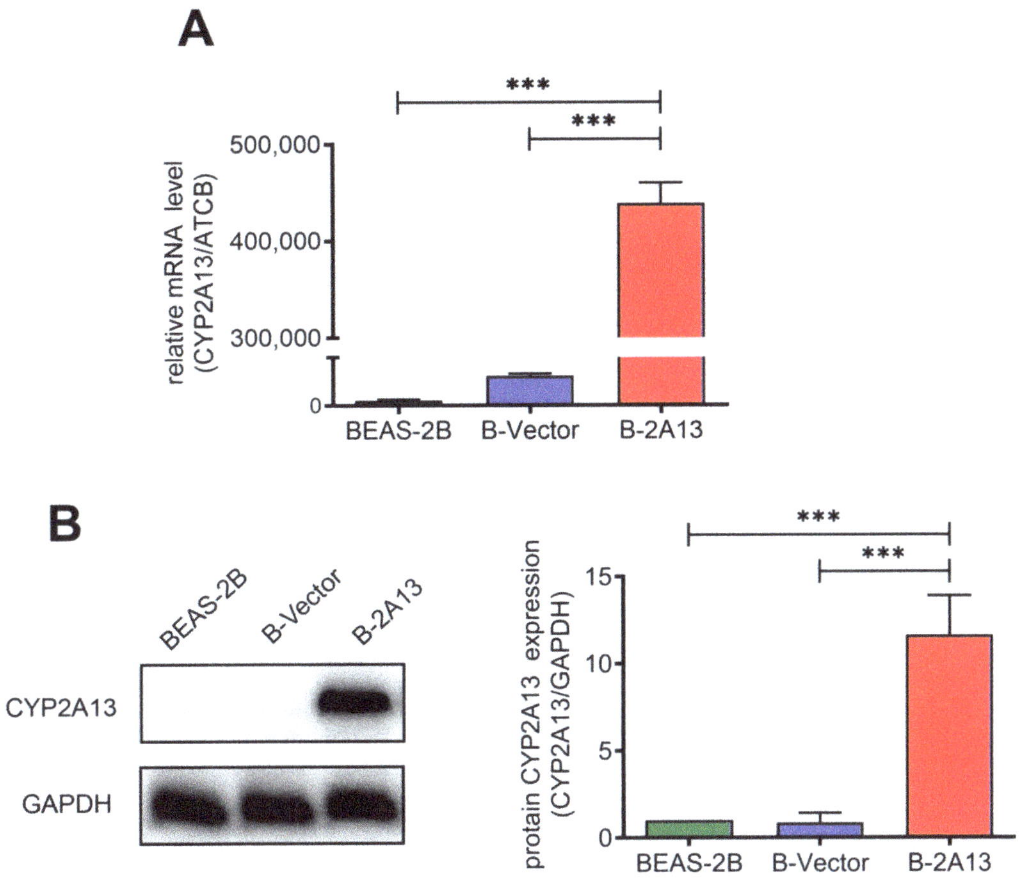


**Figure 1.** Identification of BEAS-2B cells that stably express the cytochrome P450 2A13 (CYP2A13). (**A**) mRNA expression of CYP2A13 in cells determined with quantitative Real-Time PCR (qRT-PCR). (**B**) Characterization of CYP2A13 protein expression in cell lysates. The protein levels of CYP2A13 were detected by Western blotting. Beta-actin gene (ATCB) gene served as a reference gene in the qRT-PCR experiment. Glyceraldehyde-3-phosphate dehydrogenase (GAPDH) served as a reference protein in the WB experiment. ***: $p < 0.001$.

### 2.2. *Effects of Nicotine on NNK-Induced Cytotoxicity in B-2A13 Cells*

To obtain suitable dosing concentration ranges for nicotine and NNK, cytotoxicity was first evaluated using the Cell Counting Kit-8 (CCK-8) assay. NNK induced cytotoxicity in B-2A13 cells in a dose-dependent manner, and B-2A13 cells were more sensitive to NNK than B-Vector cells and BEAS-2B cells (Figure 2A). After 24 h of treatment, the $IC_{20}$ value of NNK was 24.52 μM, 265.04 μM, and 376.14 μM for B-2A13, B-Vector, and BEAS-2B cells, respectively. These results showed that CYP2A13 played an indispensable mediating role in NNK-induced cytotoxicity.

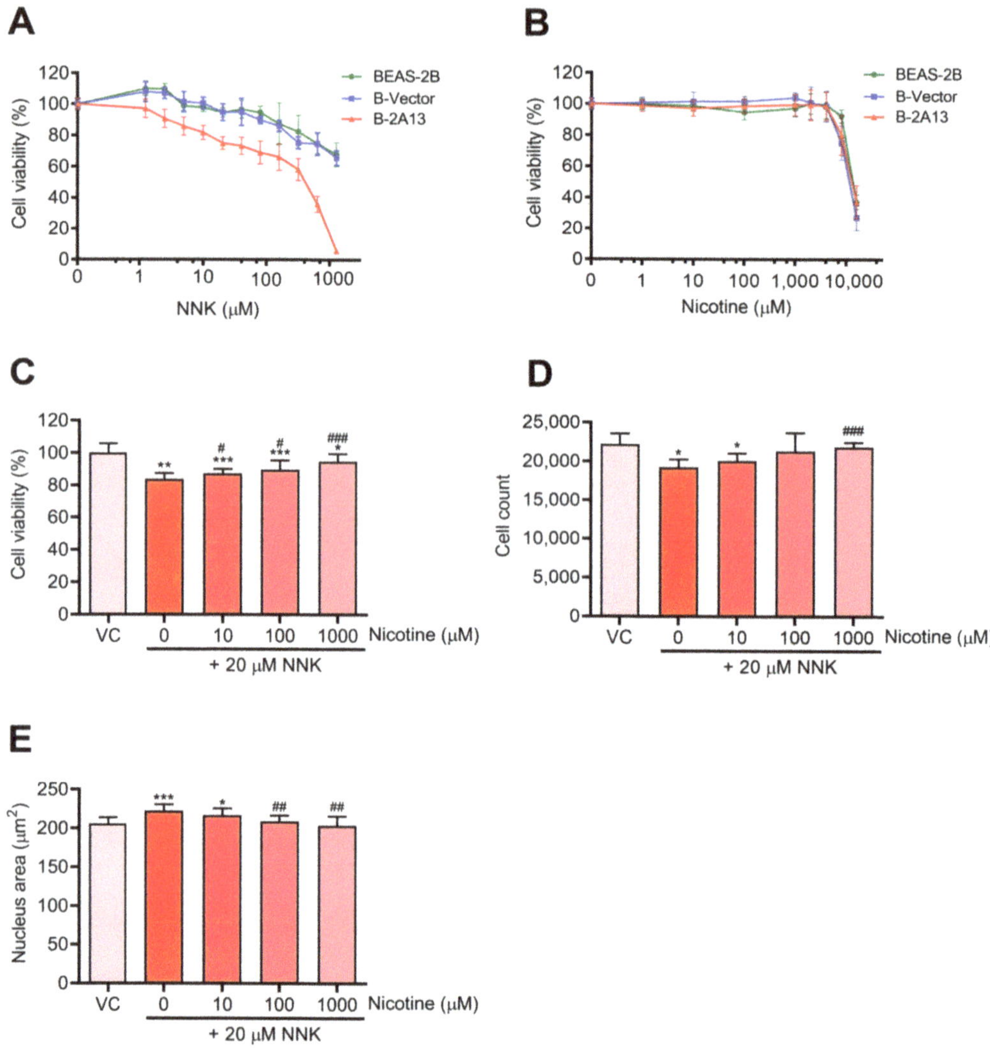


**Figure 2.** Nicotine inhibited 4-(methylnitrosamino)-1-(3-pyridyl)-1-butanone (NNK)-induced cytotoxicity in B-2A13 cells. Cells were treated with NNK and/or nicotine for 24 h. Cell viability of BEAS-2B, B-Vector, and B-2A13 cells after treatment with NNK (**A**) and nicotine (**B**). B-2A13 cells had no significant effects on the cell viability ($p > 0.05$). Cell viability (**C**), cell count (**D**), and nucleus area (**E**) of B-2A13 cells after treatment with a combination of NNK and nicotine. Vehicle control (VC) was cells treated with dissolved in dimethyl sulfoxide (DMSO) (0.004%). * Shows comparison with VC group, * $p < 0.05$, ** $p < 0.01$, and *** $p < 0.001$; # Shows comparison with 20 μM of NNK group, # $p < 0.05$, ## $p < 0.01$, and ### $p < 0.001$.

The viability of cells exposed to nicotine is shown in Figure 2B. Effects of nicotine on the viability of BEAS-2B, B-Vector, and B-2A13 cells were similar. There was no significant difference in the cell viability after 24 h of treatment with nicotine (10–1000 μM) compared to the control group ($p > 0.05$) in B-2A13 cells. Accordingly, 10–1000 μM of nicotine was selected for further experiments for the co-exposure of nicotine and NNK.

Nicotine reduced NNK-induced inhibition of cell viability in B-2A13 cells in a dose-dependent manner (Figure 2C). In the High content screening (HCS) approach, after 24 h of treatment with 20 μM of NNK in B-2A13 cells, the number of cells significantly decreased (Figure 2D, $p < 0.05$), and the area of the nucleus significantly increased (Figure 2E, $p < 0.001$). However, these adverse changes were mitigated in all the groups co-incubated with nicotine.

To avoid false positive genotoxic signals due to cell apoptosis or necrosis, experimental compounds or mixtures are generally considered to be genotoxic when the cell viability of the experimental compound or mixture is greater than 80%, and the induction of H2AX phosphorylation is at least 20% higher than that of the control [23,24]. Therefore, this study used 20 μM of NNK for the combined exposure with nicotine, and the cytotoxicity of this concentration was not higher than 20% in B-2A13 cells. To investigate the effects of nicotine with different concentration gradients on the cytotoxicity and genotoxicity of NNK, three noncytotoxic concentrations of nicotine (10, 100, and 1000 μM) were selected.

*2.3. Effects of Nicotine on NNK-Induced Genotoxicity and Cell Cycle Arrest in B-2A13 Cells*

γH2AX is widely recognized as a sensitive biomarker of DNA double-strand breaks (DSBs) [25]. DNA damage was determined using γH2AX assay. After 24 h of treatment with 1.25–20 μM of NNK, the fluorescence intensity of γH2AX dose-dependently increased in B-2A13 cells and 1.25 μM of NNK significantly induced γH2AX in B-2A13 cells ($p < 0.001$) compared with the vehicle control, while BEAS-2B cells and B-vector cells had no significant effects on NNK-induced γH2AX compared with the vehicle control (Figure 3A,B). These results demonstrated that the involvement of CYP2A13 promoted NNK-induced DNA damage. About 20 μM of NNK increased the fluorescence intensity of γH2AX in cell nuclei of B-2A13 cells, while the groups incubated with nicotine dose-dependently down-regulated the fluorescence intensity of γH2AX in B-2A13 cells (Figure 3C,D). Therefore, nicotine inhibited NNK-induced DNA damage in B-2A13 cells in a dose-dependent manner.

The cytoplasmic block micronucleus (CBMN) and non cytoplasmic block micronucleus (NCBMN) assay were used to evaluate chromosome damage. Compared with the vehicle control, CBMN frequency and NCBMN frequency of B-2A13 cells increased in a dose-dependent fashion. About 5 μM NNK significantly increased CBMN frequency ($p < 0.05$) (Figure 4A), and 2.5 μM NNK significantly increased NCBMN frequency in B-2A13 cells ($p < 0.05$) (Figure 4B), whereas no significant changes in CBMN frequency and NCBMN frequency were found in either BEAS-2B or B-Vector cells ($p > 0.05$) (Figure 4A,B). These results demonstrated that chromosomal damage induced by NNK needed mediation by CYP2A13. As expected, nicotine could reduce the CBMN frequency and the NCBMN frequency, and as the concentration increased, the CBMN frequency and the NCBMN frequency returned to the basal level (Figure 4C,D). Thus, nicotine inhibited the CYP2A13-mediated genotoxic effects (DNA damage and chromosome damage) of NNK.

The percentage of cells in the S-phase and G2-phase increased in a dose-dependent manner after 24 h of NNK treatment in B-2A13 cells (Figure 5A). However, no significant effect of NNK was found on the cell cycle arrest in BEAS-2B cells and B-vector cells ($p > 0.05$) (Figure 5B,C). In concordance with the above results, nicotine restrained NNK-induced cell cycle arrest in B-2A13 cells in a concentration-dependent manner (Figure 5D).

A systemic toxicity assessment of NNK based on normal lung cells stably expressing CYP2A13 had not been previously reported. The current study demonstrated that NNK-induced cytotoxicity was more pronounced in B-2A13 cells compared to BEAS-2B and B-Vector cells, and NNK dose-dependently increased γH2AX expression, the frequency of micronuclei, the S/G2-phase cell population, while no differences were observed in BEAS-2B and B-Vector cells between the treatment and control groups. This indicated that CYP2A13 played an indispensable mediating role in NNK-induced cytotoxicity and genotoxicity. Similar results were also found in B-2A13 cells treated with nicotine-free cigarette smoke extract and total cigarette smoke extract [6]. Previous studies demonstrated that NNK exposure to A549 and BEAS-2B cells did not have a significant dose/time-dependent effect on γH2AX at various test concentrations and exposure times [26,27]. This is because BEAS-2B and A549 cells show very little CYP2A6 and CYP2A13 activity [16], which are key enzymes in the metabolic activation of NNK [7].

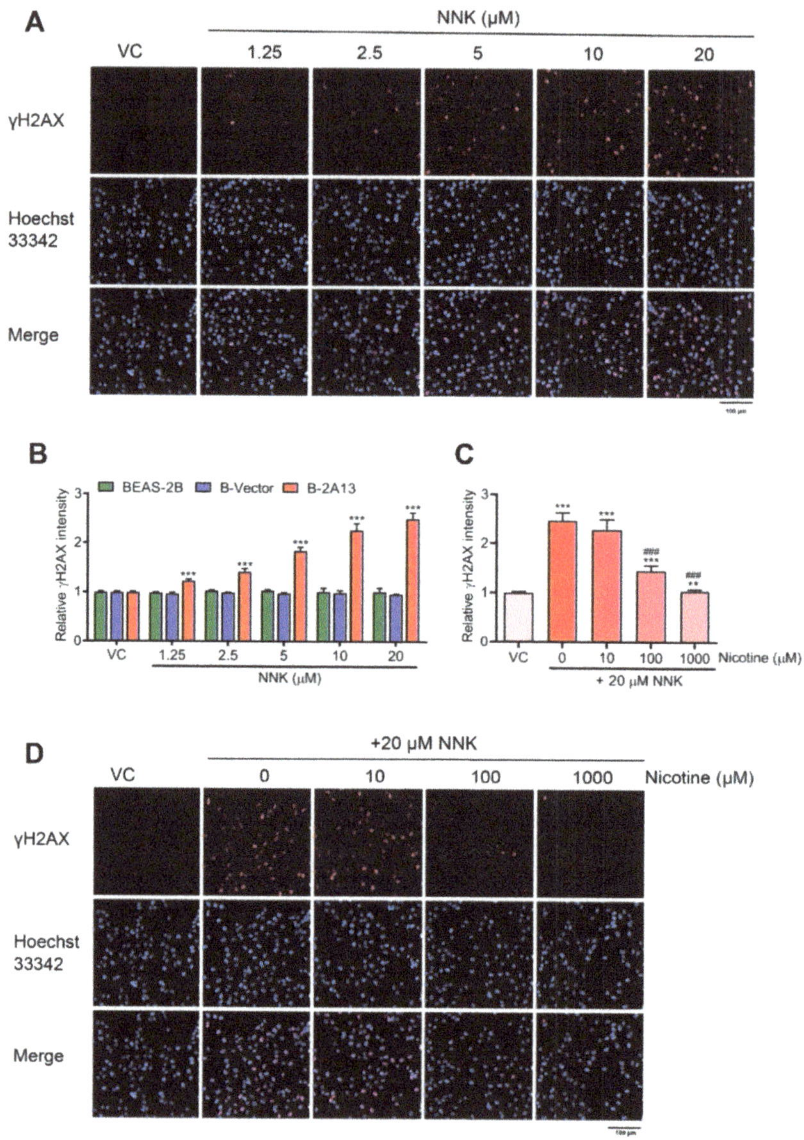


**Figure 3.** Nicotine inhibited NNK-induced DNA damage in B-2A13 cells. DNA damage was determined using γH2AX assay with high content screening. Cells were treated with NNK alone or a combination of NNK and nicotine for 24 h. (**A**) Immunostaining and imaging analysis of γH2AX in B-2A13 cells treated with NNK. (**B**) Relative γH2AX intensity in BEAS-2B, B-Vector, and B-2A13 cells treated with NNK. (**C**) Relative γH2AX intensity in B-2A13 cells treated with a combination of NNK and nicotine. (**D**) Immunostaining and imaging analysis of γH2AX in B-2A13 cells treated with a combination of NNK and nicotine. Vehicle control (VC) was cells treated with DMSO (0.004%). * Shows comparison with VC group, ** $p < 0.01$, and *** $p < 0.001$; # Shows comparison with 20 μM of NNK group, ### $p < 0.001$.

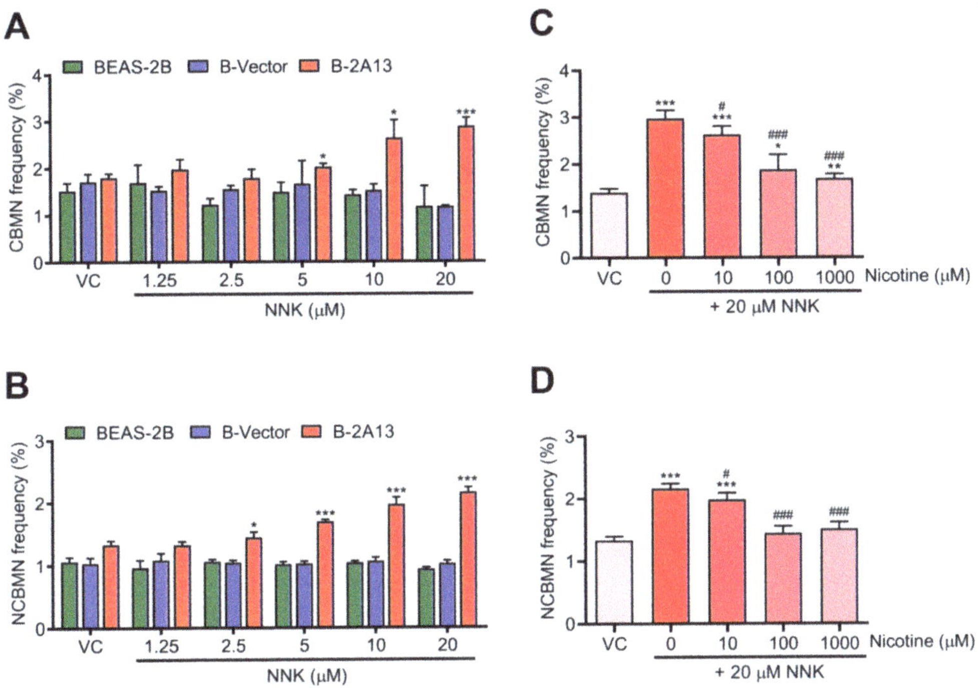


**Figure 4.** Nicotine inhibited NNK-induced chromosome damage in B-2A13 cells. Chromosome damage was detected using the cytoplasmic block micronucleus (CBMN) and non cytoplasmic block micronucleus (NCBMN) assays with high content screening. Cells were treated with NNK alone or a combination of NNK and nicotine for 24 h. CBMN frequency (**A**) and NCBMN frequency (**B**) in BEAS-2B, B-Vector, and B-2A13 cells treated with NNK. CBMN frequency (**C**) and NCBMN frequency (**D**) in B-2A13 cells treated with a combination of NNK and nicotine. Vehicle control (VC) was cells treated with DMSO (0.004%). * Shows comparison with VC group, * $p < 0.05$, ** $p < 0.01$, and *** $p < 0.001$; # Shows comparison with 20 μM of NNK group, # $p < 0.05$, and ### $p < 0.001$.

Moreover, we found that nicotine dose-dependently inhibited cytotoxicity, DNA damage, chromosome damage, and cell cycle arrest induced by NNK in B-2A13 cells. This is in agreement with previous studies in which NNK-induced DNA damage was inhibited by nicotine in hepatic-derived cell lines, possibly due to nicotine inhibiting CYP enzymes [13]. Previous researches had reported that nicotine or nicotine metabolites inhibit various CYPs involved in the bioactivation of NNK, including CYP2A13, CYP2A6, and CYP2E1 [21]. Nicotine has a competitive inhibitory effect on CYP2A13-mediated NNK metabolism with a higher affinity [21,22]. β-nicotyrine, a nicotine-related alkaloid, has also been shown to inhibit CYP2A13 and CYP2A6 in vitro [28,29]. CYP2E1 activates nitrosamines and is competitively suppressed by nicotine in the livers of both humans and rats [30]. A molecular dynamics simulation-based study showed that nicotine had a stronger affinity for CYP2A13 than NNK [31]. Therefore, CYP2A13-catalyzed NNK metabolism may be competitively inhibited by nicotine, resulting in a decrease in NNK-induced cytotoxicity and genotoxicity. Moreover, in vitro work using purified enzymes demonstrated that CYP2A6 and CYP2A13 inactivated in the process of nicotine metabolism [12]. Further investigation found that the nicotine $\Delta^{5'(1')}$ iminium ion is an inactivator of both CYP2A6 and CYP2A13 [32]. Hence another hypothesis is that nicotine metabolites inactivate CYP2A13 leading to a reduction in NNK-induced toxicity.

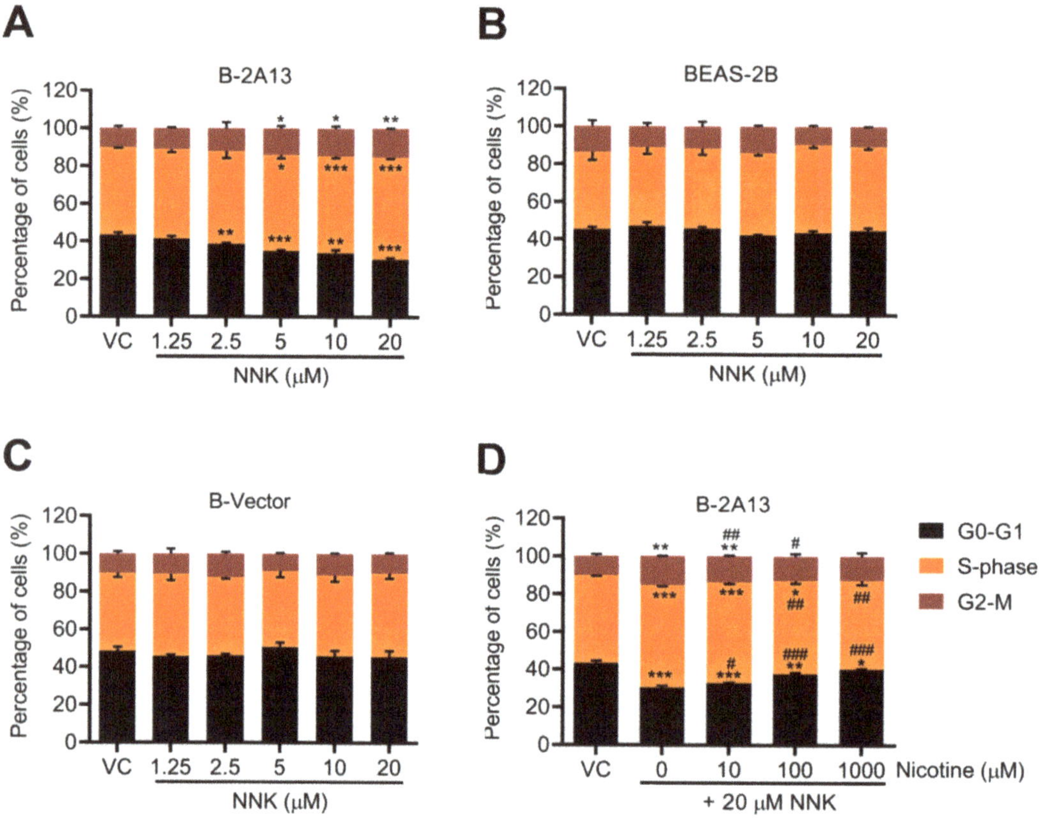


**Figure 5.** Nicotine inhibited NNK-induced cell cycle arrest in B-2A13 cells. Cells were treated with NNK alone or a combination of NNK and nicotine for 24 h. Cell cycle in B-2A13 cells (**A**), BEAS-2B cells (**B**), and B-Vector cells (**C**) treated with NNK, and cell cycle in B-2A13 cells treated with a combination of NNK and nicotine (**D**). Vehicle control (VC) was cells treated with DMSO (0.004%). * Shows comparison with VC group, * $p < 0.05$, ** $p < 0.01$, and *** $p < 0.001$; # Shows comparison with 20 μM of NNK group, # $p < 0.05$, ## $p < 0.01$, and ### $p < 0.001$.

Based on the results from the study, a putative schematic representation of the molecular mechanism involving the effects of nicotine on NNK-induced toxicity in B-2A13 cells was suggested (Figure 6). This speculates that NNK is metabolized by CYP2A13 to produce toxic substances, leading to DNA damage in cells. DNA damage can activate DNA damage checkpoints, which can slow down or arrest cell cycle progression, so that cells can repair or prevent the spread of damaged chromosomes [33,34]. If the DNA damage can be repaired, the cells return to a normal cell cycle and survive, or otherwise they will lead to further chromosomal damage [35], and eventually lead to cell death [36]. Nicotine can inhibit these toxic effects of NNK by inhibiting CYP2A13.

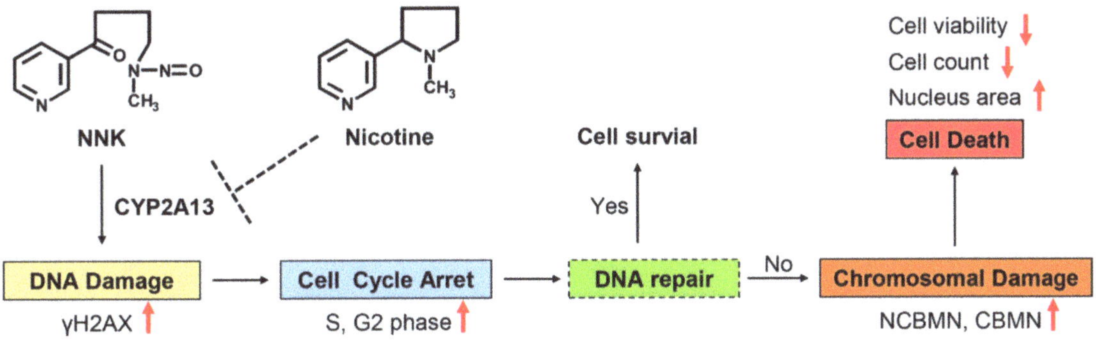


**Figure 6.** The putative schematic representation of the molecular mechanism involved in the effects of nicotine on NNK-induced toxicity in B-2A13 cells. The up and down arrows represent rising and falling respectively.

To our knowledge, this study is the first to investigate the inhibitory effects of nicotine on the cytotoxicity and genotoxicity of NNK based on normal lung cells stably expressing CYP2A13. This study clarified the protective effect of nicotine on the toxic effect of NNK. NNK is the most well-known carcinogen in cigarette smoke, its key metabolic enzyme CYP2A13 can also metabolize many other carcinogens and/or harmful components in cigarette smoke, such as benzo-pyrene, naphthalene, and 3-methylindole, polycyclic aromatic hydrocarbons (pyrene, 1-hydroxypyrene, 1-nitropyrene and 1-acetylpyrene), and heterocyclic amines [37]. They are also thought to be associated with lung cancer [38]. Therefore, nicotine has a potential protective mechanism against carcinogenesis of various cigarette smoke components. One limitation of this study is that the molecular mechanism whereby nicotine inhibits NNK toxicity has not been fully elucidated. Therefore, future studies should consider the mechanism by which nicotine inhibits NNK-induced toxicity via enzyme activity and metabolic markers with nicotine present or absent.

## 3. Materials and Methods

### *3.1. Chemicals*

NNK (>98% purity) was acquired from Toronto Research Chemicals (Toronto, ON, Canada). Nicotine (>99% purity) was obtained from the Key Laboratory of Tobacco Biological Effects (Zhengzhou, China).

### *3.2. Cell Culture and Treatment*

BEAS-2B cells were acquired from the Key Laboratory of Tobacco Biological Effects (Zhengzhou, China), and cultured in Bronchial Epithelial Cell Growth Medium (BEGM, Lonza, Walkersville, MD, USA) in a humidified incubator with 5% $CO_2$ at 37 °C. NNK was dissolved in dimethyl sulfoxide (DMSO) before it was used. The final exposure concentrations of NNK and nicotine were obtained by serial dilutions with the culture medium, and cells were exposed to NNK or/and nicotine in multi-well plates in a $CO_2$ incubator. The final concentration of DMSO was less than 0.3%.

### *3.3. Establishment of BEAS-2B Cells Expressing CYP2A13*

CYP2A13 overexpression lentivirus and empty lentivirus were purchased from GeneChem (Shanghai, China). According to the manufacturer's instructions, BEAS-2B cells ($1 \times 10^5$) were cultured in six-well plates, and when the cell density reached 20%, BEAS-2B cells were infected with lentivirus. The multiplicity of infection (MOI) value was 50. The medium in the plates was discarded and complete culture medium containing lentiviral stock solution and HitransG A (Genechem, Shanghai, China) was added into plates. The medium was replaced after 16 h and the cells were observed under fluorescence microscopy to evaluate the efficiency of the infection after 72 h. Empty lentivirus and non-infected cells were used as negative controls. About 1 μg/mL of puromycin was added to each well for screening the robust cells expressing CYP2A13.

### 3.4. qRT-PCR Analysis

qRT-PCR was performed as previously described [39]. Briefly, total RNA was extracted from cells using Trizol reagent (Pufei, Shanghai, China). The RNA was reverse-transcribed into cDNA using M-MLV reverse transcriptase (Promega, Madison, WI, USA). qRT-PCR was performed using SYBR Premix Ex Taq (Takara, Shiga, Japan). The primers used were: for CYP2A13, 5′-CGCTACGGTTTCCTGCTGCTC-3′(forward) and 5′- CATCTTGGCCCG-GTCCTCAA -3′ (reverse); and for beta-actin (ACTB), 5′- GCGTGACATTAAGGAGAAGC -3′ (forward) and 5′- CCACGTCACACT TCATGATGG -3′ (reverse). CYP2A13 mRNA expression was normalized to ACTB mRNA expression.

### 3.5. Western Blot Analysis

WB analysis was performed according to previously described procedures [40]. The primary antibodies against CYP2A13 (Invitrogen, PA5-101310) and GAPDH (Cell Signaling Technology, 97166S) were diluted at a ratio of 1:1000. The secondary antibodies, Goat anti-rabbit IgG-HRP (Cell Signaling Technology, 7074S,) and Horse anti-mouse IgG-HRP (Cell Signaling Technology, 7076S), were diluted at a ratio of 1:3000. The reference protein GAPDH was used to calibrate the errors generated by sample loading.

### 3.6. Assessment of Cytotoxicity

CCK-8 assay: Cell cytotoxicity was determined using the Cell Counting Kit-8 (CCK-8, Dojindo, Kumamoto, Japan) [41]. Approximately $1 \times 10^4$ cells were plated into each well of a 96-well microplate. After 24 h of incubation in a $CO_2$ incubator at 37 °C, the original medium was aspirated and a fresh medium containing NNK and/or nicotine was added. After 24 h of treatment, 10% ($v/v$) CCK-8 solution was added and incubated at 37 °C for 3 h. The optical density (OD value) was measured at a wavelength of 450 nm with a microplate reader (Molecular Devices LLC, San Jose, CA, USA).

HCS: The experimental protocol for HCS was previously described [42,43]. Approximately $1 \times 10^4$ cells were plated into each well of a PhenoPlate 96-well microplate (PerkinElmer, Waltham, MA, USA). After 24 h of incubation in a $CO_2$ incubator at 37 °C, the original medium was aspirated and a fresh medium containing a combination of NNK and nicotine was added. After 24 h of treatment, the B-2A13 cells were stained with Hoechst 33342 (Beyotime, Shanghai, China). Hoechst 33342 can measure features such as cell count and nuclear area by labeling DNA. Cell imaging with fluorescence analysis was performed on a PerkinElmer Operetta CLS High Content Screening platform with Harmony 4.5 software (PerkinElmer, Waltham, MA, USA).

### 3.7. γH2AX Assay

The γH2AX assay was modified on the basis of the previous experimental method [44]. Approximately $1 \times 10^4$ cells were plated into each well of a PhenoPlate 96-well microplate (PerkinElmer, Waltham, MA, USA), and 24 h later cells were continuously exposed to NNK or a combination of NNK and nicotine for 24 h. The cells were then washed with phosphate-buffered saline (PBS) fixed with 4% paraformaldehyde (Beyotime, Shanghai, China), permeabilized with immunostaining permeable fluid (Beyotime, Shanghai, China), and blocked with 5% bovine serum albumin (BSA, Sigma-Aldrich, St. Louis, MO, USA). After blocking, the cells were incubated with 100 μL of 0.1% ($v/v$) Phospho-Histone H2A.X (Ser139) Antibody (Cell Signaling Technology, Danvers, MA, USA) in 1% BSA overnight at 4 °C. After being washed three times with PBS, the cells were incubated with 100 μL of 0.1% ($v/v$) Anti-rabbit IgG (H + L), F(ab′)2 Fragment (Alexa Fluor® 647 Conjugate) (Cell Signaling Technology, Danvers, MA, USA) in 1% BSA for 2 h at room temperature in darkness. Next, 100 μL of Hoechst 33342 (Beyotime, Shanghai, China) was added to each well for 10 min. After the wells were washed with PBS, 100 μL of PBS was added per well and the plates were immediately placed on the PerkinElmer Operetta CLS High Content Screening platform for analysis.

*3.8. Micronucleus Assay*

The experimental protocol was modified according to the method previously described [43,45,46]. CBMN assay: Approximately $1 \times 10^4$ cells were plated into each well of a PhenoPlate 96-well microplate (PerkinElmer, Waltham, MA, USA). After 24 h of incubation in a $CO_2$ incubator at 37 °C, the original medium was aspirated and a fresh medium containing NNK or a combination of NNK and nicotine with cytochalasin B (4 μg/mL) was added. After 24 h of treatment, cells were washed with PBS, fixed with 4% paraformaldehyde (Beyotime, Shanghai, China), and stained with Hoechst 33342 (Beyotime, Shanghai, China) and Cell Mask Red (Life Technology, Carlsbad, CA, USA).

NCBMN assay: Approximately $1 \times 10^4$ cells were plated into each well of a PhenoPlate 96-well microplate (PerkinElmer, Waltham, MA, USA). After 24 h of incubation in a $CO_2$ incubator at 37 °C, the original medium was aspirated and a fresh medium containing NNK or a combination of NNK and nicotine was added for 24 h. The cells were then washed with PBS and stained with Hoechst 33342 for 10 min in the dark.

After being washed with PBS, 100 μL of PBS was added to each well and the plates were immediately placed on the PerkinElmer Operetta CLS High Content Screening platform for analysis. Frequencies of micronuclear cells were measured in at least 2000 binuclear cells (CBMN assay) or at least 2000 cells (NCBMN assay) per sample.

*3.9. Cell Cycle Analysis*

The cell cycle distribution was measured using the cell cycle and apoptosis analysis kit (Beyotime, Shanghai, China) [47]. Approximately $3 \times 10^5$ cells were seeded into each well of a 6-well culture plate, and 24 h later the cells were continuously exposed to NNK or a combination of NNK and nicotine for 24 h. Cells were harvested by trypsinization, centrifuged, and the pellets were resuspended in 1 mL of cold PBS. They were then centrifuged and the pellet was fixed in 75% ice-cold ethanol at 4 °C overnight. After the fixative solution was aspirated, the cells were centrifuged and the pellet was incubated in propidium iodide (PI) and RNase for 30 min. The cell cycle was analyzed by FACS flow cytometer Accuri ™ C6 Plus (BD Biosciences, San Jose, CA, USA), and the results were processed using the FlowJo (v10.8.1, BD Bioscience, San Jose, CA, USA) software program.

*3.10. Statistical Analysis*

GraphPad Prism 8.0 software (GraphPad Software, San Diego, CA, USA) was used for data analysis, the results were presented as mean ± standard deviation (SD). A two-tailed Student's *t*-test was used to determine significant differences ($p < 0.05$). All experiments were repeated at least three times.

## 4. Conclusions

To elucidate the effect of nicotine on the cytotoxicity and genotoxicity of NNK in lung cells, BEAS-2B cells stably expressing CYP2A13 were constructed and a variety of in vitro toxicological indicators were tested in this study. The results showed that CYP2A13 played an indispensable mediating role in cytotoxicity and genotoxicity of NNK. Nicotine inhibited cytotoxicity and multi-dimensional genotoxicity in NNK-induced B-2A13 cells, possibly by inhibiting CYP2A13. This study provided mechanistic insights into the protective effects of nicotine, and suggested that nicotine has a potential protective mechanism against carcinogenesis of various cigarette smoke components metabolized by CYP2A13. In addition, it suggested that nicotine may influence toxic effects of some non-carcinogenic harmful ingredients activated by CYP2A13 and CYP2A6 enzymes, and then affect the health risks of tobacco products.

**Author Contributions:** Conceptualization, Y.S. and H.W.; methodology, Y.S., H.W., H.C. and J.L.; validation, Y.S., J.L. and J.Z.; formal analysis, Y.S., H.W., H.C. and J.T.; investigation, Y.S., H.W., H.C., J.Z. and S.Z.; writing—original draft preparation, Y.S.; writing—review and editing, Y.S., H.W., S.Z. and Y.Z.; visualization, Y.S. and J.T. supervision, Y.Z., H.H. and Q.H.; project administration, Y.Z., H.H. and Q.H.; funding acquisition, H.H. and Q.H. All authors have read and agreed to the published version of the manuscript.

**Funding:** This work was supported by grants from the State Administration for Market Regulation of China (grant numbers 110202001021 (JY-14), 110202102011).

**Institutional Review Board Statement:** Not applicable.

**Informed Consent Statement:** Not applicable.

**Data Availability Statement:** The data presented in this study are available on request from the corresponding author.

**Conflicts of Interest:** The authors declare no conflict of interest.

**Sample Availability:** Samples of the compounds are not available from the authors.

## References

1. Hecht, S.S. Biochemistry, biology, and carcinogenicity of tobacco-specific N-nitrosamines. *Chem. Res. Toxicol.* **1998**, *11*, 559–603. [CrossRef] [PubMed]
2. Su, T.; Bao, Z.; Zhang, Q.Y.; Smith, T.J.; Hong, J.Y.; Ding, X. Human Cytochrome P450 CYP2A13: Predominant Expression in the Respiratory Tract and Its High Efficiency Metabolic Activation of a Tobacco-specific Carcinogen, 4-(Methylnitrosamino)-1-(3-pyridyl)-1-butanone. *Cancer Res.* **2000**, *60*, 5074–5079. [PubMed]
3. Carlson, E.S.; Upadhyaya, P.; Hecht, S.S. Evaluation of Nitrosamide Formation in the Cytochrome P450-Mediated Metabolism of Tobacco-Specific Nitrosamines. *Int. J. Electrochem. Sci.* **2011**, *29*, 2194–2205. [CrossRef] [PubMed]
4. Akopyan, G.; Bonavida, B. Understanding tobacco smoke carcinogen NNK and lung tumorigenesis. *Int. J. Oncol.* **2006**, *29*, 745–752. [CrossRef] [PubMed]
5. Peterson, L.A. Formation, Repair, and Genotoxic Properties of Bulky DNA Adducts Formed from Tobacco-Specific Nitrosamines. *J. Nucleic Acids* **2010**, *2010*, 284935. [CrossRef]
6. Ji, M.; Zhang, Y.; Li, N.; Wang, C.; Xia, R.; Zhang, Z.; Wang, S.L. Nicotine Component of Cigarette Smoke Extract (CSE) Decreases the Cytotoxicity of CSE in BEAS-2B Cells Stably Expressing Human Cytochrome P450 2A13. *Int. J. Environ. Res. Public Health.* **2017**, *14*, 1221. [CrossRef]
7. Chiang, H.C.; Wang, C.Y.; Lee, H.L.; Tsou, T.C. Metabolic effects of CYP2A6 and CYP2A13 on 4-(methylnitrosamino)-1-(3-pyridyl)-1-butanone (NNK)-induced gene mutation—A mammalian cell-based mutagenesis approach. *Toxicol. Appl. Pharmacol.* **2011**, *253*, 145–152. [CrossRef] [PubMed]
8. Peterson, L.A. Context Matters: Contribution of Specific DNA Adducts to the Genotoxic Properties of the Tobacco-Specific Nitrosamine NNK. *Chem. Res. Toxicol.* **2017**, *30*, 420–433. [CrossRef]
9. Megaraj, V.; Zhou, X.; Xie, F.; Liu, Z.; Yang, W.; Ding, X. Role of CYP2A13 in the bioactivation and lung tumorigenicity of the tobacco-specific lung procarcinogen 4-(methylnitrosamino)-1-(3-pyridyl)-1-butanone: In vivo studies using a CYP2A13-humanized mouse model. *Carcinogenesis* **2014**, *35*, 131–137. [CrossRef]
10. Sato, T.; Abe, T.; Nakamoto, N.; Tomaru, Y.; Koshikiya, N.; Nojima, J.; Kokabu, S.; Sakata, Y.; Kobayashi, A.; Yoda, T. Nicotine induces cell proliferation in association with cyclin D1 up-regulation and inhibits cell differentiation in association with p53 regulation in a murine pre-osteoblastic cell line. *Res. Commun.* **2008**, *377*, 126–130. [CrossRef]
11. Murphy, S.E.; Raulinaitis, V.; Brown, K.M. Nicotine 5′-oxidation and methyl oxidation by P450 2A enzymes. *Drug. Metab. Dispos.* **2005**, *33*, 1166. [CrossRef] [PubMed]
12. Von Weymarn, L.B.; Brown, K.M.; Murphy, S.E. Inactivation of CYP2A6 and CYP2A13 during nicotine metabolism. *J. Pharmacol. Exp. Ther.* **2006**, *316*, 295–303. [CrossRef] [PubMed]
13. Ordonez, P.; Sierra, A.B.; Camacho, O.M.; Baxter, A.; Banerjee, A.; Waters, D.; Minet, E. Nicotine, cotinine, and β-nicotyrine inhibit NNK-induced DNA-strand break in the hepatic cell line HepaRG. *Toxicol. Vitr.* **2014**, *28*, 1329–1337. [CrossRef] [PubMed]
14. Schulze, J.; Schrader, E.; Foth, H.; Kahl, G.F.; Richter, E. Effect of nicotine or cotinine on metabolism of 4-methylnitrosamino-1-(3-pyridyl)-1-butanone (NNK) in isolated rat lung and liver. *Naunyn Schmiedebergs Arch. Pharmacol.* **1998**, *357*, 344–350. [CrossRef]
15. Courcot, E.; Leclerc, J.; Lafitte, J.J.; Mensier, E.; Jaillard, S.; Gosset, P.; Shirali, P.; Pottier, N.; Broly, F.; Lo-Guidice, J.M. Xenobiotic metabolism and disposition in human lung cell models: Comparison with in vivo expression profiles. *Drug. Metab. Dispos.* **2012**, *40*, 1953–1965. [CrossRef]
16. Garcia-Canton, C.; Minet, E.; Anadon, A.; Meredith, C. Metabolic characterization of cell systems used in in vitro toxicology testing: Lung cell system BEAS-2B as a working example. *Toxicol. Vitr.* **2013**, *27*, 1719–1727. [CrossRef]

17. Newland, N.; Baxter, A.; Hewitt, K.; Minet, E. CYP1A1/1B1 and CYP2A6/2A13 activity is conserved in cultures of differentiated primary human tracheobronchial epithelial cells. *Toxicol. Vitr.* **2011**, *25*, 922–929. [CrossRef]
18. Vineis, P.M.; Alavanja, P.; Buffler, E.; Fontham, S.; Franceschi, Y.T.; Gao, P.C.; Gupta, A.; Hackshaw, E.; Matos, J.; Samet, F.; et al. Tobacco and cancer: Recent epidemiological evidence. *J. Natl. Cancer Inst.* **2004**, *96*, 99–106. [CrossRef]
19. Shen, J.; Xu, L.; Owonikoko, T.K.; Sun, S.Y.; Khuri, F.R.; Curran, W.J.; Deng, X. NNK promotes migration and invasion of lung cancer cells through activation of c-Src/PKCι/FAK loop. *Cancer Lett.* **2012**, *318*, 106–113. [CrossRef]
20. Shin, V.Y.; Jin, H.C.; Ng, E.K.; Yu, J.; Leung, W.K.; Cho, C.H.; Sung, J.J.Y. Nicotine and 4-(methylnitrosamino)-1-(3-pyridyl)-1-butanone induce cyclooxygenase-2 activity in human gastric cancer cells: Involvement of nicotinic acetylcholine receptor (nAChR) and β-adrenergic receptor signaling pathways. *Toxicol. Appl. Pharmacol.* **2008**, *233*, 254–261. [CrossRef]
21. Liu, X.; Zhang, J.; Zhang, C.; Yang, B.; Wang, L.; Zhou, J. The inhibition of cytochrome P450 2A13-catalyzed NNK metabolism by NAT, NAB and nicotine. *Toxicol. Res. (Camb)* **2016**, *5*, 1115–1121. [CrossRef] [PubMed]
22. Bao, Z.; He, X.Y.; Ding, X.; Prabhu, S.; Hong, J.Y. Metabolism of nicotine and cotinine by human cytochrome P450 2A13. *Drug. Metab. Dispos.* **2005**, *33*, 258–261. [CrossRef] [PubMed]
23. Li, X.; Chen, S.; Guo, X.; Wu, Q.; Seo, J.E.; Guo, L.; Manjanatha, M.G.; Zhou, T.; Witt, K.L.; Mei, N. Development and Application of TK6-derived Cells Expressing Human Cytochrome P450s for Genotoxicity Testing. *Toxicol. Sci.* **2020**, *175*, 251–265. [CrossRef] [PubMed]
24. Graillot, V.; Takakura, N.; Hegarat, L.L.; Fessard, V.; Audebert, M.; Cravedi, J.P. Genotoxicity of pesticide mixtures present in the diet of the French population. *Environ. Mol. Mutagen.* **2012**, *53*, 173–184. [CrossRef] [PubMed]
25. Tu, W.Z.; Li, B.; Huang, B.; Wang, Y.; Liu, X.D.; Guan, H.; Zhang, S.M.; Tang, Y.; Rang, W.Q.; Zhou, P.K. γH2AX foci formation in the absence of DNA damage: Mitotic H2AX phosphorylation is mediated by the DNA-PKcs/CHK2 pathway. *FEBS Lett.* **2013**, *587*, 3437–3443. [CrossRef]
26. Garcia-Canton, C.; Anadon, A.; Meredith, C. Genotoxicity evaluation of individual cigarette smoke toxicants using the in vitro γH2AX assay by High Content Screening. *Toxicol. Lett.* **2013**, *223*, 81–87. [CrossRef]
27. Zhang, S.; Chen, H.; Wang, A.; Liu, Y.; Hou, H.; Hu, Q. Genotoxicity analysis of five particle matter toxicants from cigarette smoke based on gamma H2AX assay combined with Hill/Two-component model. *Environ. Toxicol. Pharmacol.* **2018**, *58*, 131–140. [CrossRef] [PubMed]
28. Denton, T.T.; Zhang, X.; Cashman, J.R. Nicotine-related alkaloids and metabolites as inhibitors of human cytocrome P-450 2A6. *Biochem. Pharmacol.* **2004**, *67*, 751–756. [CrossRef]
29. Kramlinger, V.M.; Weymarn, L.B.V.; Murphy, S.E. Inhibition and inactivation of cytochrome P450 2A6 and cytochrome P450 2A13 by menthofuran, β-nicotyrine and menthol. *Chem.-Biol. Interact.* **2012**, *197*, 87–92. [CrossRef]
30. Van Vleet, T.R.; Bombick, D.W.; Coulombe Jr, R.A. Inhibition of human cytochrome P450 2E1 by nicotine, cotinine, and aqueous cigarette tar extract in vitro. *Toxicol. Sci.* **2001**, *64*, 185–191. [CrossRef] [PubMed]
31. Neves Cruz, J.; Santana de Oliveira, M.; Gomes Silva, S.; Pedro da Silva Souza Filho, A.; Santiago Pereira, D.; Lima e Lima, A.H.; de Aguiar Andrade, E.H. Insight into the interaction mechanism of nicotine, NNK, and NNN with cytochrome P450 2A13 based on molecular dynamics simulation. *J. Chem. Inf. Model.* **2019**, *60*, 766–776. [CrossRef] [PubMed]
32. Von Weymarn, L.B.; Retzlaff, C.; Murphy, S.E. CYP2A6-and CYP2A13-catalyzed metabolism of the nicotine $\Delta5'(1')$ iminium ion. *J. Pharmacol. Exp. Ther.* **2012**, *343*, 307–315. [CrossRef] [PubMed]
33. Ishikawa, K.; Ishii, H.; Saito, T. DNA damage-dependent cell cycle checkpoints and genomic stability. *DNA Cell Biol.* **2006**, *25*, 406–411. [CrossRef]
34. Tang, J.F.; Li, G.L.; Zhang, T.; Du, Y.M.; Huang, S.Y.; Ran, J.H.; Li, J.; Chen, D.L. Biophysics. Homoharringtonine inhibits melanoma cells proliferation in vitro and vivo by inducing DNA damage, apoptosis, and G2/M cell cycle arrest. *Arch. Biochem. Biophys.* **2021**, *700*, 108774. [CrossRef] [PubMed]
35. Santos, L.S.; Gil, O.M.; Silva, S.N.; Gomes, B.C.; Ferreira, T.C.; Limbert, E.; Rueff, J. Micronuclei Formation upon Radioiodine Therapy for Well-Differentiated Thyroid Cancer: The Influence of DNA Repair Genes Variants. *Genes-Basel* **2020**, *11*, 1083. [CrossRef] [PubMed]
36. Luo, T.; Yu, Q.; Zou, H.; Zhao, H.; Gu, J.; Yuan, Y.; Zhu, J.; Bian, J.; Liu, Z.P. Role of poly (ADP-ribose) polymerase-1 in cadmium-induced cellular DNA damage and cell cycle arrest in rat renal tubular epithelial cell line NRK-52E. *Environ. Pollut.* **2020**, *261*, 114149. [CrossRef] [PubMed]
37. Xiang, C.; Wang, J.; Kou, X.; Chen, X.; Qin, Z.; Jiang, Y.; Sun, C.; Xu, J.; Tan, W.; Jin, L.; et al. Pulmonary expression of CYP2A13 and ABCB1 is regulated by FOXA2, and their genetic interaction is associated with lung cancer. *FASEB J.* **2015**, *29*, 1986–1998. [CrossRef]
38. Shimada, T.; Takenaka, S.; Murayama, N.; Kramlinger, V.M.; Kim, J.H.; Kim, D.; Liu, J.; Foroozesh, M.; Yamazaki, H.; Peter Guengerich, F.; et al. Oxidation of pyrene, 1-hydroxypyrene, 1-nitropyrene and 1-acetylpyrene by human cytochrome P450 2A13. *Xenobiotica* **2016**, *46*, 211–224. [CrossRef] [PubMed]
39. Liang, H.; Ding, X.; Zhou, C.; Zhang, Y.; Xu, M.; Zhang, C.; Xu, L. Knockdown of eukaryotic translation initiation factors 3B (EIF3B) inhibits proliferation and promotes apoptosis in glioblastoma cells. *Neurol. Sci.* **2012**, *33*, 1057–1062. [CrossRef]
40. Zhang, S.; Zhang, J.; Cheng, W.; Chen, H.; Wang, A.; Liu, Y.; Hou, H.; Hu, Q. Combined cell death of co-exposure to aldehyde mixtures on human bronchial epithelial BEAS-2B cells: Molecular insights into the joint action. *Chemosphere* **2020**, *244*, 125482. [CrossRef]

41. Valashedi, M.R.; Roushandeh, A.M.; Tomita, K.; Kuwahara, Y.; Pourmohammadi-Bejarpasi, Z.; Kozani, P.S.; Sato, T.; Roudkenar, M.H. CRISPR/Cas9-mediated knockout of Lcn2 in human breast cancer cell line MDA-MB-231 ameliorates erastin-mediated ferroptosis and increases cisplatin vulnerability. *Life Sci.* **2022**, *304*, 120704. [CrossRef] [PubMed]
42. Che, S.; Yuan, J.; Chen, S.; Li, S.; Zhang, L.; Ruan, Z.; Yao, Y. Hesperidin prevents the combined toxicity of decabromodiphenyl ether and sodium nitrite in vitro. *Food Chem. Toxicol.* **2021**, *156*, 112410. [CrossRef] [PubMed]
43. Wang, H.; Chen, H.; Huang, L.; Li, X.; Wang, L.; Li, S.; Liu, M.; Zhang, M.; Han, S.; Jiang, X.; et al. In vitro toxicological evaluation of a tobacco heating product THP COO and 3R4F research reference cigarette on human lung cancer cells. *Toxicol. Vitr.* **2021**, *74*, 105173. [CrossRef] [PubMed]
44. Bleloch, J.S.; du Toit, A.; Gibhard, L.; Kimani, S.; Ballim, R.D.; Lee, M.; Blanckenberg, A.; Mapolie, S.; Wiesner, L.; Loos, B.; et al. The palladacycle complex AJ-5 induces apoptotic cell death while reducing autophagic flux in rhabdomyosarcoma cells. *Cell Death Discov.* **2019**, *5*, 60. [CrossRef] [PubMed]
45. Zhang, S.; Chen, H.; Zhang, J.; Li, J.; Hou, H.; Hu, Q. The multiplex interactions and molecular mechanism on genotoxicity induced by formaldehyde and acrolein mixtures on human bronchial epithelial BEAS-2B cells. *Environ. Int.* **2020**, *143*, 105943. [CrossRef]
46. Kim, H.R.; Kim, M.J.; Lee, S.Y.; Oh, S.M.; Chung, K.H. Genotoxic effects of silver nanoparticles stimulated by oxidative stress in human normal bronchial epithelial (BEAS-2B) cells. *Mutat. Res. Rev. Mutat. Res.* **2011**, *726*, 129–135. [CrossRef]
47. Feng, J.; Li, J.; Qie, P.; Li, Z.; Xu, Y.; Tian, Z. Long non-coding RNA (lncRNA) PGM5P4-AS1 inhibits lung cancer progression by up-regulating leucine zipper tumor suppressor (LZTS3) through sponging microRNA miR-1275. *Bioengineered* **2021**, *12*, 196–207. [CrossRef]

molecules

MDPI

Article

# Rapid Detection of *Aspergillus flavus* and Quantitative Determination of Aflatoxin $B_1$ in Grain Crops Using a Portable Raman Spectrometer Combined with Colloidal Au Nanoparticles

**Huiqin Wang *, Mengjia Liu, Yumiao Zhang, Huimin Zhao, Wenjing Lu, Taifeng Lin, Ping Zhang and Dawei Zheng**

Faculty of Environment and Life, Beijing University of Technology, Beijing 100124, China
* Correspondence: wanghuiqin@bjut.edu.cn; Tel.: +86-10-67391667

**Abstract:** *Aspergillus flavus* and Aflatoxins in grain crops give rise to a serious threat to food security and cause huge economic losses. In particular, aflatoxin $B_1$ has been identified as a Class I carcinogen to humans by the International Agency for Research on Cancer (IARC). Compared with conventional methods, Surface-Enhanced Raman Scattering (SERS) has paved the way for the detection of *Aspergillus flavus* and Aflatoxins in grain crops as it is a rapid, nondestructive, and sensitive analytical method. In this work, the rapid detection of *Aspergillus flavus* and quantification of Aflatoxin $B_1$ in grain crops were performed by using a portable Raman spectrometer combined with colloidal Au nanoparticles (AuNPs). With the increase of the concentration of *Aspergillus flavus* spore suspension in the range of $10^2$–$10^8$ CFU/mL, the better the combination of *Aspergillus flavus* spores and AuNPs, the better the enhancement effect of AuNPs solution on the *Aspergillus flavus*. A series of different concentrations of aflatoxin $B_1$ methanol solution combined with AuNPs were determined based on SERS and their spectra were similar to that of solid powder. Moreover, the characteristic peak increased gradually with the increase of concentration in the range of 0.0005–0.01 mg/L and the determination limit was 0.0005 mg/L, which was verified by HPLC in ppM concentration. This rapid detection method can greatly shorten the detection time from several hours or even tens of hours to a few minutes, which can help to take effective measures to avoid causing large economic losses.

**Keywords:** *Aspergillus flavus*; aflatoxins $B_1$; carcinogen; Surface-Enhanced Raman Scattering; colloidal Au nanoparticles; portable Raman spectrometer

**Citation:** Wang, H.; Liu, M.; Zhang, Y.; Zhao, H.; Lu, W.; Lin, T.; Zhang, P.; Zheng, D. Rapid Detection of *Aspergillus flavus* and Quantitative Determination of Aflatoxin $B_1$ in Grain Crops Using a Portable Raman Spectrometer Combined with Colloidal Au Nanoparticles. *Molecules* **2022**, *27*, 5280. https://doi.org/10.3390/molecules27165280

Academic Editors: Chongwen Wang and Guohui Sun

Received: 17 July 2022
Accepted: 15 August 2022
Published: 18 August 2022



## 1. Introduction

*Aspergillus flavus*, a common saprophytic mold widely existing all over the world, has been recognized as the main pathogenic fungus causing grain crops mildew in the process of storage. *Aspergillus flavus* and the closely related subspecies *Aspergillus parasiticus* can contaminate grain crops in a wide range [1]. According to FAO reports, the annual losses caused by fungal pollution in the world have reached tens of billions of dollars, and most of them are caused by *Aspergillus flavus* contamination [2]. They can consume a lot of nutrients, accelerate fat deterioration, and destroy protein, pantothenic acid, niacin, vitamin A, vitamin D, vitamin E, and other components, resulting in the nutritional reduction of grain crops. Moreover, 30–60% of them can produce aflatoxins under appropriate conditions [3].

Aflatoxins are cancerous secondary metabolites from *Aspergillus flavus* and *Aspergillus parasiticus* [4]. They are toxic to humans and animals causing liver damage, abnormalities, mutations, and cancer, and when in high doses, aflatoxins can be fatal [5]. Due to their high toxicity and carcinogenic potential, they are a high concern for the safety of food worldwide [6]. At present, more than 20 species of aflatoxins have been found, mainly including $B_1$, $B_2$, $G_1$, $G_2$, $M_1$, $M_2$, etc. Among them, aflatoxin $B_1$ is the most toxic and carcinogenic and has been identified as a Class I carcinogen to humans by the International Agency for Research on Cancer (IARC) [7,8]. Given these adverse effects, regulatory control limitations for aflatoxin $B_1$ in food and feed are well-established. China has strictly set limits of 20 μg/kg of aflatoxin $B_1$ in corn and corn products, according to the national food safety standard (GB 2761-2017) [9].

Several methods and techniques for the detection and quantification of *Aspergillus flavus* and Aflatoxins have been developed. DNA-based techniques have been used for the detection of aflatoxigenic strains of *Aspergillus flavus*. They mainly include the plate count method [10], polymerase chain reaction (PCR) and Quantitative PCR (qPCR) [11–14], enzyme-linked immunosorbent assay (ELISA) [15–18], and so on. High-performance liquid chromatography (HPLC) [19], thin layer chromatography (TLC) [20], enzyme-linked immunosorbent assay (ELISA) [21], and electrochemical impedance spectroscopy (EIS) [22,23] methods have been used for the detection of aflatoxins. In recent years, molecular spectrum and hyperspectral imaging were widely used to detect *Aspergillus flavus* and Aflatoxins [24]. Raman spectroscopy (RS), a modern analytical technique that provides information about molecular vibrations and consequently the structure of the analyzed specimen, has been broadly used in various research fields ranging from the medical biological field [25–28] to food safety [29,30] and electrochemistry [31].

Surface-enhanced Raman scattering (SERS) based on noble metal nanomaterials or rough surface of a metal sheet has attracted increasing attention due to its unique characteristics of high sensitivity and the capability of chemical fingerprint recognition. Caldwell et al. utilized spherical gold nanoparticles with 14 nm and 46 nm diameters to improve the scattering signal obtained during Raman spectroscopy measurements to detect small plastic particles [32]. Bharathi et al. utilized picosecond laser-ablated gold nanoparticles (Au NPs) as surface-enhanced Raman scattering (SERS) substrates to detect the dye methylene blue and a chemical warfare agent simulant (methyl salicylate) [33]. Zavyalova et al. provided a SERSaptasensor based on colloidal solutions, which combines rapidity and specificity in the quantitative determination of the SARS-CoV-2 virus [34]. Compared with the colloidal solutions, there is greater SERS signal stability and a better detection limit may be achieved that allows the detection of low concentrations up to single-molecule level based on some SERS substrates that were prepared by the template-assisted electrodeposition [35], binary-template-assisted electrodeposition [36], pulsed laser ablation [37], and other methods. The surface nanostructures of artificially roughened metal thin films display many hot spots making them excellent SERS substrates [38]. The conventional approaches have several limitations including complicated pretreatment steps, requiring expensive instruments, operational complexity, lack of instrument portability, and difficulties in real-time monitoring [39–41]. Due to the toxicity of aflatoxin and people's attention to food safety, more and more studies on using SERS technology to detect food security have appeared. The determination of aflatoxin $B_1$ in peanut based on QuEChERS purification and surface-enhanced Raman spectroscopy (SERS) was carried out by Wang et al. [42]. Yang et al. used Raman spectroscopy technology to detect zearalenone (ZEN) and aflatoxin $B_1$ in six kinds of maize samples with different mold degrees [43]. Therefore, it is of great significance to develop a fast, solvent-free, and cost-effective analytical method for noninvasive, rapid, and sensitive detection of *Aspergillus flavus* and Aflatoxin $B_1$ in grain crops to prevent potential economic losses.

An important application of SERS in pathogenic microorganisms is to rapidly detect and identify pathogenic bacteria directly isolated from samples without relying on a culture medium, so as to improve efficiency and reduce cost. In this paper, the rapid detection of *Aspergillus flavus* and quantification of Aflatoxin $B_1$ in grain crops using a portable Raman spectrometer-based colloidal Au nanoparticles (AuNPs) will be presented. Detection results of Aflatoxin $B_1$ in grain crops were verified by HPLC in ppM concentration. The characteristics of *Aspergillus flavus* and Aflatoxin $B_1$ will help to identify the degree of contamination by nondestructive testing of grain crops and gain timely control. In addition to the diagnostic application, this method is also potentially helpful for further determining the storage methods of grain crops.

## 2. Results and Discussion

### 2.1. SERS of *Aspergillus flavus* on the Medium and on Corn

A large number of biochemical components on the cell membrane surface of pathogenic microorganisms can be regarded as the characteristic signs of microorganisms. The structure and chemical composition information of the substance can be obtained based on SERS at the single-molecule level. Therefore, SERS has fingerprint recognition characteristics and high detection sensitivity of surface species, which can be used as a sign for the rapid identification and identification of fungi.

When *Aspergillus flavus* was inoculated on the culture medium, the increase of *Aspergillus flavus* on the culture medium could be clearly observed with the naked eyes with the increase of culture days. As shown in Figure 1, the surface color of the medium inoculated with *Aspergillus flavus* spores changed significantly from transparent color to black with the increase in time. The picture of day 0 showed the fresh medium, which was distinct from the following pictures of day 1, day 2, and day 3, respectively. After centrifuging the culture medium (1 mL) at 8000 *g* for 5 min, the *Aspergillus flavus* spores were collected and resuspended in 1 mL of 0.85% sterile normal saline, and then centrifuged and washed under the same conditions. The above procedure was repeated another 2–3 times to remove the culture medium and get the sample, of which 20 μL was taken and added to 500 μL of AuNPs solution and mixed for SERS detection. The SERS of *Aspergillus flavus* had fingerprints at 400–1800 $cm^{-1}$, as shown in Figure 2a, which were mainly reflected in the characteristic peak absorption and spectral shape, especially at the range of 600–800 $cm^{-1}$, 1200–1400 $cm^{-1}$, and 1500–1600 $cm^{-1}$. The peak of the surface-enhanced Raman spectra of *Aspergillus flavus* was analyzed and explained. The peak at 1605–1615 $cm^{-1}$ represents C=O stretching in proteins [44], 1343–1346 $cm^{-1}$ represents DNA base [45], 1315–1317 $cm^{-1}$ represents the vibration of (–C=C–) conjugated of Amine III [26], 1302–1306 $cm^{-1}$ represents carbohydrates [44], and 805–825 $cm^{-1}$ represents protein, respectively [46]. The intensity of the Raman characteristic peak at 1535–1537 $cm^{-1}$ increased obviously with the growth of culture time of *Aspergillus flavus* spore suspension. The experimental results shown from the SERS of *Aspergillus flavus* were consistent with the color changes observed directly on the culture medium. The comparison of non-SERS and SERS of *Aspergillus flavus* was shown in Figure 2b. The results showed that the SERS signals of *Aspergillus flavus* would increase with the coupling with AuNPs, while under non-SERS conditions, the Raman signals of normal *Aspergillus flavus* could hardly be detected by a portable Raman spectrometer.

**Figure 1.** Growth of *Aspergillus flavus* on the medium.

When *Aspergillus flavus* was inoculated on corn grains, the amount of *Aspergillus flavus* gradually increased and the color of corn changed significantly with the increase in time, as shown in Figure 3. The SERS of *Aspergillus flavus* inoculated on corn grains shown in Figure 4 was similar to that inoculated on the medium, and they had the same fingerprints at 400–1800 $cm^{-1}$. Moreover, the intensity of Raman characteristic peak at 1536–1537 $cm^{-1}$ increased obviously with the growth of culture time of Aspergillus flavus on corn in different culture periods. The experimental results shown from the SERS of Aspergillus flavus were consistent with the color changes observed directly on corn.

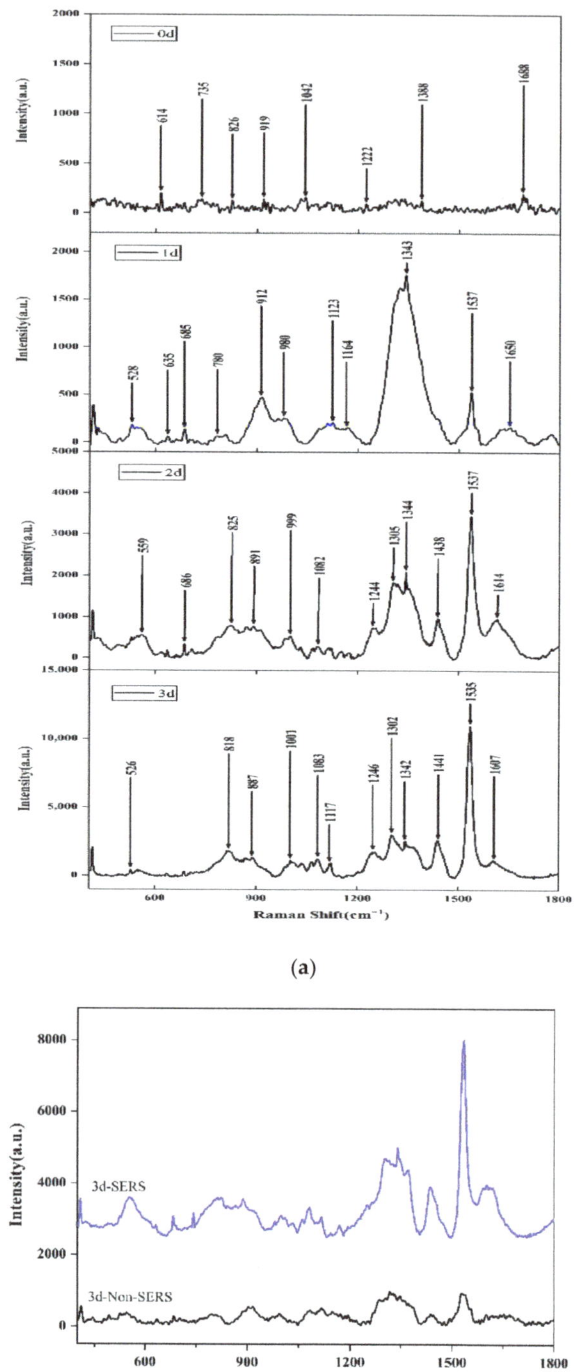


(a)

(b)

**Figure 2.** (**a**) SERS of *Aspergillus flavus* on the medium in different culture periods. (**b**) The comparison of non-SERS and SERS of *Aspergillus flavus*.

**Figure 3.** Growth of *Aspergillus flavus* on Corn.

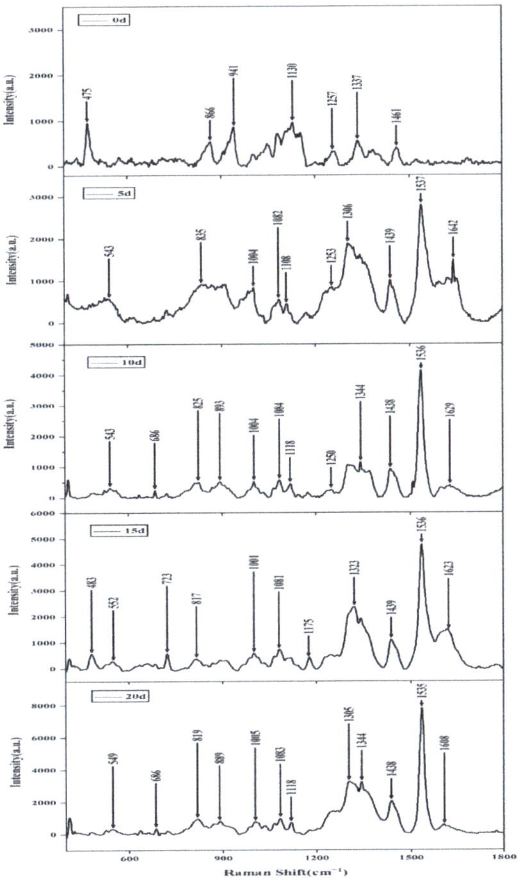


**Figure 4.** SERS of *Aspergillus flavus* on corn in different culture periods.

*2.2. SERS of Aspergillus flavus with Different Concentrations*

Colloidal Au nanoparticles (AuNPs) with a particle size of about 60 nm, prepared according to a previously published procedure [47], served as SERS substrates in this work. *Aspergillus flavus* with different concentrations were added to a certain amount of AuNPs solution and mixed, then different colors of the mixture appeared.

As shown in Figure 5, $10^2$–$10^8$ CFU/mL *Aspergillus flavus* spore suspension and AuNPs solution were mixed, with the concentration increase of *Aspergillus flavus* spore suspension, the mixture color changed from pink to gray gradually. Since the color of AuNPs solution will change after *Aspergillus flavus* is combined with the AuNPs, the greater the color change, the better the combination between the *Aspergillus flavus* and the AuNPs. The result showed that with the increase of the concentration of *Aspergillus flavus* spore suspension, the better the combination of *Aspergillus flavus* spores and AuNPs, the better the enhancement effect of AuNPs solution on the *Aspergillus flavus*.

**Figure 5.** Color change of *Aspergillus flavus* spore suspension mixed with nano gold sol.

A significant challenge for many applications of Raman spectroscopy is that the spectra are often accompanied by a strong fluorescence background, especially for biological samples. This background is generally dominated by intrinsic fluorescence from the sample. There is no doubt that the existence of the resonance effect depends on the wavelength of the excitation laser. If the excited photon cannot provide enough energy for the molecule to be in the excited state, the corresponding fluorescence transition will not occur. However, once the fluorescence is generated, its intensity will be much greater than the Raman scattering light, thus masking the characteristics of the Raman signal. Therefore, choosing laser wavelength is an effective way to avoid fluorescent radiation.

The SERS spectra of *Aspergillus flavus* with different concentrations were determined by using a portable Raman spectrometer with a 785 nm laser under the conditions of 300 mW of laser power, 20 s of integration time, and three integration times. According to flocculation theory and hot-spot effect [48,49], when the AuNPs are close to a certain distance, in the gap between particles a highly enhanced electromagnetic field will be formed, and hot spots with excellent enhancement effects will be formed, resulting in strong Raman enhancement signal with the maximum enhancement factor possibly up to $10^{14}$–$10^{15}$. When the analyte concentration is high, the high concentration ratio AuNPs provides sufficient adsorption sites for the analyte molecules and flocculates to obtain a strong SERS signal. As shown in Figure 6, the average SERS spectra of *Aspergillus flavus* with different concentrations were obtained after baseline correction, normalization and smoothing. The results showed that *Aspergillus flavus* with different concentrations had similar SERS fingerprints at 400–1800 $cm^{-1}$, which were mainly reflected in the characteristic peak absorption and spectral shape, especially at the range of 600–800 $cm^{-1}$, 1200–1400 $cm^{-1}$, and 1500–1600 $cm^{-1}$. Moreover, the intensity of the Raman characteristic peak at 1536 $cm^{-1}$ and in the range of 1200–1400 $cm^{-1}$ increased gradually with the increasing concentration of *Aspergillus flavus* spore suspension. In addition, it also showed that the combination of *Aspergillus flavus* spores and AuNPs was more sufficient with the increase of *Aspergillus flavus* spore concentration and the enhancement effect was better, which was consistent with the results shown in Figure 6. Probably due to the limited hot-spot, the Raman signal does not increase with the concentration of analyte molecules. Even if the concentration ratio of *Aspergillus flavus* is further increased, the AgNPs will not combine with more *Aspergillus flavus*, and the SERS signal will no longer increase. Therefore, $10^2$–$10^8$ CFU/mL *Aspergillus flavus* spore suspension should be selected to enable the AgNPs to absorb the abundant *Aspergillus flavus* to generate the Raman signals for quantitative analysis.

The dynamic three-dimensional Raman spectrum revealed a dynamic result of the determination of *Aspergillus flavus* spore suspension added to the AuNPs solution and combined with the AuNPs, as shown in Figure 7. The results showed that the combination state of *Aspergillus flavus* and AuNPs tended to be stable after *Aspergillus flavus* was added to the

AuNPs solution for about 10 min. Therefore, the determination effect of the SERS spectrum would be better after 10 min of sufficient combination of *Aspergillus flavus* and AuNPs.

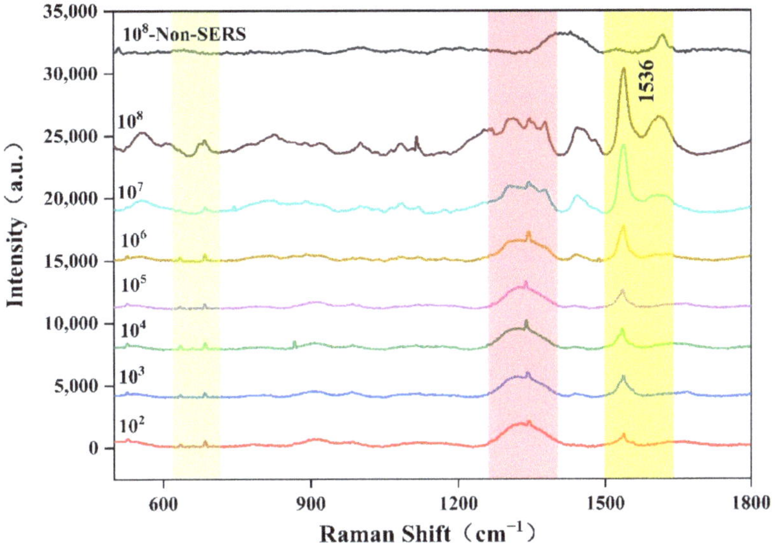


**Figure 6.** SERS spectra of different concentrations of *Aspergillus flavus* spore suspension.

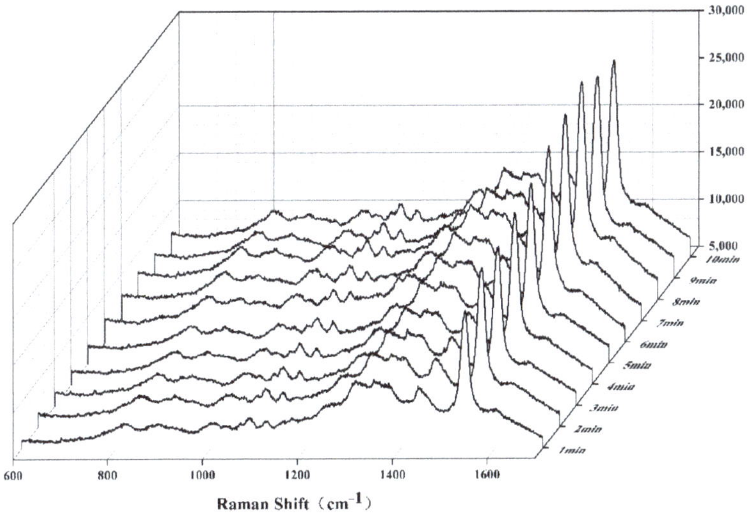


**Figure 7.** SERS spectrum of dynamic determination of *Aspergillus flavus*.

The Raman peak at 1536 $cm^{-1}$ in the SERS diagram of *Aspergillus flavus* was selected as the characteristic peak to carry on the semi-quantitative analysis. According to the signal intensity, the relationship between different concentrations of *Aspergillus flavus* spore suspension and Raman peak intensity at 1536 $cm^{-1}$ was drawn by using the least square fitting method, as shown in Figure 8, indicating that the method could be used for the determination of *Aspergillus flavus*. According to the experimental results, when *Aspergillus flavus* was between $10^2$–$10^5$ cfu/mL, the SERS signal intensity did not change significantly with the increase in concentration. Therefore, it could be considered that $10^2$ cfu/mL was the lowest detectable concentration in this detection. Considering the dilution effect of the

AuNPs solution on the solution of *Aspergillus flavus*, 20 μL of *Aspergillus flavus* was diluted in 500 μL of AuNPs solution. Therefore, the limit of detection (LOD) was 3.85 cfu/mL.

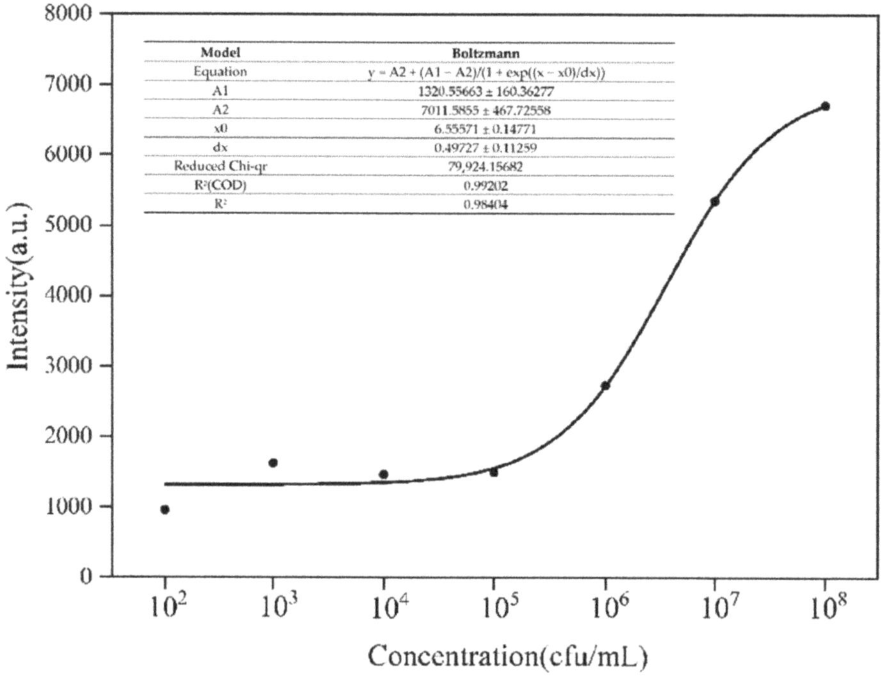


**Figure 8.** Fitting analysis of SERS test results of different concentrations of *Aspergillus flavus* spore Suspension.

## 2.3. SERS of Aflatoxin $B_1$

The Raman spectra of solid aflatoxin $B_1$ were determined so as to reduce the interference of solvents and other factors. It can be seen from Figure 9 that there were many Raman peaks of solid aflatoxin $B_1$, and the most obvious characteristic peaks were produced by inelastic scattering between incident laser and aflatoxin $B_1$. The aflatoxin $B_1$ molecule contains an oxanaphthalene *o*-ketone, a difuran ring, and a cyclopentene ring. The SERS characteristic peaks of aflatoxin $B_1$ are 662, 686, 717, 776, 829, 927, 1007, 1082, 1135, 1247, 1309, 1366, 1481, 1554, 1559, 1628, 1691, and 1760 $cm^{-1}$. Among them, the pyran ring respiratory vibration is at 686 $cm^{-1}$, the C–O–C Tensile vibration is at 1247 $cm^{-1}$, and the C–O–C Tensile vibration is at 1309 $cm^{-1}$; 1554 $cm^{-1}$ is the C–C Tensile vibration, and 1599 $cm^{-1}$ is the C–H plane vibration [50,51].

However, due to the large interference of grain surface and interior in the determination of aflatoxin in the process of practical application, as aflatoxin exists in the interior of grain, a certain pretreatment was needed to extract aflatoxin $B_1$. Therefore, in the later experiments, aflatoxin $B_1$ was dissolved in methanol for SERS determination.

A series of different concentrations of aflatoxin $B_1$ methanol solution (0.01 mg/L, 0.005 mg/L, 0.003 mg/L, 0.001 mg/L, and 0.0005 mg/L) were prepared for SERS detection. The non-SERS and SERS of aflatoxin $B_1$ were shown in Figure 10, The results showed that the SERS signals of aflatoxin $B_1$ would increase with the coupling with AuNPs, while under non-SERS conditions, the Raman signals of normal aflatoxin $B_1$ could hardly be detected by portable Raman spectrometer. However, the SERS of aflatoxin $B_1$ had fingerprints at 400–1800 $cm^{-1}$, which were mainly reflected in the characteristic peak absorption and spectral shape. In addition, the SERS spectra of aflatoxin $B_1$ were similar to that of solid powder by laser confocal micro Raman spectrometer and the characteristic peak increased gradually with the increase of concentration when the standard concentration was in the range of 0.0005–0.01 mg/L. Compared with other peaks, the characteristic peak intensity at 1556 $cm^{-1}$ was more linear with the change of concentrations, so the characteristic peak at 1556 $cm^{-1}$ was selected for the data calculation of detection limit and repeatability. The fitting equation of the curve was shown in Figure 11. Compared with the results of HPLC determination in the published paper carried out by the same research group [52], SERS determination showed the liner range was 0.0005–0.01 mg/L and the limit of detection was 0.0005 mg/L.

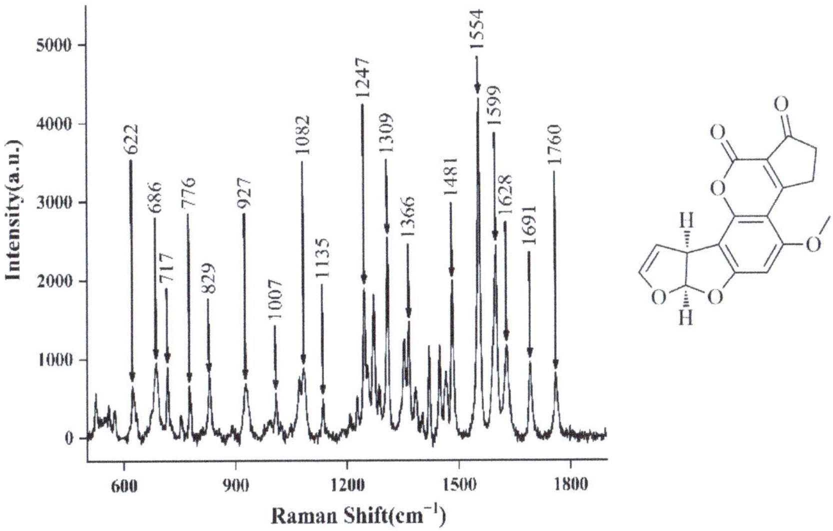


**Figure 9.** Raman spectrum of aflatoxin $B_1$ solid powder determined by laser confocal micro Raman spectrometer.

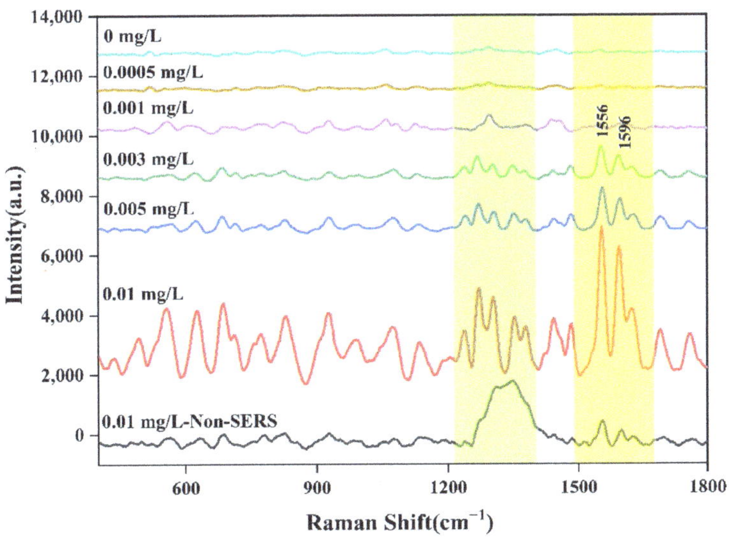


**Figure 10.** SERS spectra of aflatoxin $B_1$ with different concentrations.

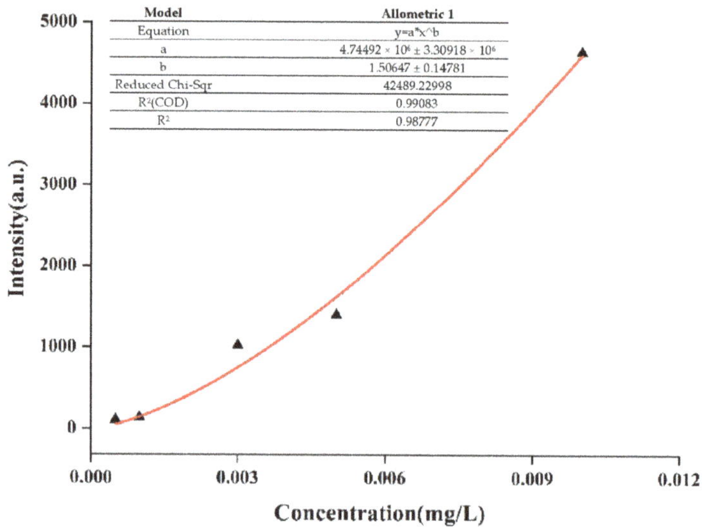


**Figure 11.** Fitting analysis of SERS test results of different concentrations of aflatoxin $B_1$ methanol solution.

## 3. Materials and Methods

### 3.1. Reagents and Materials

Colloidal AuNPs mainly at 60 nm particle size, prepared and provided by our research group according to the primary work published article, were used as the enhanced substrate to magnify the Raman signals [47]. Grain crops (corn) were purchased from the Songyuli farmers' market of Beijing (China). *Aspergillus flavus* was provided by the Yanjing beer company of Beijing (China). Aflatoxin $B_1$ was purchased from Sigma-Aldrich (Shanghai) Trading Co. Ltd. (Shanghai, China).Potato glucose agar medium (PDA) was purchased from Beijing Luqiao Technology Co., Ltd. (Beijing, China). Ultrapure water (18.2 MΩ•cm) was prepared by Millipore (Direct-Q 8 UV-R) and used to prepare all aqueous solutions. The glassware used in the experiment was cleaned with aqua regia ($HCl:HNO_3$ = 3:1, $v/v$), thoroughly rinsed in water, and dried in an oven at 100 °C prior to use.

### 3.2. Culture of Aspergillus flavus

Next, 8.2 g of PDA medium was placed into a 500 mL conical flask with 200 mL of ultrapure water to obtain the PDA medium solution, which was then heated in a water bath until the medium was clear and transparent. The solution was then placed in a sterilization pot at 121 °C for 20 min. Further, 15 mL of sterilized PDA medium solution was transferred to a disposable dish, then *Aspergillus flavus* was inoculated on the medium and cultured in a constant temperature incubator at 30 °C. After culturing in the incubator for 3–5 days, the mature standard strain was removed and placed in an ultra-clean workbench. The spores on the culture medium were washed with sterile water and filtered to obtain a spore suspension, which was inoculated on corns and cultured at 30 °C.

### 3.3. Preparation of Aflatoxin $B_1$ Samples

For the preparation of aflatoxin $B_1$ samples, 1.0 mg of Aflatoxin $B_1$ powder was accurately weighed with an analytical balance and was transferred to a 1000 mL volumetric flask to obtain a 1.0-mg/L Aflatoxin $B_1$ Standard solution after being fixed to the scale with chromatographic methanol. A series of different concentrations of aflatoxin $B_1$ methanol solution (0.01 mg/L, 0.005 mg/L, 0.003 mg/L, 0.001 mg/L, and 0.0005 mg/L) was prepared by diluting 1.0 mg/L aflatoxin $B_1$ solution for SERS determination.

### 3.4. SERS Measurement and Spectra Preprocessing

The SERS measurement was conducted by using a portable Raman Spectrometer (RamTracer–200, Suzhou OptoTrace Technologies Co. Ltd., Suzhou, China) equipped with a diode laser with a wavelength of 785 nm and power of 300 mW. The wave number range was 250–1800 $cm^{-1}$, the resolution was 2 $cm^{-1}$, and the laser bandwidth was 0.2 nm. The laser power was 300 mW with an integration time of 20 s and was integrated three times.

Before each measurement, X-axis calibration was carried out with acetonitrile (Spectrally Pure) to ensure that the experimental conditions were consistent. Then, 500 μL colloidal AuNPs solution was transferred with a micropipette and placed into a 2.0 mL glass bottle, and 20 μL of *Aspergillus flavus* solution was added and mixed with a vibrator for Raman spectrum measurement immediately. The original spectra were sequentially pre-processed by removing cosmic rays and by baseline correction using Wire 4.1 software (in Via, Renishaw, Gloucestershire, London UK). All of the figures were plotted with Origin software (version 8.0, OriginLab, Northampton, MA, USA).

## 4. Conclusions

In this work, the rapid detection of *Aspergillus flavus* and quantification of Aflatoxin $B_1$ in grain crops were presented by using a portable Raman spectrometer combined with AuNPs. The SERS of *Aspergillus flavus* had fingerprints at 400–1800 $cm^{-1}$, which were mainly reflected in the characteristic peak absorption and spectral shape, especially at the range of 600–800 $cm^{-1}$, 1200–1400 $cm^{-1}$, and 1500–1600 $cm^{-1}$. However, the intensity of the Raman characteristic peak at 1536 $cm^{-1}$ increased obviously with the growth of culture time of *Aspergillus flavus* spore suspension. When *Aspergillus flavus* was inoculated on corn grains, the SERS of *Aspergillus flavus* was similar to that inoculated on the medium. With the increase of the concentration of *Aspergillus flavus* spore suspension in the range of $10^2$–$10^8$ CFU/mL, the better the combination of *Aspergillus flavus* spores and AuNPs, the better the enhancement effect of AuNPs solution on the *Aspergillus flavus*. A series of different concentrations of aflatoxin $B_1$ methanol solution combined with AuNPs were determined based on SERS and their spectra were similar to that of solid powder. Moreover, the characteristic peak increased gradually with the increase of concentration in the range of 0.0005–0.01 mg/L.

The results showed that rapid detection of *Aspergillus flavus* and quantitative determination of aflatoxin $B_1$ by using a portable Raman spectrometer combined with colloidal Au nanoparticles based on SERS was reliable and could be used for the assessment of *Aspergillus flavus* and Aflatoxin $B_1$ contaminants in grain crops during storage conditions. Compared with the conventional methods, this rapid detection method can greatly shorten the detection time from several hours or even tens of hours to a few minutes, which is very important in the detection of grain crops. Once the grain crops are contaminated by *Aspergillus flavus* and aflatoxin $B_1$, we can quickly take effective measures according to the detection results to avoid causing large economic losses.

**Author Contributions:** Conceptualization, H.W.; methodology, H.W., Y.Z. and D.Z.; software, M.L. and W.L.; validation, M.L. and H.Z.; data curation, M.L.; writing—original draft preparation, H.W.; writing—review and editing, H.W. and M.L.; supervision, P.Z. and T.L. All authors have read and agreed to the published version of the manuscript.

**Funding:** This research received no external funding.

**Institutional Review Board Statement:** Not applicable.

**Informed Consent Statement:** Not applicable.

**Data Availability Statement:** All data are contained within the article.

**Acknowledgments:** The authors all thank the Yanjing beer company of Beijing for providing *Aspergillus flavus*.

**Conflicts of Interest:** The authors declare no conflict of interest.

**Sample Availability:** Samples of the compounds are available from the authors.

## References


1. Perrone, G.; Susca, A.; Cozzi, G.; Ehrlich, K.; Varga, J.; Frisvad, J.C.; Meijer, M.; Noonim, P.; Mahakarnchanakul, W.; Samson, R.A. Biodiversity of Aspergillus species in some important agricultural products. *Stud. Mycol.* **2007**, *59*, 53–66. [CrossRef] [PubMed]
2. Lai, X.W.; Zhang, H.; Liu, R.C.; Liu, C.L. Potential for aflatoxin $B_1$ and $B_2$ production by *Aspergillus flavus* strains isolated from rice samples. *Saudi J. Biol. Sci.* **2015**, *22*, 176–180. [CrossRef]
3. Kosegarten, C.E.; Ramírez-Corona, N.; Mani-López, E.; Palou, E.; López-Malo, A. Description of *Aspergillus flavus* growth under the influence of different factors (water activity, incubation temperature, protein and fat concentration, pH, and cinnamon essential oil concentration) by kinetic, probability of growth, and time-to-detection models. *Int. J. Food Microbiol.* **2017**, *240*, 115–123. [PubMed]
4. Nakai, V.K.; Rocha, L.O.; Goncalez, E.; Fonseca, H.; Ortega, E.M.M.; Correa, B. Distribution of fungi and aflatoxins in a stored peanut variety. *Food Chem.* **2008**, *106*, 285–290. [CrossRef]
5. Bbosa, G.S.; Kitya, D.; Lubega, A.; Ogwal-Okeng, J.; Anokbonggo, W.W.; Kyegombe, D.B. Review of the Biological and Health Efects of Aflatoxins on Body Organs and Body Systems. In *Aflatoxins-Recent Advances and Future Prospects*, 1st ed.; Razzaghi-Abyaneh, M., Ed.; InTech: Rijeka, Croatia, 2013; pp. 239–265.
6. Nguyen, T.B.; Vu, T.B.; Pham, H.M.; Tran, C.S.; Le, H.H.; Vo, N.T. Detection of Aflatoxins $B_1$ in Maize Grains Using Fluorescence Resonance Energy Transfer. *Appl. Sci.* **2020**, *10*, 1578–1586. [CrossRef]
7. IARC. *IARC Monographs on the Evaluation of Carcinogenic Risks to Humans: Some Naturally Occurring Substances: Food Items and Constituents, Heterocyclic Aromatic Amines and Mycotoxins*; IARC Monographs: Lyon, France, 1993; pp. 359–362.
8. Piermarini, S.; Micheli, L.; Ammida, N.H.S.; Palleschi, G.; Moscone, D. Electrochemical immunosensor array using a 96-well screen-printed microplate for Aflatoxin $B_1$ detection. *Biosens. Bioelectron.* **2007**, *22*, 1434–1440. [CrossRef]
9. National Health Commission of the People's Republic of China, State Administration for Market Regulation. *GB 2761-2017[S]*; National Food Safety Standard Limit of Mycotoxins in Food. Quality and Standards Publishing & Media Co., Ltd.: Beijing, China, 2017.
10. Tournas, V.H.; Niazi, N.S. Potentially toxigenic fungi from selected grains and grain products. *J. Food Saf.* **2018**, *38*, 12422–12427. [CrossRef]
11. Aboutalebin, S.; Ahmadikia, K.; Fakhim, H.; Chabavizadeh, J.; Okhovat, A.; Nikaeen, M.; Mirhendi, H. Direct detection and identification of the most common bacteria and fungi causing otitis externa by a stepwise multiplex PCR. *Front. Cell. Infect. Microbiol.* **2021**, *11*, 1–11. [CrossRef]
12. Feng, Y.Y.; Yang, A.F.; Zheng, Q.Y.; Wan, C.; Cao, J.J. Establishment of a real-time PCR method for detection of main fungus in grain. *Cereal Feed Ind.* **2020**, *4*, 1–3.
13. Priyanka, S.R.; Venkatramana, M.; Balakrishna, K.; Murali, H.S.; Batra, H.V. Development and evaluation of a multiplex PCR assay for simultaneous detection of major mycotoxigenic fungi from cereals. *J. Food Sci. Tech.* **2015**, *52*, 486–492. [CrossRef]
14. Sardiñas, N.; Vázquez, C.; Gil-Serna, J.; González-Jaén, T.; Patiño, B. Specific detection and quantification of *Aspergillus flavus* and *Aspergillus parasiticus* inwheat flour by SYBR®Green quantitative PCR. *Int. J. Food Microbiol.* **2011**, *145*, 121–125. [CrossRef] [PubMed]
15. Dong, G.; Pan, Y.; Wang, Y.; Ahmed, S.; Liu, Z.; Peng, D.; Yuan, Z. Preparation of a broad-spectrum anti-zearalenone and its primary analogues antibody and its application in an indirect competitive enzyme-linked immunosorbent assay. *Food Chem.* **2018**, *247*, 8–15. [CrossRef] [PubMed]
16. Gabriel, M.F.; Uriel, N.; Teifoori, F.; Postigo, I.; Suñén, E.; Martínez, J. The major Alternaria alternata allergen, Alt a 1: A reliable and specific marker of fungal contamination in citrus fruits. *Int. J. Food Microbiol.* **2017**, *257*, 26–30. [CrossRef]
17. Henri, A.; Antti, T.; Alexis, N.; Tarja, N. A simple and specific noncompetitive ELISA method for HT-2 toxin detection. *Toxins* **2017**, *9*, 145–154.
18. Jayaratne, W.M.S.C.; Abeyratne, A.H.M.A.K.; Zoysa, H.K.S.; Dissanayake, D.M.R.B.N.; Bamunuarachchige, T.C.; Waisundara, V.Y.; Chang, S. Detection and quantification of Aflatoxin $B_1$ in corn and corn-grown soils in the district of Anuradhapura, Sri Lanka. *Heliyon* **2020**, *6*, 5319–5324. [CrossRef]
19. Hu, Y.Y.; Zheng, P.; Zhang, Z.X.; He, Y.Z. Determination of aflatoxins in high-pigment content samples by matrix solid-phase dispersion and high-performance liquid chromatography. *J. Agric. Food Chem.* **2006**, *54*, 4126–4130. [CrossRef]
20. Lin, L.; Zhang, J.; Wang, P.; Wang, Y.; Chen, J. Thin-layer chromatography of mycotoxins and comparison with other chromatographic methods. *J. Chromatogr. A* **1998**, *815*, 3–20. [CrossRef]
21. Rossi, C.N.; Takabayashi, C.R.; Ono, M.A.; Saito, G.H.; Itano, E.N.; Kawamura, O.K.; Hirooka, E.Y.; Ono, E.Y.S. Immunoassay based on monoclonal antibody for aflatoxin detection in poultry feed. *Food Chem.* **2012**, *132*, 2211–2216. [CrossRef]
22. Vig, A.; Radoi, A.; Munoz-Berbel, X.; Gyemant, G.; Marty, J.L. Impedimetric aflatoxin $M_1$ immunosensor based on colloidal gold and silver electrodeposition. *Sens. Actuators B Chem.* **2009**, *138*, 214–220. [CrossRef]
23. Yu, L.L.; Zhang, Y.; Hu, C.Y.; Wu, H.; Yang, Y.Y.; Huang, C.S.; Jia, N.Q. Highly sensitive electrochemical impedance spectroscopy immunosensor for the detection of $AFB_1$ in olive oil. *Food Chem.* **2015**, *176*, 22–26. [CrossRef]
24. Chu, X.; Wang, W.; Ni, X.Z.; Zheng, H.T.; Zhao, X.; Zhang, R.; Li, Y.F. Growth Identification of *Aspergillus flavus* and *Aspergillus parasiticus* by Visible/Near-Infrared Hyperspectral Imaging. *Appl. Sci.* **2018**, *8*, 513–524. [CrossRef]

25. Zhang, P.; Wang, L.M.; Fang, Y.P.; Zheng, D.W.; Lin, T.F.; Wang, H.Q. Label-free exosomal detection and classification in rapid discriminating different cancer types based on specific Raman phenotypes and multivariate statistical analysis. *Molecules* **2019**, *24*, 2947–2961. [CrossRef]
26. Zheng, D.W.; Liu, X.Y.; Zhang, P.; Su, L.; Wang, L.M.; Wei, X.D.; Wang, H.Q.; Lin, T.F. Rapid identification of mixed enteropathogenic bacteria by means of Au nanoparticles@bacteria using portable Raman spectrometer. *J. Nanosci. Nanotechnol.* **2018**, *18*, 6776–6785. [CrossRef]
27. Fang, Y.P.; Lin, T.F.; Zheng, D.W.; Zhu, Y.W.; Wang, L.M.; Fu, Y.Y.; Wang, H.Q.; Wu, X.H.; Zhang, P. Rapid and label-free identification of different cancer types based on surface-enhanced Raman scattering profiles and multivariate statistical analysis. *J. Cell. Biochem.* **2020**, *122*, 277–289. [CrossRef]
28. Cui, X.Y.; Liu, T.; Xu, X.S.; Zhao, Z.Y.; Tian, Y.; Zhao, Y.; Chen, S.; Wang, Z.; Wang, Y.D.; Hu, D.Y.; et al. Label-free detection of multiple genitourinary cancers from urine by surface-enhanced Raman spectroscopy. *Spectrochim. Acta A Mol. Biomol. Spectrosc.* **2020**, *240*, 118543–118549. [CrossRef] [PubMed]
29. Wang, T.; Liu, M.H.; Huang, S.G.; Yuan, H.C.; Zhao, J.H.; Chen, J. Surface-enhanced Raman spectroscopy method for classification of doxycycline hydrochloride and tylosin in duck meat using gold nanoparticles. *Poult. Sci.* **2021**, *100*, 101165–101173. [CrossRef] [PubMed]
30. Hua, B.X.; Sun, D.W.; Pu, H.B.; Wei, Q.Y. Rapid nondestructive detection of mixed pesticides residues on fruit surface using SERS combined with self-modeling mixture analysis method. *Talanta* **2020**, *217*, 120998–121009. [CrossRef]
31. Audilakshmi, S.; Stenhouse, J.W.; Reddy, T.P. Genetic analysis of grain mold resistance in white seed sorghum genotypes. *Euphytica* **2005**, *145*, 95–101. [CrossRef]
32. Caldwell, J.; Taladriz-Blanco, P.; Rothen-Rutishauser, B.; Petri-Fink, A. Detection of Sub-Micro- and Nanoplastic Particles on Gold Nanoparticle-Based Substrates through Surface-Enhanced Raman Scattering (SERS) Spectroscopy. *Nanomaterials* **2021**, *11*, 1149–1166. [CrossRef]
33. Bharathi, M.S.S.; Byram, C.; Banerjee, D.; Sarma, D.; Barkakaty, B.; Soma, V.R. Gold nanoparticle nanofibres as SERS substrate for detection of methylene blue and a chemical warfare simulant (methyl salicylate). *Bull. Mater. Sci.* **2021**, *44*, 103–110. [CrossRef]
34. Zavyalova, E.; Ambartsumyan, O.; Zhdanov, G.; Gribanyov, D.; Gushchin, V.; Tkachuk, A.; Rudakova, E.; Nikiforova, M.; Kuznetsova, N.; Popova, L.; et al. SERS-Based Aptasensor for Rapid Quantitative Detection of SARS-CoV-2. *Nanomaterials* **2021**, *11*, 1394–1406. [CrossRef] [PubMed]
35. Kozhina, E.P.; Bedin, S.A.; Nechaeva, N.L.; Podoynitsyn, S.N.; Tarakanov, V.P.; Andreev, S.N.; Grigoriev, Y.V.; Naumov, A.V. Ag-Nanowire Bundles with Gap Hot Spots Synthesized in Track-Etched Membranes as Effective SERS-Substrates. *Appl. Sci.* **2021**, *11*, 1375–1387. [CrossRef]
36. Zhu, C.H.; Meng, G.W.; Zheng, P.; Huang, Q.; Li, Z.B.; Hu, X.Y.; Wang, X.J.; Huang, Z.L.; Li, F.D.; Wu, N.Q. A Hierarchically Ordered Array of Silver-Nanorod Bundles for Surface-Enhanced Raman Scattering Detection of Phenolic Pollutants. *Adv. Mater.* **2016**, *28*, 4871–4876. [CrossRef] [PubMed]
37. Tommasini, M.; Zanchi, C.; Lucotti, A.; Bombelli, A.; Villa, N.S.; Casazza, M.; Ciusani, E.; Grazia, U.; Santoro, M.; Fazio, E.; et al. Laser-Synthesized SERS Substrates as Sensors toward Therapeutic Drug Monitoring. *Nanomaterials* **2019**, *9*, 677–691. [CrossRef]
38. Halas, N.J.; Moskovits, M. Surface-enhanced Raman spectroscopy: Substrates and materials for research and applications. *MRS Bull.* **2013**, *38*, 607–611. [CrossRef]
39. Alsammarraie, F.K.; Lin, M.S.; Mustapha, A.; Lin, H.T.; Chen, X.; Chen, Y.H.; Wang, H.; Huang, M.Z. Rapid determination of thiabendazole in juice by SERS coupled with novel gold nanosubstrates. *Food Chem.* **2018**, *259*, 219–225. [CrossRef]
40. Wang, K.Q.; Sun, D.W.; Pu, H.B.; Wei, Q.Y. Shell thickness-dependent Au@Ag nanoparticles aggregates for high-performance SERS applications. *Talanta* **2019**, *195*, 506–515. [CrossRef]
41. Hu, B.X.; Sun, D.W.; Pu, H.B.; Wei, Q.Y. Recent advances in detecting and regulating ethylene concentrations for shelf-life extension and maturity control of fruit: A review. *Trends Food Sci. Technol.* **2019**, *91*, 66–82. [CrossRef]
42. Wang, Y.L.; Zhai, W.L.; Han, C.R.; Wang, H.; Zhao, Z.L.; Wang, M. Determination of aflatoxin $B_1$ in peanut by QuEChERS-surface-enhanced Raman spectroscopy. *J. Food Saf. Qual.* **2022**, *13*, 1502–1508.
43. Yang, X.Q.; Yu, H.C.; Yin, Y.; Yuan, Y.X.; Wu, H.; Li, X. Detection of Aflatoxin $B_1$ and Zearalenone in maize by Raman Spectroscopy. *J. Nucl. Agric. Sci.* **2021**, *35*, 159–166.
44. Egging, V.; Nguyen, J.; Kurouski, D. Detection and Identification of Fungal Infections in Intact Wheat and Sorghum Grain Using a Hand-Held Raman Spectrometer. *Anal. Chem.* **2018**, *90*, 8616–8621. [CrossRef] [PubMed]
45. Shetty, G.; Kendall, C.; Shepherd, N.; Stone, N.; Barr, H. Raman spectroscopy: Elucidation of biochemical changes in carcinogenesis of oesophagus. *Br. J. Cancer* **2006**, *94*, 1460–1464. [CrossRef]
46. Bittel, M.; Assaf, A.; Cordella, C.B.Y.; Durand, M.J.; Thouand, G. Potential of Raman Spectroscopy to Monitor Arsenic Toxicity on Bacteria: Insights toward Multiparametric Bioassays. *Environ. Sci. Technol.* **2015**, *49*, 12324–12332. [CrossRef] [PubMed]
47. Wei, X.D.; Zheng, D.W.; Zhang, P.; Lin, T.F.; Wang, H.Q.; Zhu, Y.W. Surface enhanced Raman scattering investigation of bovine albumin with different size of Au nanoparticles. *J. Appl. Biomater. Func.* **2018**, *16*, 157–162.
48. Shi, Y.Y.; Liu, W.F.; Chen, C.P. Two-step centrifugation method for subpicomolar surface-enhanced Raman scattering detection. *Anal. Chem.* **2016**, *88*, 5009–5015. [CrossRef] [PubMed]
49. Futamata, M.; Ishikura, M.; Iida, C.; Handa, S. The critical importance of gap modes in surface enhanced Raman scattering. *Faraday Discuss.* **2015**, *178*, 203–220. [CrossRef]

50. Wu, X.; Gao, S.; Wang, J.S.; Wang, H.; Huang, Y.W.; Zhao, Y. The surface-enhanced Raman spectra of aflatoxins: Spectral analysis, density functional theory calculation. *Detect Differ. Anal.* **2012**, *137*, 4226–4234. [CrossRef]
51. Gao, S.M.; Wang, H.Y.; Lin, Y.X.; Li, R.H. Surface-enhanced Raman spectra of aflatoxin $B_1$ adsorbed on silver clusters. *Acta Phys. -Chim. Sin.* **2012**, *28*, 2044–2050.
52. Liu, M.J.; Ren, X.F.; Zheng, D.W.; Lin, T.F.; Zhang, P.; Wang, H.Q. Determination of aflatoxin $B_1$ in corn or corn products by post-column photochemical derivatization. *China Feed* **2022**, *9*, 92–95.

molecules

MDPI

Article

# BDE-47 Induces Immunotoxicity in RAW264.7 Macrophages through the Reactive Oxygen Species-Mediated Mitochondrial Apoptotic Pathway

Qian Gao [1,2,†], Zhong-Yuan Zhou [3,†], Ya-Ning He [1,2], Ming-Hui Dong [1,2], Zhao-Ning Wang [3] and Hong-Mei Chen [1,2,*]

1 Key Laboratory of Xinjiang Phytomedicine Resources and Utilization, Ministry of Education, School of Pharmacy, Shihezi University, Shihezi 832002, China
2 Pharmacology Department, School of Pharmacy, Shihezi University, Shihezi 832002, China
3 Department of Marine Ecology, College of Marine Life Science, Ocean University of China, Qingdao 266003, China
* Correspondence: hongmeichen2017@shzu.edu.cn
† These authors contributed equally to this work.

**Abstract:** Polybrominated diphenyl ethers (PBDEs) are classic and emerging pollutants that are potentially harmful to the human immune system. Research on their immunotoxicity and mechanisms suggests that they play an important role in the resulting pernicious effects of PBDEs. 2,2′,4,4′-Tetrabrominated biphenyl ether (BDE-47) is the most biotoxic PBDE congener, and, in this study, we evaluated its toxicity toward RAW264.7 cells of mouse macrophages. The results show that exposure to BDE-47 led to a significant decrease in cell viability and a prominent increase in apoptosis. A decrease in mitochondrial membrane potential (MMP) and an increase in cytochrome C release and caspase cascade activation thus demonstrate that cell apoptosis induced by BDE-47 occurs via the mitochondrial pathway. In addition, BDE-47 inhibits phagocytosis in RAW264.7 cells, changes the related immune factor index, and causes immune function damage. Furthermore, we discovered a significant increase in the level of cellular reactive oxygen species (ROS), and the regulation of genes linked to oxidative stress was also demonstrated using transcriptome sequencing. The degree of apoptosis and immune function impairment caused by BDE-47 could be reversed after treatment with the antioxidant NAC and, conversely, exacerbated by treatment with the ROS-inducer BSO. These findings indicate that oxidative damage caused by BDE-47 is a critical event that leads to mitochondrial apoptosis in RAW264.7 macrophages, ultimately resulting in the suppression of immune function.

**Keywords:** PBDE; immunotoxicity; macrophage; reactive oxygen species; apoptosis; immune function

**Citation:** Gao, Q.; Zhou, Z.-Y.; He, Y.-N.; Dong, M.-H.; Wang, Z.-N.; Chen, H.-M. BDE-47 Induces Immunotoxicity in RAW264.7 Macrophages through the Reactive Oxygen Species-Mediated Mitochondrial Apoptotic Pathway. *Molecules* **2023**, *28*, 2036. https://doi.org/10.3390/molecules28052036

Academic Editor: Giovanni Li Volti

Received: 16 January 2023
Revised: 17 February 2023
Accepted: 20 February 2023
Published: 21 February 2023



## 1. Introduction

As a kind of flame retardant, polybrominated diphenyl ethers (PBDEs) are widely utilized in both residential and commercial items [1,2]. They are highly fat soluble and difficult to break down. They can be transported in diverse environmental media following discharge into the environment [3] and reach the human body following the bioconcentration and amplification effect of the food chain [4]. In 2004, analysts noted that the total amount of PBDEs in human blood, milk, and tissue had increased by nearly 100 times in the past three decades [5]. Due to their toxicity and persistence, some PBDEs were restricted by the Stockholm Convention in 2009 and 2017 [6], although the use of PBDEs has been considerably reduced in recent years. However, affected by electronics recycling and time lag effects, PBDEs in the environment will continue to be detected in the coming decades [7,8]. From 2000 to 2019, a meta-analysis of the global distribution and trend of PBDEs in human blood and milk indicated no significant declining trend [9]. It is nevertheless critical to pay attention to the research on the toxicity of PBDEs and other chemicals, particularly 2,2′,4,4′-tetrabrominated biphenyl ether (BDE-47) and 2,2′,4,4′,5-pentabromodiphenyl ether (BDE-99), which are relatively more hazardous to human health and the environment [10].

Many studies have recently revealed the systemic toxic effect of PBDEs on reproduction [11], nerves [12,13], and development [14,15]. PBDEs also damage target organs such as the liver [16,17] and kidneys [18]. The immune system is an important defense system of biological organisms, and functional changes in this system can serve as a sensitive index for evaluating the safety of exogenous chemicals. Although data on PBDEs immunotoxicity are limited, some studies on rodents have shown that long-term exposure to PBDEs damages components of the immune systems of mice and rats [19–22]. In studies of other animals, young salmon exposed to PBDEs (270 ng/g) in the lower sections of the Willamette River demonstrated higher susceptibility to illness, which may be connected to immune function impairment [23]. Environmentally realistic concentrations (0−634 ng/g wet diet) of PBDEs are able to alter the immune function of frogs [24], and similar results were presented for in vitro toxicity research. Exposure to some brominated flame retardants (DE-71, DE-79, etc.) in a concentration range of 0.01–10 mg/mL for 24 h may stimulate primary splenocytes by boosting antigen-presenting-related molecule expression and IL-4 production [25]. Finally, concentrations above 0.01 μg/mL of DE-71 can enhance the production of normal human PBMC cytokines stimulated by LPS or PHA-L in vitro [26]. Exposure to BDE-47 at concentrations above 5 μM or BDE-209 at concentrations above 20 μM for 24 h can increase intracellular reactive oxygen species (ROS) formation, activate apoptosis pathways, and damage immune function, such as cell phagocytosis, in peritoneal macrophages [27]. PBDEs have features of low bioaccessibility [28], easy compound adsorption onto plastic [29], and binding to serum proteins [30], meaning that not all PBDEs can enter the cells [31]. Thus, the concentration of PBDEs examined in experiments is higher than the environmental concentration. In addition, the effect of PBDEs is influenced by the concentration of exposure, wherein higher concentrations that exceed the defense ability can cause toxic damage. Depending on the concentration, PBDEs can either stimulate or inhibit the autoimmune response of various immune cells to some extent. Different immune response indicators and response trends also complicate the investigation of immunotoxicity mechanisms and the development of effective defense pathways.

Research on immunotoxicity mechanisms currently focuses on mitochondrial apoptosis and the control of ROS. Apoptosis mediated by the mitochondrial pathway is frequently triggered by ROS, which causes structural changes, rupture, and fragmentation of the mitochondrial membrane system, with subsequent programmed cell death [32]. ROS and nitric oxide (NO) generated by cell oxidative stress can disrupt redox equilibrium, thus resulting in damage to proteins, DNA, and membranes, and finally promoting cell death [33]. Ivermectin (IVM) has previously been demonstrated to induce the overproduction of macrophage ROS, cytochrome C release, and damage to double-stranded DNA, indicating that it might cause immunotoxicity via immunological malfunction and cytotoxicity [34]. Exposure to microplastics and two persistent organic contaminants can considerably affect immune responses in clams and dramatically increase their intracellular ROS levels [35]. The immune response induced by the fungicides azoxystrobin (AZO) and iprodione (IPR) can be partially reversed by the antioxidant N-acetylcysteine (NAC) [36]. Similar findings have been shown in related studies of PBDE immunotoxicity in recent years. When seals were exposed to PBDE homologs at a higher concentration, the ROS levels and phagocytosis were affected [37].

Macrophages are a type of immune cell that is used to study cell phagocytosis, cellular immunity, and molecular immunology. They are essential for the host to defend itself against foreign substances. On the one hand, macrophages can induce inflammation via the use of reactive nitrogen (RNS) and ROS receptors to mediate phagocytosis and particle clearance, activating intracellular inflammatory bodies, caspase processes, and the release of inflammatory mediators [38]. On the other hand, they present antigens by expressing co-stimulatory molecules and specific cytokines, thereby initiating and controlling specific

T cell responses. As a result, macrophages are critical for both the innate and adaptive immune systems [39]. The RAW264.7 cell line is an immortal mouse macrophage line that is frequently used as an in vitro model to study the immunotoxicity of environmental pollutants. RAW264.7 cells were chosen to investigate the immunotoxicity and potential mechanisms of BDE-47 in this study, with a focus on the mitochondrial apoptotic pathway and phagocytosis, antigen presentation, inflammatory response, phenotypic activation, and other immune responses. Furthermore, the antioxidant N-acetyl-L-cysteine (NAC) and ROS-inducer L-buthionine-(S,R)-sulfoximine(BSO) were applied to study the correlation between immunotoxicity and intracellular ROS levels induced by BDE-47 in RAW264.7 cells.

## 2. Results

### 2.1. Effects of BDE-47 on Cell Viability and Morphology

The results of an MTT assay showed that, with an increase in BDE-47 concentration, the proliferation viability of RAW264.7 cells presented a trend of decreasing with increasing time and concentration (Figure 1A). This difference was significant at concentrations above 10 μM at 24 h ($p < 0.05$), and, when beyond 48 h, the results exhibited a very significant difference for concentrations above 5 μM ($p < 0.01$). The rate of cell lethality also exhibited time and concentration-dependent increases (Figure 1B). The cell lethality rate at 24 h markedly increased in the groups treated with concentrations above 10 μM ($p < 0.01$). The obtained results are consistent with the findings of the MTT assay.

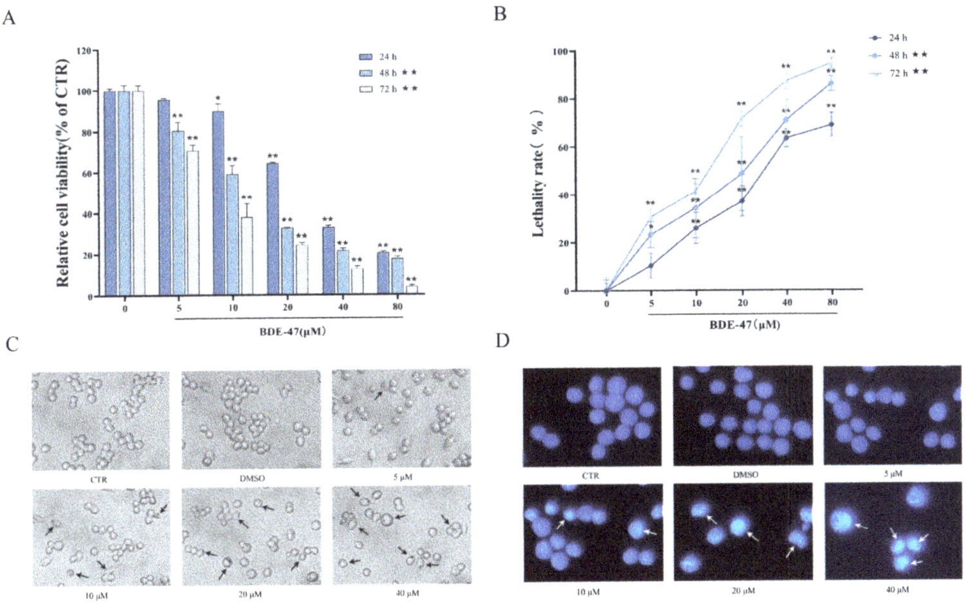


**Figure 1.** Effect of BDE-47 on proliferative viability and morphology in RAW264.7 cells. (**A**) Cell proliferation viability characterized by MTT assay ($n = 6$). (**B**) Cell mortality determined using trypan blue staining ($n = 6$). (**C**) Cells photographed under light microscopy (200×). Arrows indicate cells exhibiting cell shrinkage and cytoplasmic membrane blebbing. (**D**) Cells stained with DAPI and photographed under inverted fluorescence microscopy (400×). Arrows indicate condensed nuclei. * Significant difference compared with the control (* $p < 0.05$; ** $p < 0.01$). ★ Significant difference compared with BDE-47-treated 6 h group (★★ $p < 0.01$).

The number of cells decreased with the increase in the concentration of BDE-47 used in the treatment for 24 h (Figure 1C). For normal RAW264.7 cells, the edges are clear, the refractive index of the cells is higher, and a small number of cells become spindle-shaped in the DMSO solvent control and 5 μM BDE-47 groups. At concentrations greater than 10 μM, the number of cells is significantly reduced, and the cell edges appear blurry. When the concentration of BDE-47 is 40 μM, the microscopic cells become blurred and black, and a large number of cells are broken. When analyzed by DAPI nuclear staining, treatment with BDE-47 results in the nuclei becoming brighter, solidified, and concentrated, and partially fragmented forms appeared compared with the CTR group (Figure 1D).

### *2.2. BDE-47-Induced Apoptosis in RAW264.7 Cells Is Mediated by the Mitochondrial Pathway*

The BDE-47-induced RAW264.7 cell apoptosis in a concentration-dependent manner. The cell apoptotic rate increased steadily (from 11.34 ± 0.12% to 24.26 ± 0.17%) after 24 h exposure, and the difference between the treated groups and the control group was statistically significant ($p < 0.05$) (Figure 2A,B). The intensity of rhodamine 123 fluorescence in the cells gradually decreased with the increase in the BDE-47 concentration, which indicates a drop in MMP levels. A significant decrease in the MMP levels was observed for 5, 10, 20, and 40 μM treatments compared with the control ($p < 0.01$). For instance, the intensity of rhodamine 123 fluorescence decreased to 38.41 ± 1.40% in the treated group of 40 μM at 24 h, which was much lower than that of the CTR group (Figure 2C,D). According to the Western blot results, the expression of pro-apoptotic protein BAX in RAW274.7 cells exposed to BDE-47 showed an increasing trend. The expression level of anti-apoptotic protein Bcl-2 showed the opposite trend, with its expression level significantly reduced after treatment with BDE-47 at concentrations above 20 μM ($p < 0.01$). The protein expression of cytochrome C was also examined in cells; it increased in the cytoplasm after BDE-47 exposure, and the differences werE significant at concentrations above 10 μM ($p < 0.01$) (Figure 2E,F). The results indicate that cytochrome C is released from the mitochondria into the cytoplasm. No significant changes in the activities of Caspase-3 and Caspase-9 enzymes were observed between the solvent control group and the control group, while their activities showed an increasing trend with the increase in the BDE-47 concentration, which was significant in the 40 μM treatment group ($p < 0.01$) (Figure 2G).

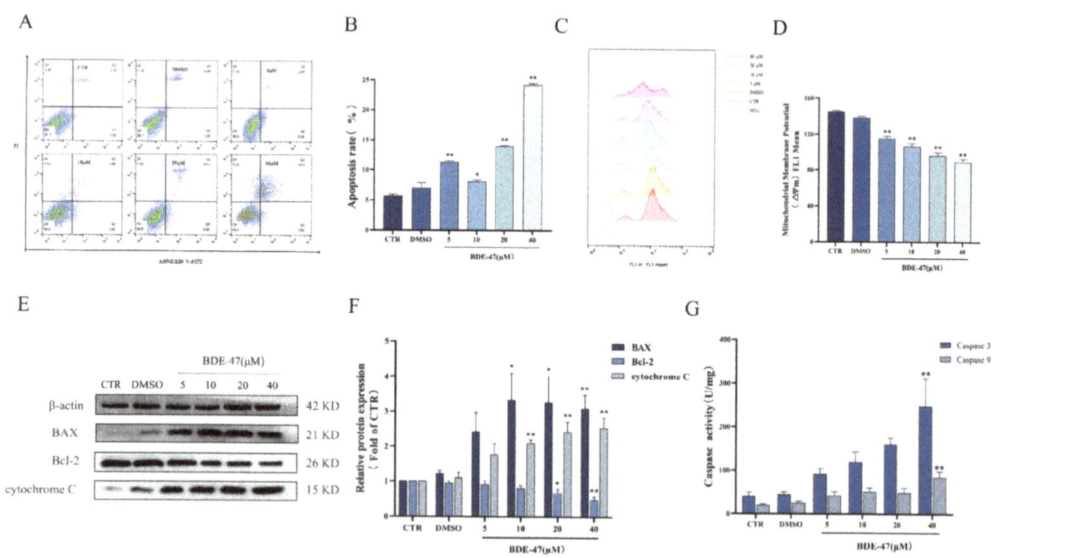


**Figure 2.** Effect of BDE-47 on apoptosis in RAW264.7 via the mitochondrial pathway. (**A**,**B**) Representative graphs and quantitative data of cell apoptosis obtained from flow cytometry analysis. (**C**,**D**) Representative graphs and quantification of mean fluorescence value of MMP changes obtained from flow cytometry analysis. (**E**,**F**) Protein expression level and quantitative statistics of BAX, Bcl-2, and cytochrome C. (**G**) Changes in Caspase-3 and Caspase-9 enzyme activities. $n = 3$; * $p < 0.01$; ** $p < 0.05$.

### 2.3. BDE-47 Alters Immune Function of RAW264.7 Cells

The capacity of RAW264.7 cells to phagocytize *E. coli* gradually decreases with the treatment of BDE-47 for 24 h, and this was significant for the 20 and 40 μM treatments compared with the CTR ($p < 0.05$) (Figure 3A,B). BDE-47 treatment could decrease the gene expression of exogenous antigen-presenting molecule MCH-II, which was significant for the 5 μM treatment ($p < 0.05$). The gene expression levels of co-stimulators CD40, CD80, and CD86 were significantly increased after exposure to BDE-47. The expression level of the CD80 gene was significantly increased at 10 μM ($p < 0.05$), and the expression level of CD86 was obviously increased at 20 μM ($p < 0.05$) (Figure 3C). Moreover, NO production was steadily elevated with the increase in the BDE-47 concentration, being significant at concentrations greater than 10 μM compared with the CTR group (Figure 3D). BDE-47 also increased the secretion of inflammatory factors IL-1β, IL-6, and TNF-α. Compared with the CTR group, the cytokine level of IL-6 was greatly increased, and significance was observed at a concentration of 40 μM ($p < 0.01$); IL-1β was significantly elevated at 20 and 40 μM ($p < 0.05$). The cytokine levels of TNF-α were greatly elevated with the increase in the BDE-47 concentration compared with the CTR group ($p < 0.01$). However, there was no significant change in the anti-inflammatory factor IL-10 after exposure to BDE-47 (Figure 3E). We further examined the level of intracellular TNF-α protein expression using Western blotting experiments. The results are shown in Figure 3F,G. After BDE-47 exposure, the expression level of TNF-α protein in RAW264.7 cells was significantly increased. The markers of different phenotypes of RAW264.7 macrophages were detected by flow cytometry. The expression level of iNOS, which is an M1 pro-inflammatory phenotype marker, was elevated with the increase in the BDE-47 concentration (Figure 3H,I). However, the expression level of M2 anti-inflammatory phenotypic marker Arg-1 exhibited no significant change compared with the CTR group after BDE-47 exposure (Figure 3J,K).

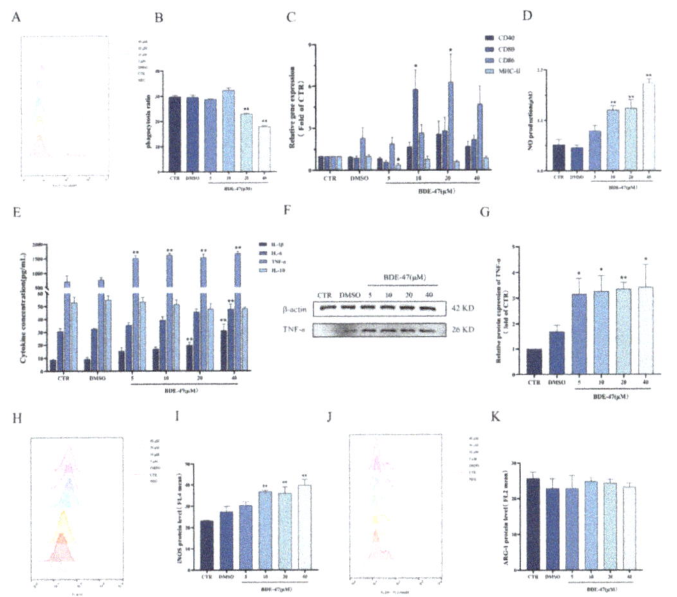


**Figure 3.** Effect of BDE-47 on immune function and polarization phenotype of RAW264.7 cells. (**A**,**B**) Representative graphs and quantitative data of cell phagocytosis ratios obtained from flow cytometry analysis. (**C**) Gene expression levels of presented antigens. (**D**) NO release level. (**E**) Levels of IL-1β, IL-6, TNF-α, and IL-10 secretion. (**F**,**G**) Protein expression level and quantitative statistics of TNF-α. (**H**,**I**) Representative graphs and quantitative data of iNOS expression levels obtained from flow cytometry analysis. (**J**,**K**) Representative graphs and quantitative data of Arg-1 expression levels obtained from flow cytometry analysis. $n = 3$; * $p < 0.01$, ** $p < 0.05$.

*2.4. BDE-47 Increases the Level of Reactive Oxygen Species in RAW264.7 Cells and the Regulation of Oxidative Stress Genes*

DCFH-DA was used to detect the intracellular ROS levels, and the results show a trend of concentration-dependent increases with the BDE-47 exposure. A significant increase was observed for the concentrations greater than 10 μM BDE-47 ($p < 0.05$) (Figure 4A,B). In the Pearson correlation matrix, a strong negative correlation was observed between ROS levels and cell proliferative viability, MMP, and cell phagocytosis capacity (r = 0.8–1, $p < 0.01$). However, a strong positive correlation was observed between ROS levels and cell mortality, apoptotic rate, Caspase-3 and Caspase-9 enzyme activity, cytochrome C protein expression, CD40, CD86, IL-1β, IL-6, TNF-α, and iNOS gene expression, and the capacity to generate NO (Figure 4C). The 440 oxidative stress genes obtained in the CTD database were mapped to the differentially expressed genes (1113 upregulated, 696 downregulated) obtained by transcriptome sequencing, yielding 39 genes differentially regulated by oxidative stress, of which 23 are upregulated and 16 are downregulated (Figure 4D). The biological processes found, by Metascape analysis, to be enriched were mainly related to "response to oxidative stress response", "response to reactive oxygen species", and "positive regulation of apoptotic process" (Figure 4E). The protein–protein interaction network is shown in Figure 4F. The modular analysis identified two core MCODEs, a cellular response to oxidative stress, and aerobic electron transport. Increased ROS and modulation of genes associated with oxidative stress in the transcriptome provide further evidence that ROS is a critical regulator of BDE-47 stress in RAW264.7 cells.

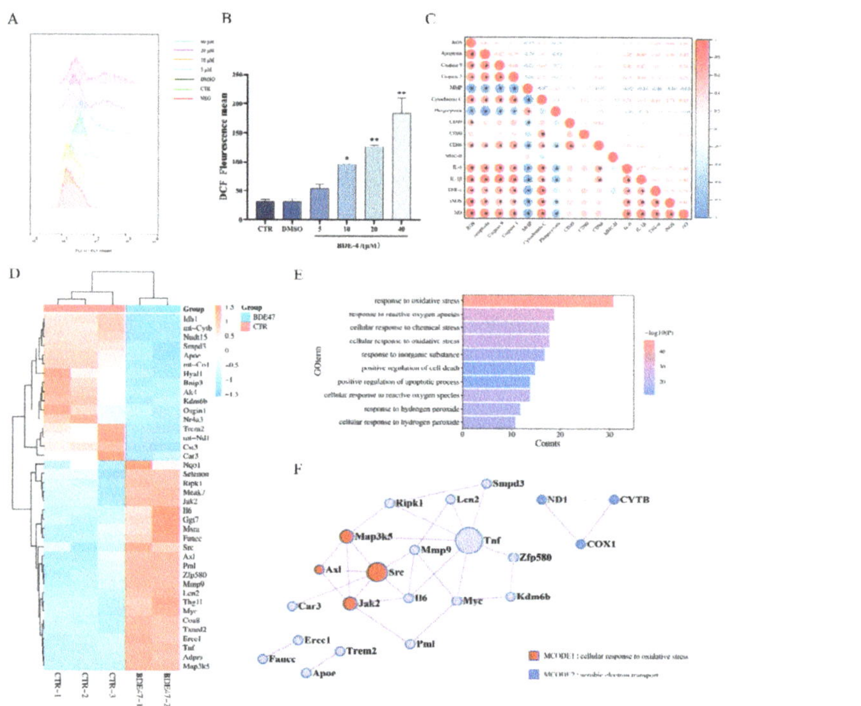


**Figure 4.** Effect of BDE-47 treatment on intracellular level of ROS and oxidative stress regulatory network in RAW264.7 cells. (**A**,**B**) Representative graphs and quantitative data of ROS levels obtained from flow cytometry analysis after BDE-47 treatment. (**C**) Pearson correlation analysis of the relevant metrics. (**D**) Differentially expressed genes associated with oxidative stress in transcriptome sequencing. (**E**) Biological process analysis of genes differentially expressed under oxidative stress. (**F**) Protein–protein network interaction analysis of differentially expressed oxidative stress genes. $n = 3$; * $p < 0.01$, ** $p < 0.05$.

## 2.5. Role of ROS in BDE-47-Induced Apoptosis and Immune Impairment in RAW264.7 Cells

### 2.5.1. Effect of Antioxidant NAC Pretreatment on BDE-47-Induced Apoptosis and Cellular Immune Function on RAW264.7 Cells

The antioxidant NAC was employed in the pretreatment to more thoroughly characterize the role of ROS in the process of stress induced by BDE-47 on RAW264.7. The results indicate that NAC could reduce the rate of BDE-47-induced apoptosis, while there was no significant effect on the apoptotic rate due to NAC treatment alone. Analyses of the apoptotic rate revealed that treatment with a combination of NAC and BDE-47 reduced the apoptotic rate (from 24.00 ± 0.20% to 9.76 ± 0.18%) compared with BDE-47 alone (Figure 5A,B). Moreover, the NAC measurably alleviated the effects of BDE-47, causing a drop in MMP and an increase in the intensity of rhodamine 123 fluorescence (from 38.41 ± 1.40% to 15.39 ± 1.04%) compared with BDE-47 only (Figure 5C,D). BDE-47 caused an elevation in the levels of cytochrome C protein in the cytoplasm, which was inhibited by the NAC. Treatment with NAC in combination with BDE-47 decreased the expression of cytochrome C compared with the BDE-47 group ($p < 0.01$) (Figure 5E,F). The increase in Caspase-3 and Caspase-9 enzyme activities due to BDE-47 were also inhibited by the NAC (Figure 5G).

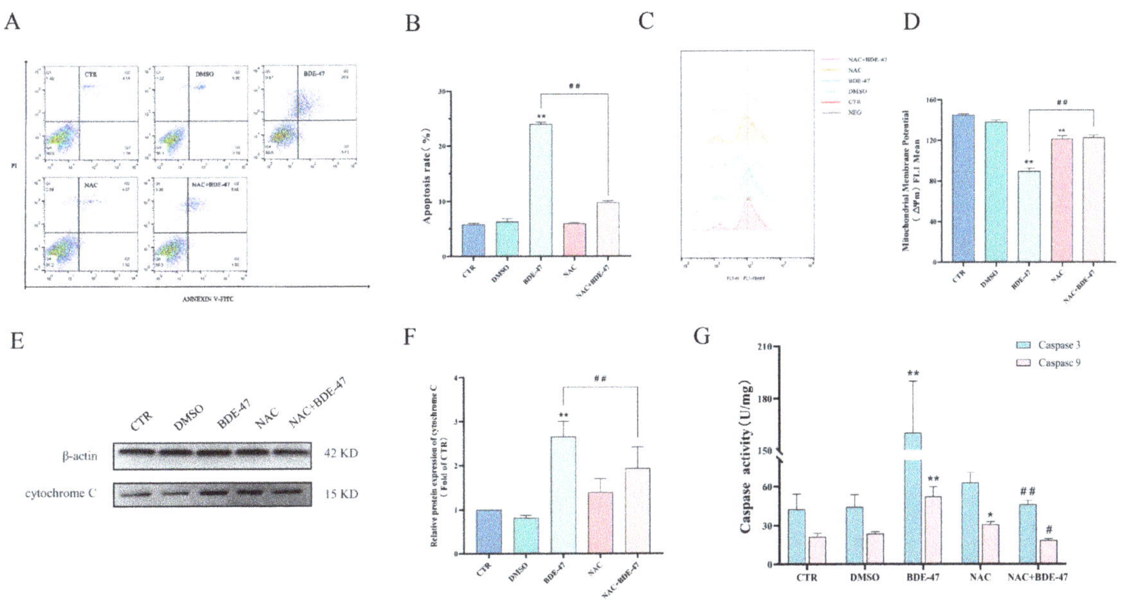


**Figure 5.** Effect of NAC pretreatment on BDE-47-induced apoptosis in RAW264.7 cells. Cells were pre-incubated with or without 100 μM NAC and then treated with BDE-47. (**A**,**B**) Cell apoptosis is shown in flow cytometric graphs and histogram. (**C**,**D**) Changes in MMP are shown in flow cytometric graphs and histogram. (**E**,**F**) Protein expression level and quantitative statistics of cytochrome C. (**G**) Changes in Caspase-3 and Caspase-9 enzyme activities. $n = 3$; * Significant difference compared with the untreated controls (* $p < 0.05$; ** $p < 0.01$). # Significant difference compared with the BDE-47-treated group (# $p < 0.05$; ## $p < 0.01$).

NAC-alleviated BDE-47 caused a decrease in phagocytosis capacity. The capacity oF phagocytosis increased (from 17.93 ± 0.25% to 30.37 ± 1.12%) compared with the BDE-47 treatment (Figure 6A,B). BDE-47 caused elevated levels of CD40, CD80, IL-1β, and TNF-α, which were inhibited by the NAC pretreatment ($p < 0.05$). Similar results were obtained in Western blotting experiments, where NAC pretreatment reduced the degree of increase in TNF-α protein expression caused by BDE-47 exposure (Figure 6F,G). Furthermore, the

NAC and BDE-47 combined-treatment group showed a lower protein expression of NO release level and iNOS protein level compared with the BDE-47 group ($p < 0.01$). The effect of NAC pretreatment on the expression of CD86, IL-6, MHC-II, IL-10, and Arg-1 had no significant change after the BDE-47 exposure (Figure 6C–E,H–K).

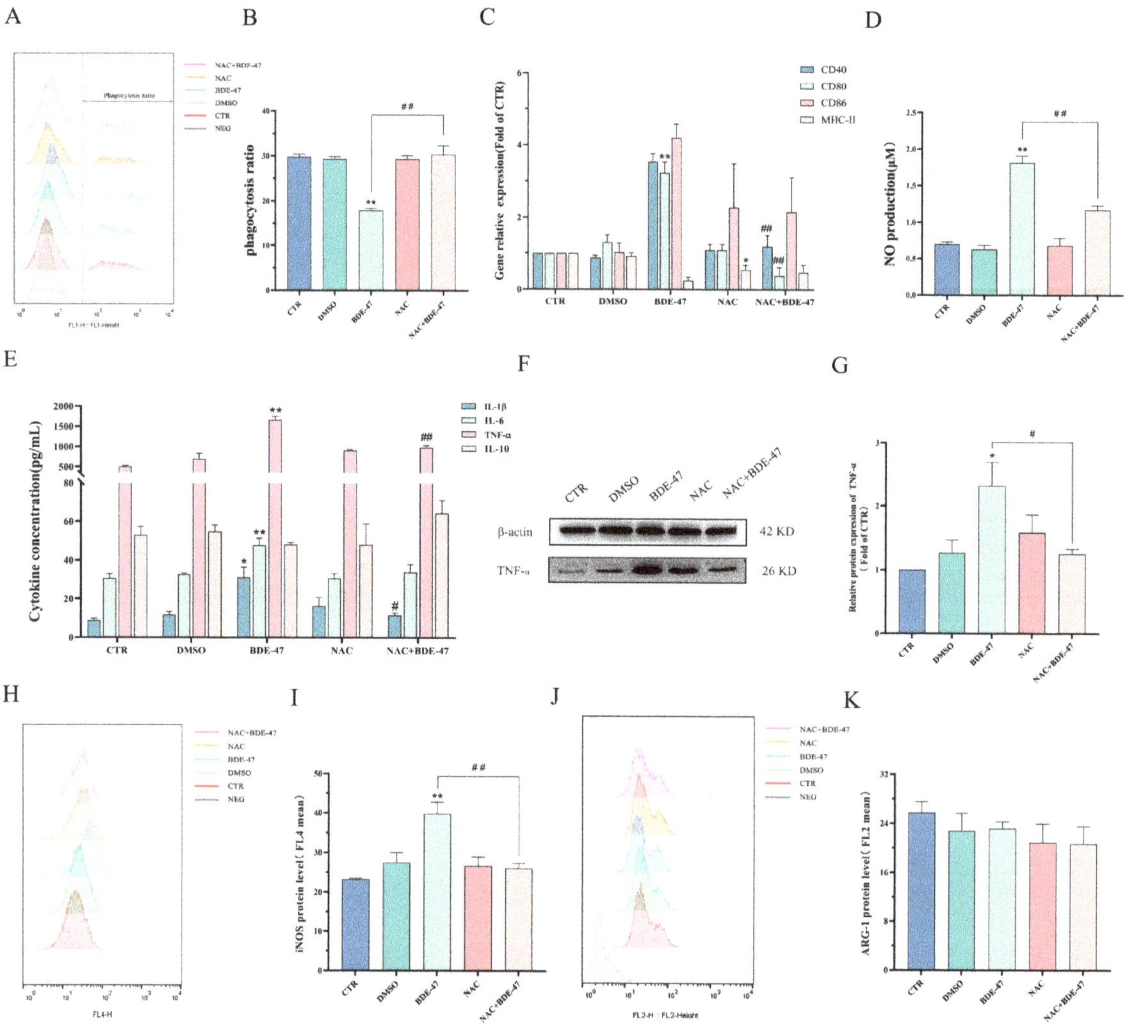


**Figure 6.** Effect of NAC pretreatment on immune function and polarization phenotype after BDE-47 exposure of RAW264.7 cells. Cells were pre-incubated with or without 100 μM NAC and then treated with BDE-47. (**A**,**B**) Cell phagocytosis function shown in flow cytometric graphs and histogram. (**C**) Gene expression levels of antigen presentation. (**D**) NO release levels. (**E**) Levels of IL-1β, IL-6, TNF-α, and IL-10 secretion. (**F**,**G**) Protein expression level and quantitative statistics of TNF-α. Flow cytometric graphs and histogram of (**H**,**I**) iNOS, and (**J**,**K**) Arg-1 protein levels. $n = 3$. * Significant difference compared with the untreated controls (* $p < 0.05$; ** $p < 0.01$). # Significant difference compared with the BDE-47-treated group (# $p < 0.05$; ## $p < 0.01$).

2.5.2. Effect of Inducer BSO Pretreatment on Apoptosis and Cellular Immune Function after BDE-47 Exposure

BSO pretreatment increased the rates of BDE-47-induced apoptosis of RAW264.7 cells from 24.00 ± 0.20% to 60.86 ± 0.27% (Figure 7A,B), and the degree of apoptosis was also intensified, with a significance level of $p < 0.01$. In addition, we also measured the level of NO production and inflammatory factor secretion after the BSO pretreatment, and the results (Figure 7C,D) were the same as those for the BDE-47-treated group. BSO pretreatment leads to NO release and increased levels of secretion of the inflammatory factors IL-1β, IL-6, and TNF-α ($p < 0.01$). Finally, we used flow cytometry to detect the macrophage M1-type marker, iNOS, at the protein level, revealing that pretreatment with the inducer BSO enables increased polarization of RAW264.7 M1 pro-inflammatory typing after BDE-47 induction ($p < 0.05$).

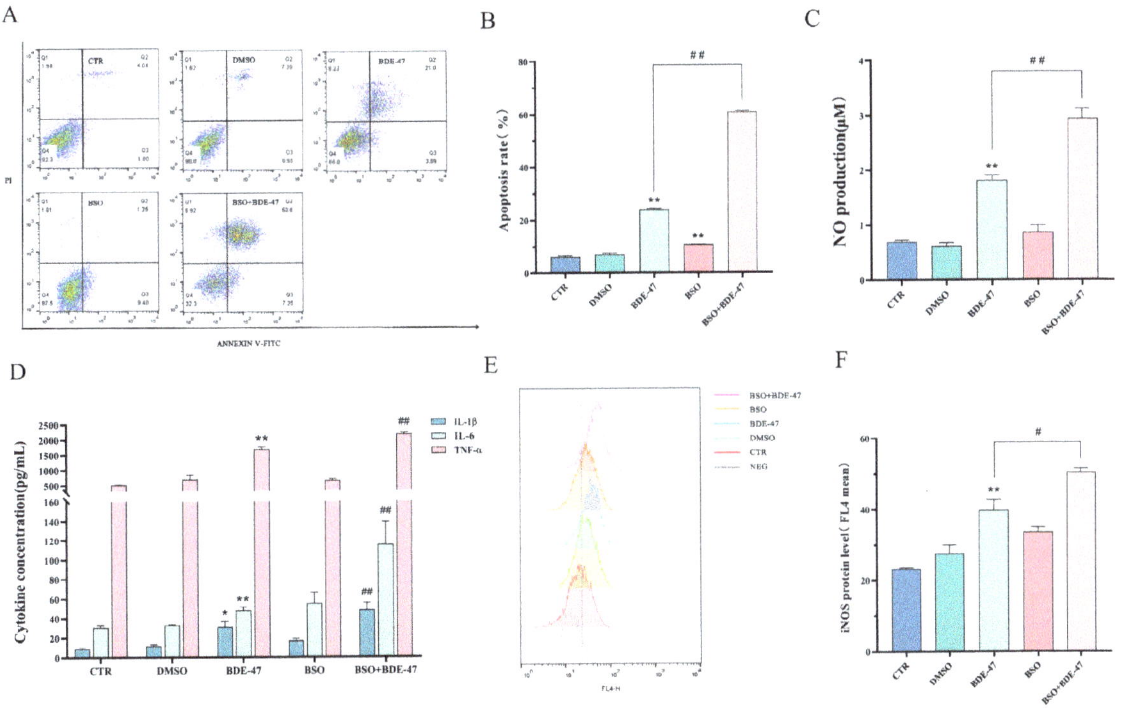


**Figure 7.** Effect of BSO pretreatment on apoptosis and immune function after BDE-47 exposure of RAW264.7 cells. Cells were pre-incubated, with or without 500 μM BSO, and then treated with BDE-47. (**A**,**B**) Cell apoptosis is shown in flow cytometric graphs and histogram. (**C**) NO release levels. (**D**) Levels of IL-1β, IL-6, and TNF-α secretion. (**E**,**F**) Expression of iNOS protein level is shown in flow cytometric graphs and histogram. $n = 3$. * Significant difference compared with the untreated controls (* $p < 0.05$; ** $p < 0.01$). # Significant difference compared with the BDE-47-treated group (# $p < 0.05$; ## $p < 0.01$).

## 3. Discussion

Exposure to BDE-47 can affect the immune system, manifesting as susceptibility to diseases in the body, abnormal immune responses, immune dysfunction, and cellular damage [27,40,41]. Macrophages have an innate immune defense mechanism and are crucial to many pathophysiological processes. This type of immune cell is found in many tissues of the human body, characterized by its powerful phagocytic activity, and it is highly active and sensitive. In this study, we report for the first time that BDE-47 can significantly damage the activity and immune function of RAW264.7 mouse macrophages in vitro, and we explore the role of ROS in immunotoxicity caused by BDE-47.

In the present study, a concentration range of 5 to 80 μM BDE-47 was chosen to investigate cell toxicity, according to our preliminary study and other previous studies [42,43]. With the increase in concentration, the cell viability was significantly reduced for 24, 48, and 72 h. However, the cell mortality increased with the increase in BDE-47 concentration, indicating that BDE-47 inhibits the proliferation of RAW264.7 mouse macrophages. Similar results were observed in other cells exposed to BDE-47 [16,44–49]. BDE-47-induced immunotoxicity is attributed to a decrease in the number of macrophages in various tissues in the body and weakened proliferative activity, which interferes with responses of the immune system, similar to the perfluorinated compounds PFOA and PFOS [50].

Current research on the immunotoxicity mechanism of exogenous compounds is mainly focused on autophagy [51], apoptosis [27], oxidative damage [52], and inflammatory response [53]. Mitochondrial pathway-mediated apoptosis is a process of cell death that is highly regulated by apoptosis genes. In our findings, the protein expression of Bcl-2 decreased and BAX significantly increased. Furthermore, the MMP levels decreased, cytochrome C release to the cytoplasm was increased, and the enzyme activities of Caspase-3 and Caspase-9 were upregulated, suggesting that the endogenous mitochondrial apoptotic pathway was triggered. These phenomena are consistent with our previous studies on Neuro-2a cells [13]. There was also a similar result in a study on acrylamide, which was found to induce caspase-dependent apoptosis in mouse splenocytes through mitochondrial-dependent signaling [54].

Macrophages play an extremely important role in inflammation, an innate immune response [55], and they exert defensive functions through phenotypic polarization, cytokines release, and phagocytosis of foreign particles to maintain the stability of the internal environment [38]. To further characterize the damage to the macrophage immune function resulting from BDE-47, we examined changes in phagocytosis, antigen presentation, cytokines secretion levels, and phenotypic markers. The results show that BDE-47 reduces the phagocytic capacity of macrophages and the expression of antigen-presenting molecules but increases the expression of co-stimulating factors and the secretion of inflammatory factors. Macrophages are divided into at least two major polarized phenotypes: M1-polarized macrophages and M2-activated macrophages [56]. We found that BDE-47 significantly promotes macrophage polarization to type M1, manifested by the elevated expression of the pro-inflammatory cytokines TNF-α, IL-6, and IL-1β, increased expression of TNF-α protein in cells, and increased NO release levels, triggering an inflammatory response in the body that is accompanied by an elevated expression of the macrophage-type M1 polarization marker, iNOS. Nevertheless, BDE-47 has no significant effect on macrophage polarization to the M2 type. The above data show that BDE-47 can significantly affect the macrophage immune function, such as through decreased phagocytosis, blocking of antigen presentation, and increased secretion of inflammatory factors, promoting NO production. Moreover, BDE-47 also has a certain impact on the phenotype of macrophages, inducing macrophages to M1-type polarization. Bisphenol A, one of the most widely used industrial compounds for synthesizing materials such as polycarbonate (PC) and epoxy resin, can promote mouse intraperitoneal macrophage differentiation from the pro-inflammatory M1-subtype but inhibits differentiation from the anti-inflammatory M2-subtype macrophages [57]. PM2.5 in the air can significantly enhance the polarization of macrophage inflammation M1-type, interfering with the balance of macrophage inflammation M1 and anti-inflammatory M2 polarization [58]. The above findings are consistent with our results.

ROS is often considered an essential regulatory factor when macrophages defend against inflammation and other stress [59]. In our study, ROS levels significantly increased with the concentration of BDE-47 used in exposure to RAW264.7 cells. The ROS that accumulates in macrophages can not only serve as an upstream factor to trigger apoptosis [60] but also directly damage biological macromolecules, eliciting various inflammatory responses [61]. ROS plays a mediating role between innate and adaptive immune cells and further influences immune processes, such as T cell activation, by antigen delivery [62]. A Pearson correlation analysis of biological indicators shows a significant correlation with ROS levels in most cells (Figure 4C). The transcriptome sequencing results further supported our previous results, such as for IL-6; TNF upregulated oxidative stress-associated genes also showed the same trend in protein expression in our ELISA and Western blotting results. To further explore the effects of ROS on BDE-47-induced immunotoxicity, the antioxidant N-acetylcysteine (NAC) was utilized to study the relationship between the immunotoxicity and ROS level caused by BDE-47 in RAW264.7 cells. The cytotoxicity results show that pretreatment with NAC could attenuate BDE-47-caused MMP damage, cytochrome C release, and the increased enzyme activity of Caspase-3 and Caspase-9, eventually reversing apoptosis. The biological process enrichment and protein interaction network modular analysis of the differentially transcriptionally expressed oxidative stress genes revealed that they were closely related to apoptosis and the transfer of aerobic electron chains in mitochondria, which, to some extent, supports our research finding that ROS can regulate the mitochondrial apoptotic pathway.

For the immune function, NAC can inhibit a degree of decreased cell phagocytosis function after exposure to BDE-47. It has been demonstrated that an increase in ROS is linked with M1-type activation in macrophages [63], which further stimulates the release of pro-inflammatory cytokines. NAC also alleviates the production level of NO, the elevated secretion of inflammatory factors, gene expression of co-stimulating factors, and protein expression of TNF-$\alpha$ and the M1 macrophage marker, iNOS. NAC was found to mitigate immunotoxicity in other studies, where NAC pretreatment could alleviate the immunotoxic effects of azolinone and isomethoxamine fungicides in mice in vitro [36]. Hou et al. showed that the immunotoxicity of aflatoxin B1 and ochratoxin A is associated with ROS regulation, and the antioxidant NAC can mitigate the combined toxicity in porcine alveolar macrophages [64]. However, we also found that gene expression levels of the exogenous antigen-presenting complex, MHC-II, did not increase significantly after NAC pretreatment, and minimal correlation was also observed in the Pearson correlation analysis, suggesting that it may have little involvement in the regulation of ROS. More research is still required to explain the observed phenomena. In addition, we also used pretreatment with the ROS-inducer BSO to explore the effect of ROS elevation on BDE-47-induced immunotoxicity [65]. Compared with the NAC pretreatment, diametrically opposite trends are shown: after the BSO pretreatment, several key indicators, such as apoptosis, NO production, inflammatory factor level, and M1 polarization of RAW264.7 cells, were significantly increased. This indicates that the BDE-47-induced immunotoxicity was intensified.

Therefore, our results suggest that the environmental pollutant BDE-47 can cause an increase in intracellular ROS levels in macrophages. Genes related to oxidative stress showed transcriptional differences, and the intracellular antioxidant protection mechanism was affected by the increase in BDE-47 concentration. Moreover, BDE-47 further induces apoptosis of cells through the mitochondrial pathway, damaging immune function and generating immunotoxic effects on macrophages (Figure 8).

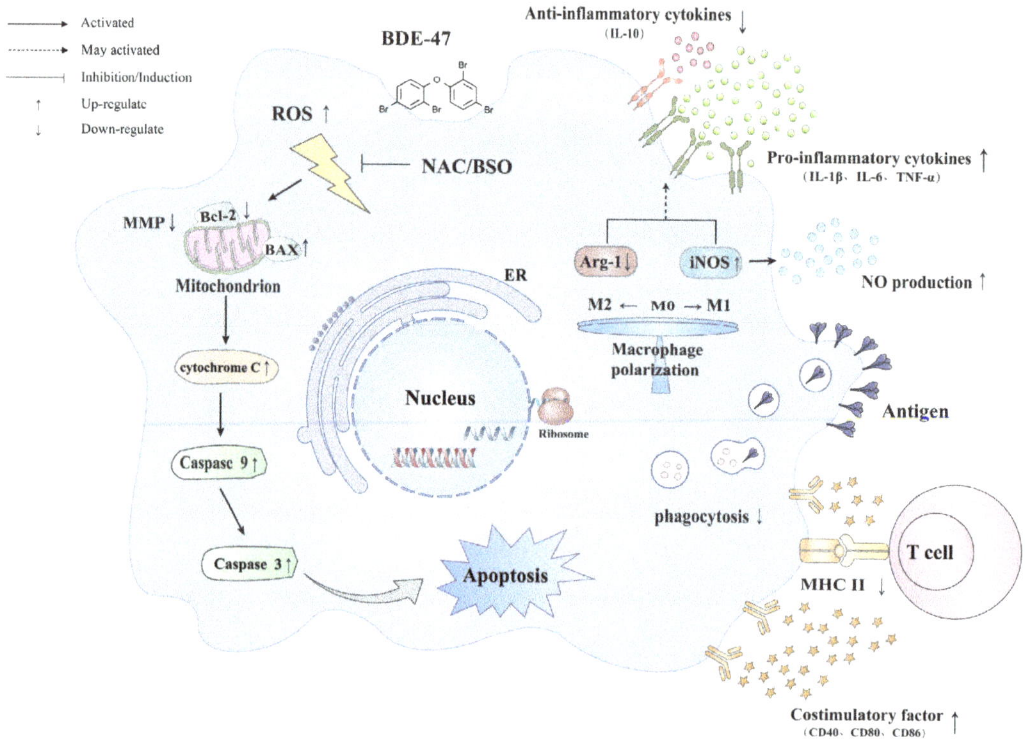


**Figure 8.** The mechanism of BDE-47-induced immunotoxicity in RAW264.7 cells.

## 4. Materials and Methods

### 4.1. Chemicals and Reagents

BDE-47 was purchased from AccuStandard (New Haven, CT, USA). DCFH2-DA, thiazole blue, and dimethyl sulfoxide (DMSO) were purchased from Sigma (St. Louis, MO, USA). The Annexin V-FITC/PI apoptosis kit was purchased from BD Bioscience (Franklin Lake, NJ, USA). Arg-1 and iNOS antibodies were purchased from eBioscience (San Diego, CA, USA). *E. coli* (strain K-12) was purchased from Invitrogen (Carlsbad, CA, USA). The IL-10 enzyme-linked immunosorbent assay (ELISA) kit was purchased from MultiSciences (Hangzhou, China). Cytochrome C antibodies, IL-6, IL-1β, TNF-α ELISA kits, DAPI dye, rhodamine 123, and Caspase-3 and Caspase-9 activity assay kits were purchased from Beyotime Biotechnology (Shanghai, China). BAX, Bcl-2, and TNF-α antibodies were purchased from WanLeiBio Co., Ltd. (Shenyang, China). ECL luminescent reagent was purchased from Sangon Biotech Co., Ltd. (Shanghai, China). Real-time fluorescent quantitative reagents were purchased from Takara Biomedical Technology Co., Ltd. (Beijing, China). N-acetylcysteine (NAC) and L-buthionine-(S,R)-sulfoximine (BSO) were purchased from Yuanye Bio-Technology Co., Ltd. (Shanghai, China).

### 4.2. Cell Culture and Experimental Design

The RAW264.7 mouse macrophage line was obtained from the Shanghai Institute of Cell Biology (Chinese Academy of Sciences, Shanghai, China). The cells were cultured in DMEM (Gibco, Grand Island, NE, USA), containing 10% fetal bovine serum (Biological Industries, Kibbutz Beit Haemek, Israel) and 1% penicillin–streptomycin antibiotics (Gibco, Grand Island, NE, USA), and then cultured in an incubator at 37 °C and 5% $CO_2$.

Logarithmically growing RAW264.7 cells were inoculated in 96-well and 6-well plates at a concentration of $2.5 \times 10^4$ and $2.5 \times 10^5$ $mL^{-1}$, respectively, and cultured overnight.

The cells were then treated with BDE-47 at various concentrations (0, 5, 10, 20, and 40 μM) for 24 h for the determination of cell viability, fluorescence level, protein, and mRNA expression. Untreated cells were used as the control (CTR); cells treated with 0.1% DMSO (*v*/*v*) were used as the DMSO solvent control (DMSO). In addition, RAW264.7 cells were precultured in 100 μM NAC or 500 μM BSO for 4 h, and then 40 μM of BDE-47 was added to the cells for 24 h, followed by the calculation of indexes related to apoptosis and immune function, and for studying the relationship between ROS and immunotoxicity.

### 4.3. Cytotoxicity Evaluation

The RAW264.7 cell proliferation activity was measured using the MTT assay. In brief, the cells were exposed to various concentrations of BDE-47 for 24, 48, and 72 h; then, 10 μL of MTT solution was added to each well and incubated at 37 °C for 4 h. After that, 150 μL DMSO was added to each well to dissolve the methylzan; then, a microplate reader was used to determine the absorbance (OD) value at 490 nm. Cell viability was calculated as the percentage of live cells relative to the total number of cells. Trypan blue staining was used to determine the rate of cell death, as in previous studies [66,67].

### 4.4. Morphology Observation

After the exposure to BDE-47, the cell surface morphological changes were directly visualized through an inverted microscope. Additionally, the cells were stained with 0.1 μg/mL DAPI staining solution for 30 min, followed by washing with PBS twice. Then, the cells were placed under a fluorescence microscope to observe cell nucleus pyknosis.

### 4.5. Apoptosis Detection

Cell apoptotic rates were detected using an Annexin V-FITC/PI staining kit. The cells were digested and collected following EDTA-free trypsinization and washed twice with pre-chilled PBS before being resuspended in Annexin-V binding buffer. The cells were stained with PI and Annexin V-FITC for 10 min in the dark. The stained cells were assessed using a BD flow cytometer, and FlowJo software was used to analyze the flow cytometry data. The apoptotic rate was expressed as the sum of the Q2 quadrant (the percentage of late apoptotic cells) and Q3 quadrant (the percentage of early apoptotic cells).

### 4.6. Detection of Mitochondrial Membrane Potential (MMP)

Rhodamine 123 chemical dye was used to evaluate the changes in the MMP. The treated cells were incubated with rhodamine 123 (5 mg/mL) for 30 min. The cells were collected by centrifugation and resuspended in PBS. The average FL1 channel fluorescence intensity, as assessed using the BD flow cytometer, was taken to represent the MMP in the cells.

### 4.7. Examination of ROS Level

Intracellular ROS formation was determined using the DCFH-DA probe. Briefly, after the exposure treatment of the cells, the supernatant was aspirated and washed before adding the DCFH2-DA probe (10 μM), then incubating for 30 min. The fluorescence intensity in the FL1 channel was then measured using the BD flow cytometer.

### 4.8. Caspase Enzymatic Activity

The Caspase-3 and Caspase-9 activities were determined using commercial Caspase-3 and Caspase-9 activity assay kits. First, the treated cell samples were collected, the protein extracted, and the concentration was then determined using the Bradford method. Subsequently, 50 μL of the sample was added to the assay buffer and catalytic substrate for reaction and incubated at 37 °C overnight. The absorbance at 405 nm was detected using a microplate reader, and the units of enzyme activity were calculated, according to the standard curve, as well as the sample protein concentration.

### 4.9. Western Blotting

The protein expression levels were determined by Western blot analysis, according to a previous study [68]. Briefly, protein samples were extracted using RIPA buffer, separated by SDS-PAGE, transferred to PVDF membranes, incubated for 12 h with antibodies diluted at the appropriate proportion (BAX antibody 1:1000; Bcl-2 antibody 1:1000; cytochrome C antibody: 1:200), and, for 1 h, with horseradish peroxidase-conjugated antibodies. The membrane was washed with TBST three times for 5 min each. The immunoreactive bands were detected with ECL reagents, according to the manufacturer's instructions. ImageJ software was used to quantify the bands, and the protein expression levels of BAX, Bcl-2, and cytochrome C were standardized, based on comparison with β-actin protein.

### 4.10. Phagocytic Capacity

The treated cells were cultured in a serum-free fresh medium containing the biotinylated particles of FITC-labeled *Escherichia coli* (K-12 strain) for 1 h. After that, the cells were collected and washed three times with PBS to eliminate unincorporated *E. coli*, and the cell fluorescence was then detected by using flow cytometry, according to the method referred to in a previous study [69].

### 4.11. Quantitative Real-Time Polymerase Chain Reaction (QRT-PCR) Analysis

Total RNA was extracted from the treated RAW264.7 cells with TRIizol, according to the manufacturer's protocol. Then, the cDNA was synthesized from the total RNA using a first-strand cDNA synthesis kit. The real-time PCR was performed using SYBR Green Supermix and PCR primers in a cycler. The sequences of primers used in the real-time PCR are listed in Table 1. The amplification was performed in 20 μL of the total mixture volume. The amplification program was as follows: predenaturation at 95 °C for 30 s, followed by 40 cycles of 95 °C for 5 s, and 60 °C for 34 s. GAPDH was used for normalization in the relative quantitative analysis, and the standard $2^{-\Delta\Delta Ct}$ method was used in the calculation.

**Table 1.** Specific primer sequences for *qRT-PCR*.

| Target Gene | Forward (5′-3′) | Reverse (5′-3′) |
|---|---|---|
| MHC-II | GTGTGCAGACACAACTACGAGG | CTGTCACTGAGCAGACCAGAGT |
| CD40 | ACCAGCAAGGATTGCGAGGCAT | GGATGACAGACGGTATCAGTGG |
| CD80 | CCTCAAGTTTCCATGTCCAAGGC | GAGGAGAGTTGTAACGGCAAGG |
| CD86 | ACGTATTGGAAGGAGATTACAGCT | TCTGTCAGCGTTACTATCCCGC |
| GAPDH | CATCACTGCCACCCAGAAGACTG | ATGCCAGTGAGCTTCCCGTTCAG |

### 4.12. Cytokine Secretion Levels and NO Production

The supernatant of the treated cells was collected, and the cytokine levels of IL-10, IL-6, IL-1β, and TNF-α were detected by ELISA, according to the manufacturer's instructions. The NO production in each group was determined using the Griess kit, with the OD determined at 550 nm wavelength.

### 4.13. Cell Phenotype

INOS and Arg-1 are classic markers of the M1 pro-inflammatory phenotype and M2 anti-inflammatory phenotype, respectively. The detection of polarization marker protein expression levels was carried out using flow cytometry, according to a method referred to in previous studies [70,71]. The treated cells were fixed with a fixative solution for 20 min at 25 °C, and, after being washed with PBS twice, the cells were treated with a permeabilization solution and then incubated with PE-labeled Arg-1 and APC-labeled iNOS antibodies for 30 min at 4 °C in the dark. Then, the samples were analyzed by BD flow cytometry after washing.

### 4.14. RNA-Seq and Analysis of the Oxidative Stress Regulatory Network

RAW264.7 cells were treated with BDE-47 at a concentration of 20 μM (BDE-47) for 24 h and then harvested. The RNA extraction, purity, and quantification, and the transcriptome sequencing and analysis were performed by Personalbio Technology Co., Ltd. (Shanghai, China). After the sequencing, clean reads were obtained and aligned to the mouse genome, and the differentially expressed genes (DEGs) between the two groups were analyzed using DESeq (2012) R software. The significance of the gene expression differences was assessed with the $p$-value < 0.05 and log2 (fold change) > 1 as the threshold value. Relevant entries were retrieved from the CTD database (https://ctdbase.org, accessed on 15 October 2022) with the keyword "oxidative stress" to obtain associated genes, which were mapped to the differentially expressed genes obtained from the transcriptome sequencing. Gene ontology and protein–protein interaction analyses were performed using Metascape (https://metascape.org, accessed on 5 November 2022).

### 4.15. Data Analysis

For each experiment, we analyzed three or more replicates in a completely random manner. The experimental data are the average ± standard error (mean ± SEM), with GraphPad Prism 8.4 software for chart visualization. SPSS 26 software was used for the statistical analysis of the data. The HSD Tukey test was used to assess the significance between the control and experimental groups, and $p < 0.05$ indicates significance. Pearson correlation was used to analyze the interaction between the variables, for which visualization was realized using the ggplot2 package.

## 5. Conclusions

We found that BDE-47 can induce apoptosis of RAW264.7 cells via the mitochondrial pathway, causing immune function damage. The ROS inhibitor NAC could reverse apoptosis and alleviate the immune impairment caused by BDE-47, which could be conversely exacerbated by the inducer BSO. Intracellular ROS levels were found to be one of the regulators leading to BDE-47 immunotoxicity.

**Author Contributions:** Q.G.: formal analysis, validation, writing—original draft. Z.-Y.Z.: investigation, writing—review and editing. Y.-N.H., M.-H.D. and Z.-N.W.: formal analysis, methodology, investigation. H.-M.C.: conceptualization, funding acquisition, project administration, supervision, writing—review and editing. All authors have read and agreed to the published version of the manuscript.

**Funding:** This study was financially supported by the Natural Science Foundation of China (NSFC) (No. 21806114, 42106139).

**Institutional Review Board Statement:** Not applicable.

**Informed Consent Statement:** Not applicable.

**Data Availability Statement:** The data are available on request due to privacy. The data presented in this study are available on request from the corresponding author.

**Acknowledgments:** The authors are thankful to all the members in the lab for their assistance.

**Conflicts of Interest:** The authors declare no conflict of interest.

**Sample Availability:** Samples of the compounds are not available.

## References

1. Costa, L.G.; Giordano, G. Polybrominated Diphenyl Ethers. In *Encyclopedia of Toxicology*, 3rd ed.; Wexler, P., Ed.; Academic Press: Oxford, UK, 2014; pp. 1032–1034.
2. Li, K.; Fu, S. Polybrominated Diphenyl Ethers (PBDEs) in House Dust in Beijing, China. *Bull. Environ. Contam. Toxicol.* **2013**, *91*, 382–385. [CrossRef]
3. Ohoro, C.R.; Adeniji, A.O.; Okoh, A.I.; Okoh, O.O. Polybrominated diphenyl ethers in the environmental systems: A review. *J. Environ. Health Sci. Eng.* **2021**, *19*, 1229–1247. [CrossRef]

4. Zhang, Y.; Wang, W.; Song, J.; Ren, Z.; Yuan, H.; Yan, H.; Zhang, J.; Pei, Z.; He, Z. Environmental Characteristics of Polybrominated Diphenyl Ethers in Marine System, with Emphasis on Marine Organisms and Sediments. *Biomed. Res. Int.* **2016**, *2016*, 1317232. [CrossRef] [PubMed]
5. RA, H. Polybrominated diphenyl ethers in the environment and in people: A meta-analysis of concentrations. *Environ. Sci. Technol.* **2004**, *38*, 945–956.
6. Sharkey, M.; Harrad, S.; Abou-Elwafa Abdallah, M.; Drage, D.S.; Berresheim, H. Phasing-out of legacy brominated flame retardants: The UNEP Stockholm Convention and other legislative action worldwide. *Environ. Int.* **2020**, *144*, 106041. [CrossRef] [PubMed]
7. Maddela, N.R.; Venkateswarlu, K.; Kakarla, D.; Megharaj, M. Inevitable human exposure to emissions of polybrominated diphenyl ethers: A perspective on potential health risks. *Environ. Pollut.* **2020**, *266*, 115240. [CrossRef] [PubMed]
8. Hites, R.A.; Lehman, D.C.; Salamova, A.; Venier, M. Temporal environmental hysteresis: A definition and implications for polybrominated diphenyl ethers. *Sci. Total Environ.* **2021**, *753*, 141849. [CrossRef] [PubMed]
9. Meng, T.; Cheng, J.; Tang, Z.; Yin, H.; Zhang, M. Global distribution and trends of polybrominated diphenyl ethers in human blood and breast milk: A quantitative meta-analysis of studies published in the period 2000–2019. *J. Environ. Manag.* **2021**, *280*, 111696. [CrossRef] [PubMed]
10. Bramwell, L.; Glinianaia, S.V.; Rankin, J.; Rose, M.; Fernandes, A.; Harrad, S.; Pless-Mulolli, T. Associations between human exposure to polybrominated diphenyl ether flame retardants via diet and indoor dust, and internal dose: A systematic review. *Environ. Int.* **2016**, *92–93*, 680–694. [CrossRef]
11. Sun, M.H.; Li, X.H.; Xu, Y.; Xu, Y.; Sun, S.C. Exposure to PBDE47 affects mouse oocyte quality via mitochondria dysfunction-induced oxidative stress and apoptosis. *Ecotoxicol. Environ. Saf.* **2020**, *198*, 110662. [CrossRef]
12. Chen, H.; Streifel, K.M.; Singh, V.; Yang, D.; Mangini, L.; Wulff, H.; Lein, P.J. BDE-47 and BDE-49 Inhibit Axonal Growth in Primary Rat Hippocampal Neuron-Glia Co-Cultures via Ryanodine Receptor-Dependent Mechanisms. *Toxicol. Sci.* **2017**, *156*, 375–386. [CrossRef] [PubMed]
13. Chen, H.; Tang, X.; Zhou, B.; Xu, N.; Wang, Y. Mechanism of Deca-BDE-induced apoptosis in Neuro-2a cells: Role of death-receptor pathway and reactive oxygen species-mediated mitochondrial pathway. *J. Environ. Sci.* **2016**, *46*, 241–251. [CrossRef] [PubMed]
14. Pallocca, G.; Grinberg, M.; Henry, M.; Frickey, T.; Hengstler, J.G.; Waldmann, T.; Sachinidis, A.; Rahnenfuhrer, J.; Leist, M. Identification of transcriptome signatures and biomarkers specific for potential developmental toxicants inhibiting human neural crest cell migration. *Arch. Toxicol.* **2016**, *90*, 159–180. [CrossRef] [PubMed]
15. Khalil, A.; Parker, M.; Brown, S.E.; Cevik, S.E.; Guo, L.W.; Jensen, J.; Olmsted, A.; Portman, D.; Wu, H.; Suvorov, A. Perinatal exposure to 2,2′,4′4′ -Tetrabromodiphenyl ether induces testicular toxicity in adult rats. *Toxicology* **2017**, *389*, 21–30. [CrossRef] [PubMed]
16. Liu, X.; Wang, J.; Lu, C.; Zhu, C.; Qian, B.; Li, Z.; Liu, C.; Shao, J.; Yan, J. The role of lysosomes in BDE 47-mediated activation of mitochondrial apoptotic pathway in HepG2 cells. *Chemosphere* **2015**, *124*, 10–21. [CrossRef]
17. Tang, S.; Liu, H.; Yin, H.; Liu, X.; Peng, H.; Lu, G.; Dang, Z.; He, C. Effect of 2, 2′, 4, 4′-tetrabromodiphenyl ether (BDE-47) and its metabolites on cell viability, oxidative stress, and apoptosis of HepG2. *Chemosphere* **2018**, *193*, 978–988. [CrossRef]
18. Shan, Q.; Zheng, G.H.; Han, X.R.; Wen, X.; Wang, S.; Li, M.Q.; Zhuang, J.; Zhang, Z.F.; Hu, B.; Zhang, Y.; et al. Troxerutin Protects Kidney Tissue against BDE-47-Induced Inflammatory Damage through CXCR4-TXNIP/NLRP3 Signaling. *Oxid. Med. Cell. Longevity.* **2018**, *2018*, 9865495. [CrossRef]
19. Liu, X.; Zhan, H.; Zeng, X.; Zhang, C.; Chen, D. The PBDE-209 exposure during pregnancy and lactation impairs immune function in rats. *Mediat. Inflamm.* **2012**, *2012*, 692467. [CrossRef]
20. Fair, P.A.; Stavros, H.C.; Mollenhauer, M.A.; DeWitt, J.C.; Henry, N.; Kannan, K.; Yun, S.H.; Bossart, G.D.; Keil, D.E.; Peden-Adams, M.M. Immune function in female B(6)C(3)F(1) mice is modulated by DE-71, a commercial polybrominated diphenyl ether mixture. *J. Immunotoxicol.* **2012**, *9*, 96–107. [CrossRef]
21. Zeng, W.; Wang, Y.; Liu, Z.; Khanniche, A.; Hu, Q.; Feng, Y.; Ye, W.; Yang, J.; Wang, S.; Zhou, L.; et al. Long-term exposure to decabrominated diphenyl ether impairs CD8 T-cell function in adult mice. *Cell. Mol. Immunol.* **2014**, *11*, 367–376. [CrossRef]
22. Fowles, J.; Fairbrother, A.; Baecher-Steppan, L.; Kerkvliet, N. Immunologic and endocrine effects of the flame-retardant pentabromodiphenyl ether (DE-71) in C57BL/6J mice. *Toxicology* **1994**, *86*, 49–61. [CrossRef]
23. Arkoosh, M.R.; Boylen, D.; Dietrich, J.; Anulacion, B.F.; Bravo, C.F.; Johnson, L.L.; Loge, F.J.; Collier, T.K. Disease susceptibility of salmon exposed to polybrominated diphenyl ethers (PBDEs). *Aquat. Toxicol.* **2010**, *98*, 51–59. [CrossRef]
24. Cary, T.L.; Ortiz-Santaliestra, M.E.; Karasov, W.H. Immunomodulation in post-metamorphic northern leopard frogs, Lithobates pipiens, following larval exposure to polybrominated diphenyl ether. *Environ. Sci. Technol.* **2014**, *48*, 5910–5919. [CrossRef]
25. Koike, E.; Yanagisawa, R.; Takigami, H.; Takano, H. Brominated flame retardants stimulate mouse immune cells in vitro. *J. Appl. Toxicol.* **2013**, *33*, 1451–1459. [CrossRef] [PubMed]
26. Mynster Kronborg, T.; Frohnert Hansen, J.; Nielsen, C.H.; Ramhoj, L.; Frederiksen, M.; Vorkamp, K.; Feldt-Rasmussen, U. Effects of the Commercial Flame Retardant Mixture DE-71 on Cytokine Production by Human Immune Cells. *PLoS ONE* **2016**, *11*, e0154621. [CrossRef] [PubMed]
27. Lv, Q.Y.; Wan, B.; Guo, L.H.; Zhao, L.; Yang, Y. In vitro immune toxicity of polybrominated diphenyl ethers on murine peritoneal macrophages: Apoptosis and immune cell dysfunction. *Chemosphere* **2015**, *120*, 621–630. [CrossRef] [PubMed]

28. Chen, G.; Jiang, X.; Gu, C.; Sun, C.; Li, M. Bioaccessibility of BDE 47 in a simulated gastrointestinal system and its metabolic transformation mechanisms in Caco-2 cells. *Chemosphere* **2019**, *214*, 408–417. [CrossRef] [PubMed]
29. Neumann, S.; Harju, M.; Herzke, D.; Anker-Nilssen, T.; Christensen-Dalsgaard, S.; Langset, M.; Gabrielsen, G.W. Ingested plastics in northern fulmars (Fulmarus glacialis): A pathway for polybrominated diphenyl ether (PBDE) exposure? *Sci. Total Environ.* **2021**, *778*, 146313. [CrossRef] [PubMed]
30. Salam, K.A.; Furuta, A.; Noda, N.; Tsuneda, S.; Sekiguchi, Y.; Yamashita, A.; Moriishi, K.; Nakakoshi, M.; Tani, H.; Roy, S.R.; et al. PBDE: Structure-activity studies for the inhibition of hepatitis C virus NS3 helicase. *Molecules* **2014**, *19*, 4006–4020. [CrossRef]
31. Wei, R.G.; Zhao, Y.X.; Liu, P.Y.; Qin, Z.F.; Yan, S.S.; Li, Y.; Qin, X.F.; Xia, X.J.; Xu, X.B.; Yan, M.C. Determination of environmentally relevant exposure concentrations of polybrominated diphenyl ethers for in vitro toxicological studies. *Toxicol. In Vitro* **2010**, *24*, 1078–1085. [CrossRef]
32. Chen, Y.; Zhou, Z.; Min, W. Mitochondria, Oxidative Stress and Innate Immunity. *Front. Physiol.* **2018**, *9*, 1487. [CrossRef]
33. Scherlinger, M.; Tsokos, G.C. Reactive oxygen species: The Yin and Yang in (auto-)immunity. *Autoimmun. Rev.* **2021**, *20*, 102869. [CrossRef]
34. Zhang, P.; Li, Y.; Xu, W.; Cheng, J.; Zhang, C.; Gao, J.; Li, Z.; Tao, L.; Zhang, Y. Immunotoxicity induced by Ivermectin is associated with NF-kappaB signaling pathway on macrophages. *Chemosphere* **2022**, *289*, 133087. [CrossRef]
35. Sun, S.; Shi, W.; Tang, Y.; Han, Y.; Du, X.; Zhou, W.; Hu, Y.; Zhou, C.; Liu, G. Immunotoxicity of petroleum hydrocarbons and microplastics alone or in combination to a bivalve species: Synergic impacts and potential toxication mechanisms. *Sci. Total Environ.* **2020**, *728*, 138852. [CrossRef] [PubMed]
36. Naasri, S.; Helali, I.; Aouni, M.; Mastouri, M.; Harizi, H. N-acetylcysteine reduced the immunotoxicity effects induced in vitro by azoxystrobin and iprodione fungicides in mice. *Environ. Toxicol.* **2021**, *36*, 562–571. [CrossRef] [PubMed]
37. Frouin, H.; Lebeuf, M.; Hammill, M.; Masson, S.; Fournier, M. Effects of individual polybrominated diphenyl ether (PBDE) congeners on harbour seal immune cells in vitro. *Mar. Pollut. Bull.* **2010**, *60*, 291–298. [CrossRef] [PubMed]
38. Gordon, S.; Martinez-Pomares, L. Physiological roles of macrophages. *Pflugers. Arch.* **2017**, *469*, 365–374. [CrossRef] [PubMed]
39. Arts, R.J.W.; Netea, M.G. Adaptive Characteristics of Innate Immune Responses in Macrophages. *Microbiol. Spectr.* **2016**, *4*, 4. [CrossRef]
40. Lundgren, M.; Darnerud, P.O.; Blomberg, J.; Friman, G.; Ilback, N.G. Polybrominated diphenyl ether exposure suppresses cytokines important in the defence to coxsackievirus B3 infection in mice. *Toxicol. Lett.* **2009**, *184*, 107–113. [CrossRef] [PubMed]
41. Beineke, A.; Siebert, U.; McLachlan, M.; Bruhn, R.; Thron, K.; Failing, K.; Müller, G.; Baumgärtner, W. Investigations of the potential influence of environmental contaminants on the thymus and spleen of harbor porpoises (*Phocoena phocoena*). *Environ. Sci. Technol.* **2005**, *39*, 3933–3938. [CrossRef] [PubMed]
42. Zhou, Z.; Chen, H.; Li, Y.; Liu, Q.; Lu, K.; Zhu, X.; Wang, Y. Transcriptome and biochemical analyses of rainbow trout (*Oncorhynchus mykiss*) RTG-2 gonadal cells in response to BDE-47 stress indicates effects on cell proliferation. *Aquat. Toxicol.* **2022**, *245*, 106108. [CrossRef] [PubMed]
43. Chen, H.; Tang, X.; Zhou, B.; Zhou, Z.; Xu, N.; Wang, Y. A ROS-mediated mitochondrial pathway and Nrf2 pathway activation are involved in BDE-47 induced apoptosis in Neuro-2a cells. *Chemosphere* **2017**, *184*, 679–686. [CrossRef] [PubMed]
44. Hu, X.; Zhang, J.; Jiang, Y.; Lei, Y.; Lu, L.; Zhou, J.; Huang, H.; Fang, D.; Tao, G. Effect on metabolic enzymes and thyroid receptors induced by BDE-47 by activation the pregnane X receptor in HepG2, a human hepatoma cell line. *Toxicol. In Vitro* **2014**, *28*, 1377–1385. [CrossRef]
45. Yang, C.; Wong, C.M.; Wei, J.; Chung, A.C.K.; Cai, Z. The brominated flame retardant BDE 47 upregulates purine metabolism and mitochondrial respiration to promote adipocyte differentiation. *Sci. Total Environ.* **2018**, *644*, 1312–1322. [CrossRef]
46. Dong, L.; Li, P.; Yang, K.; Liu, L.; Gao, H.; Zhou, G.; Zhao, Q.; Xia, T.; Wang, A.; Zhang, S. Promotion of mitochondrial fusion protects against developmental PBDE-47 neurotoxicity by restoring mitochondrial homeostasis and suppressing excessive apoptosis. *Theranostics* **2020**, *10*, 1245–1261. [CrossRef]
47. Song, J.; Li, Y.; Zhao, C.; Zhou, Q.; Zhang, J. Interaction of BDE-47 with nuclear receptors (NRs) based on the cytotoxicity: In vitro investigation and molecular interaction. *Ecotoxicol. Environ. Saf.* **2021**, *208*, 111390. [CrossRef]
48. Sun, S.; Zhao, Z.; Rao, Q.; Li, X.; Ruan, Z.; Yang, J. BDE-47 induces nephrotoxicity through ROS-dependent pathways of mitochondrial dynamics in PK15 cells. *Ecotoxicol. Environ. Saf.* **2021**, *222*, 112549. [CrossRef]
49. Park, H.R.; Loch-Caruso, R. Protective effect of nuclear factor E2-related factor 2 on inflammatory cytokine response to brominated diphenyl ether-47 in the HTR-8/SVneo human first trimester extravillous trophoblast cell line. *Toxicol. Appl. Pharmacol.* **2014**, *281*, 67–77. [CrossRef]
50. Liang, L.; Pan, Y.; Bin, L.; Liu, Y.; Huang, W.; Li, R.; Lai, K.P. Immunotoxicity mechanisms of perfluorinated compounds PFOA and PFOS. *Chemosphere* **2022**, *291 Pt 2*, 132892. [CrossRef]
51. Su, R.; Jin, X.; Zhang, W.; Li, Z.; Liu, X.; Ren, J. Particulate matter exposure induces the autophagy of macrophages via oxidative stress-mediated PI3K/AKT/mTOR pathway. *Chemosphere* **2017**, *167*, 444–453. [CrossRef]
52. Huang, F.; Liu, Q.; Xie, S.; Xu, J.; Huang, B.; Wu, Y.; Xia, D. Cypermethrin Induces Macrophages Death through Cell Cycle Arrest and Oxidative Stress-Mediated JNK/ERK Signaling Regulated Apoptosis. *Int. J. Mol. Sci.* **2016**, *17*, 885. [CrossRef]
53. Zhang, H.; Shen, L.; Fang, W.; Zhang, X.; Zhong, Y. Perfluorooctanoic acid-induced immunotoxicity via NF-kappa B pathway in zebrafish (*Danio rerio*) kidney. *Fish Shellfish Immunol.* **2021**, *113*, 9–19. [CrossRef]

54. Zamani, E.; Shaki, F.; AbedianKenari, S.; Shokrzadeh, M. Acrylamide induces immunotoxicity through reactive oxygen species production and caspase-dependent apoptosis in mice splenocytes via the mitochondria-dependent signaling pathways. *Biomed. Pharmacother.* **2017**, *94*, 523–530. [CrossRef]
55. Faas, M.M.; de Vos, P. Mitochondrial function in immune cells in health and disease. *Biochim. Biophys. Acta Mol. Basis Dis.* **2020**, *1866*, 165845. [CrossRef]
56. Murray, P.J. Macrophage Polarization. *Annu. Rev. Physiol.* **2017**, *79*, 541–566. [CrossRef]
57. Lu, X.; Li, M.; Wu, C.; Zhou, C.; Zhang, J.; Zhu, Q.; Shen, T. Bisphenol A promotes macrophage proinflammatory subtype polarization via upregulation of IRF5 expression in vitro. *Toxicol. In Vitro* **2019**, *60*, 97–106. [CrossRef]
58. Zhao, Q.; Chen, H.; Yang, T.; Rui, W.; Liu, F.; Zhang, F.; Zhao, Y.; Ding, W. Direct effects of airborne PM2.5 exposure on macrophage polarizations. *Biochim. Biophys. Acta* **2016**, *1860*, 2835–2843. [CrossRef]
59. Virag, L.; Jaen, R.I.; Regdon, Z.; Bosca, L.; Prieto, P. Self-defense of macrophages against oxidative injury: Fighting for their own survival. *Redox Biol.* **2019**, *26*, 101261. [CrossRef]
60. Dupré-Crochet, S.; Erard, M.; Nüβe, O. ROS production in phagocytes: Why, when, and where? *J. Leukocyte Biol.* **2013**, *94*, 657–670. [CrossRef]
61. Morgan, M.J.; Liu, Z.G. Crosstalk of reactive oxygen species and NF-κB signaling. *Cell Res.* **2011**, *21*, 103–115. [CrossRef]
62. Herb, M.; Schramm, M. Functions of ROS in Macrophages and Antimicrobial Immunity. *Antioxidants* **2021**, *10*, 313. [CrossRef]
63. Costa Carvalho, J.; Brito, A. The M1/M2 Pattern and the Oxidative Stress are Modulated by Low-Level Laser in Human Macrophage. *J. Clin. Cell. Immunol.* **2016**, *7*, 387.
64. Hou, L.; Zhou, X.; Gan, F.; Liu, Z.; Zhou, Y.; Qian, G.; Huang, K. Combination of Selenomethionine and N-Acetylcysteine Alleviates the Joint Toxicities of Aflatoxin B1 and Ochratoxin A by ERK MAPK Signal Pathway in Porcine Alveolar Macrophages. *J. Agric. Food. Chem.* **2018**, *66*, 5913–5923. [CrossRef]
65. Schweikl, H.; Birke, M.; Gallorini, M.; Petzel, C.; Bolay, C.; Waha, C.; Hiller, K.A.; Buchalla, W. HEMA-induced oxidative stress inhibits NF-kappaB nuclear translocation and TNF release from LTA- and LPS-stimulated immunocompetent cells. *Dent. Mater.* **2021**, *37*, 175–190. [CrossRef]
66. Chen, H.; Tang, X.; Zhou, B.; Xu, N.; Zhou, Z.; Fang, K.; Wang, Y. BDE-47 and BDE-209 inhibit proliferation of Neuro-2a cells via inducing G1-phase arrest. *Environ. Toxicol. Pharmacol.* **2017**, *50*, 76–82. [CrossRef]
67. Zhou, Z.; Tang, X.; Chen, H.; Wang, Y. Comparative studies of saxitoxin (STX)-induced cytotoxicity in Neuro-2a and RTG-2 cell lines: An explanation with respect to changes in ROS. *Chemosphere* **2018**, *192*, 66–74. [CrossRef]
68. Zhou, Z.; Zhou, B.; Chen, H.; Tang, X.; Wang, Y. Reactive oxygen species (ROS) and the calcium-($Ca^{2+}$) mediated extrinsic and intrinsic pathways underlying BDE-47-induced apoptosis in rainbow trout (Oncorhynchus mykiss) gonadal cells. *Sci. Total Environ.* **2019**, *656*, 778–788. [CrossRef]
69. Santos, S.S.; Brunialti, M.K.C.; Rodrigues, L.; Liberatore, A.M.A.; Koh, I.H.J.; Martins, V.; Soriano, F.G.; Szabo, C.; Salomao, R. Effects of the PARP Inhibitor Olaparib on the Response of Human Peripheral Blood Leukocytes to Bacterial Challenge or Oxidative Stress. *Biomolecules* **2022**, *12*, 788. [CrossRef]
70. Zhao, C.; Tang, Z.; Xie, P.; Lin, K.; Chung, A.C.K.; Cai, Z. Immunotoxic Potential of Bisphenol F Mediated through Lipid Signaling Pathways on Macrophages. *Environ. Sci. Technol.* **2019**, *53*, 11420–11428. [CrossRef]
71. Ma, J.; Liu, R.; Wang, X.; Liu, Q.; Chen, Y.; Valle, R.P.; Zuo, Y.Y.; Xia, T.; Liu, S. Crucial Role of Lateral Size for Graphene Oxide in Activating Macrophages and Stimulating Pro-inflammatory Responses in Cells and Animals. *ACS Nano* **2015**, *9*, 10498–10515. [CrossRef]

*molecules*

MDPI

*Article*

# Integrated Analysis of Transcriptome and microRNA Profile Reveals the Toxicity of *Euphorbia* Factors toward Human Colon Adenocarcinoma Cell Line Caco-2

**Lingyue Zou [1,†], Wenqiang Bao [1,†], Yadong Gao [2,3], Mengting Chen [1], Yajiao Wu [1], Shuo Wang [2], Chutao Li [1], Jian Zhang [1], Dongcheng Zhang [1], Qi Wang [2,4,5,*] and An Zhu [1,*]**

1 Key Laboratory of Ministry of Education for Gastrointestinal Cancer, School of Basic Medical Sciences, Fujian Medical University, Fuzhou 350108, China
2 Department of Toxicology, School of Public Health, Peking University, Beijing 100191, China
3 Fujian Provincial Key Laboratory of Zoonosis Research, Fujian Center for Disease Control and Prevention, Fuzhou 350001, China
4 Key Laboratory of State Administration of Traditional Chinese Medicine for Compatibility Toxicology, Beijing 100191, China
5 Beijing Key Laboratory of Toxicological Research and Risk Assessment for Food Safety, Beijing 100191, China
* Correspondence: wangqi@bjmu.edu.cn (Q.W.); zhuan@fjmu.edu.cn (A.Z.)
† These authors contributed equally to this work.

**Abstract:** *Euphorbia* factors, lathyrane-type diterpenoids isolated from the medical herb *Euphorbia lathyris* L. (*Euphorbiaceae*), have been associated with intestinal irritation toxicity, but the mechanisms underlying this phenomenon are still unknown. The objective of this study was to evaluate the transcriptome and miRNA profiles of human colon adenocarcinoma Caco-2 cells in response to *Euphorbia* factors L1 (EFL1) and EFL2. Whole transcriptomes of mRNA and microRNA (miRNA) were obtained using second generation high-throughput sequencing technology in response to 200 μM EFL treatment for 72 h, and the differentially expressed genes and metabolism pathway were enriched. Gene structure changes were analyzed by comparing them with reference genome sequences. After 72 h of treatment, 16 miRNAs and 154 mRNAs were differently expressed between the EFL1 group and the control group, and 47 miRNAs and 1101 mRNAs were differentially expressed between the EFL2 group and the control. Using clusters of orthologous protein enrichment, the sequenced mRNAs were shown to be mainly involved in transcription, post-translational modification, protein turnover, chaperones, signal transduction mechanisms, intracellular trafficking, secretion, vesicular transport, and the cytoskeleton. The differentially expressed mRNA functions and pathways were enriched in transmembrane transport, T cell extravasation, the IL-17 signaling pathway, apoptosis, and the cell cycle. The differentially expressed miRNA EFLs caused changes in the structure of the gene, including alternative splicing, insertion and deletion, and single nucleotide polymorphisms. This study reveals the underlying mechanism responsible for the toxicity of EFLs in intestinal cells based on transcriptome and miRNA profiles of gene expression and structure.

**Keywords:** *Euphorbia* semen; *Euphorbia* factors; transcriptome; RNA sequencing; bioinformatics; gene structure

**Citation:** Zou, L.; Bao, W.; Gao, Y.; Chen, M.; Wu, Y.; Wang, S.; Li, C.; Zhang, J.; Zhang, D.; Wang, Q.; et al. Integrated Analysis of Transcriptome and microRNA Profile Reveals the Toxicity of *Euphorbia* Factors toward Human Colon Adenocarcinoma Cell Line Caco-2. *Molecules* **2022**, *27*, 6931. https://doi.org/10.3390/molecules27206931

Academic Editors: Chongwen Wang and Guohui Sun

Received: 20 September 2022
Accepted: 14 October 2022
Published: 16 October 2022



## 1. Introduction

*Euphorbia* semen is the ripe and dry seeds of *Euphorbia lathyris* Linnaeus (*Euphorbiaceae*), which are widely distributed in East Asia, South Europe, Russia, Australia, and America [1]. In China, *Euphorbia semen* has been used as a traditional medicine for thousands of years, since the Song dynasty. It is used for the treatment of amenorrhea, edema, abdominal distension, constipation, oliguria, and anuria by oral administration, as well as for therapy for stubborn dermatophytoses and warts by external use on the skin [2,3]. In recent years, *Euphorbia* semen has been reported to have excellent pharmacological actions allowing it to inhibit adipogenesis, reverse tumor multidrug resistance, and treat terminal schistosomiasis, giving it the potential to be developed as a weight-loss and oncotherapy drug [4–6].

Safety should be a matter of concern during drug use. In the Compendium of Materia Medica, *Euphorbia* semen was classified as a toxic medicinal herb. In the clinical setting, the recommended oral dosage of *Euphorbia* semen is 1–2 g in accordance with the 2020 edition of the Chinese Pharmacopoeia [7]. At an overdose of 9–15 g, the poisoning symptoms are generally manifested as dizziness, nausea, vomiting, palpitations, spontaneous cold sweating, and a pale complexion. However, blood pressure drops, profuse sweating, cold limbs, shallow and thick breathing, and a weak pulse can occur in serious illness cases [8]. After intragastric administration of an aqueous extract of *Euphorbia* semen, the $LD_{50}$ was 1.7950 g/kg (95% confidence interval of 1.6211–1.9879) in mice [9]. So far, more than 15 types of 240 compounds have been isolated and identified from *Euphorbia* semen, and *Euphorbia* factors belonging to the lathyrane-type diterpenoids family are the main toxic components [10]. *Euphorbia* factors L1 (EFL1) and L2 (EFL2) can accelerate the peristalsis rate of the small intestine and thus contribute to the induction of intestinal tract irritation and diarrhea [11]. The intense intestinal toxicity of *Euphorbia* semen has limited its safe clinical use.

As a classical toxic herb that is used to target to the intestine in traditional Chinese medicine, there is an urgent need to clarify the intestinal toxicological mechanism associated with *Euphorbia* semen, especially regarding *Euphorbia* factors. Next generation high-throughput RNA sequencing (RNA-seq) provides methods with which to find the gene targets of toxic xenobiotics. With the advantages of low cost and time, ultra-high data output and high accuracy, RNA-seq has been introduced into the pharmacological and toxicological studies of traditional medicines such as Fuzi decoction [12], lycorine [13], Shuxinyin formula [14], and so forth. Messenger RNA (mRNA) is formed by transcription and executes its biological functions after being translated into proteins. However, the translation of mRNA into functional proteins can be silenced by microRNA (miRNA) at the post-transcriptional level [15]. miRNA, a kind of small non-coding RNA, is about 21–23 nucleobases in length, and widely exists in viruses, plants, and animals with high evolutionary conservation [16]. miRNA is transcribed from DNA sequences into pri-miRNA, before being processed into pre-miRNA and exported into the cytoplasm. After being processed by RNase III Dicer, the mature miRNA duplex is produced [17]. To form the RNA-induced silencing complex (RISC), the miRNA duplex is loaded into the Argonaut (Ago) family of proteins. The 5′ ends of Ago proteins from the RISCs of miRNA will pair with 3′ untranslated regions (3′-UTRs) of cytoplasmic mRNAs, inhibiting translation or promoting the degradation of the mRNAs [18]. The 3′-UTR is the main region of binding, and other regions of gene promoters, the 5′UTR, and the coding sequence have also been reported as binding locations [19].

To provide evidence for safe clinical drug use, in the present study, we comprehensively used mRNA and miRNA sequencing technologies to explore the mechanisms underlying EFL-induced intestinal toxicity in the human colon adenocarcinoma cell line Caco-2. The gene expression pattern and structural modification were analyzed, and the gene function was enriched by bioinformatics analysis.

## 2. Results

### 2.1. *Effects of EFLs on the Survival of Caco-2 Cells*

The cytotoxic effects of EFL1 and EFL2 on Caco-2 were assessed at different time intervals using the MTT assay. EFL1 and EFL2 significantly inhibited cell viability in a concentration-dependent manner (Figure 1). After being exposed to 200 μM EFL1 for 72 h, the cell viability of Caco-2 cells was 77.9%; after being exposed to EFL2 concentrations of 50, 100, and 200 μM, the cell viability decreased to 91.4%, 80.2%, and 73.1%, respectively.

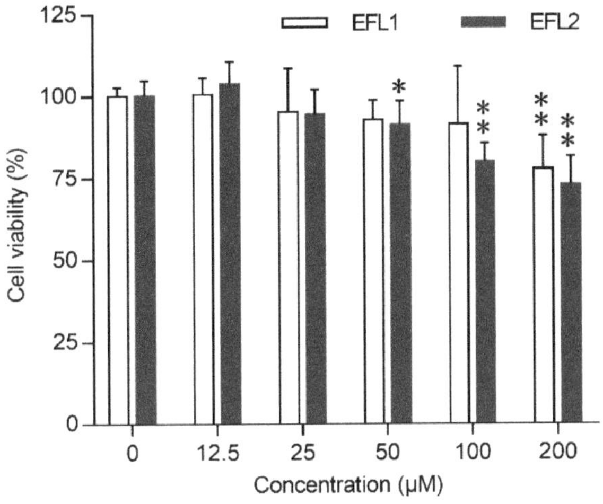


**Figure 1.** Cytotoxicity of EFLs in Caco-2 cells. Concentration effect of EFLs on cell viability was determined by the MTT assay, the cells viability of each EFLs group was expressed in percentage in comparison with vehicle control group. Data were presented as mean ± standard deviation from three independent experiments. * $p < 0.05$ or ** $p < 0.01$ compared with vehicle control of the same compound.

### 2.2. Quality Control and Mapping of RNA Sequencing Data

Based on the cell viability of Caco-2 cells after EFL exposure, a concentration of 200 μM was chosen for further study. High-throughput sequencing was performed for the analysis of mRNA and miRNA expression in Caco-2 cells with or without EFLs exposure. As shown in Table S3, average quantities of 69.4, 74.7, and 71.0 million raw reads of mRNA were obtained from the control, EFL1, and EFL2 groups, respectively. After checking the read quality, the Q20 was 98.3% and Q30 was 94.5% in all samples. In the control, EFL1, and EFL2 groups, 68.4, 73.6, and 70.1 million clean reads were chosen to map to the human reference genome, and the unique mapping ratios were 90.04%, 89.74%, and 89.27%, respectively.

The miRNA quality control is shown in Table S4. The Q20 was 99.43% and Q30 was 98.2% in all the samples. In the control, EFL1, and EFL2 groups, average quantities of 19.3, 18.3, and 18.9 million raw reads were obtained with 16.9, 15.4, and 16.6 million clean reads, respectively, and 8.5, 7.3, and 8.2 million reads were mapped to the positive and negative RNAs of the human reference genome.

### 2.3. Assembly of mRNA Transcripts

The mRNA transcripts were assembled in StringTie software. Based on the network flow algorithms and optional de novo assembly, the complex datasets were assembled into transcripts. The number of transcripts contained in most genes was <4, and the number of genes consisting of only one transcript reached 34,965 (Figure S1A in Supplementary Materials). All samples contained a total of 221,912 transcripts, including 66,911 transcripts > 1800 bp, followed by 49,384 transcripts with a length of 401–600 bp (Figure S1B). Most of the transcripts contained exons with fewer than 10 bp, and the number of transcripts consisting of 2 exons reached 32,846 (Figure S1C).

### 2.4. Functional Annotation of mRNA and miRNA in Databases

The mRNA functional annotation was performed using the COG, GO, KEGG, NR, Swiss-Prot, and Pfam databases (Table S5). In total, 203,268 (97.48%) transcripts and 46,195 (78.45%) genes were successfully annotated. Most of the transcripts and genes were annotated in NR (89.39% and 77.46%, respectively), followed by COG (83.44% and 68.78%, respectively), Swiss-Prot (82.16% and 66.89%, respectively), GO (70.57% and 53.47%, respectively), KEGG (67.14% and 48.15%, respectively), and Pfam (41.17% and 28.11%, respectively).

The known and novel miRNA functional annotation were performed in miRBase and miRDeep2, respectively. An average of 1153, 1109, and 1130 known miRNAs, as well as 356, 355, and 356 novel miRNAs were reported in the control, EFL1, and EFL2 groups, respectively (Table S6).

### 2.5. The Fold Change in mRNA and miRNA Quantified by RT-qPCR

In the differentially expressed genes, a total of 14 mRNAs and 10 miRNAs were detected using RT-qPCR, and the correlation between the fold change in RT-qPCR and mRNA-seq was analyzed using linear regression. In the EFL1 group, the mRNA had a regression equation of $Y = 0.892X - 2.761$, $R^2 = 0.957$ (Figure S2A), with miRNA having $Y = 1.108X - 0.025$, $R^2 = 0.967$ (Figure S2B). In the EFL2 group, the mRNA had a regression equation of $Y = 0.924X - 3.759$, $R^2 = 0.9903$ (Figure S2C), with miRNA having $Y = 0.941X + 0.079$, $R^2 = 0.9987$ (Figure S2D). The gene expression levels detected by RT-qPCR and mRNA-seq were consistent.

### 2.6. Differentially Expressed Genes and miRNA-mRNA Crosstalk

Standards of fold change (FC) $\geq 2$ and $p$ value $\leq 0.05$ were set as significant up- or downregulated differentially expressed genes (DEGs) for mRNA, and FC $\geq 1.5$ and $p$ value $\leq 0.05$ were used for miRNA. After treatment of EFL1, a total of 16 miRNAs were differentially expressed compared with the control group, of which 6 miRNAs were downregulated and 10 miRNAs were upregulated (Figure 2A); 154 mRNAs were differentially expressed, of which 66 mRNAs were downregulated and 88 miRNAs were upregulated (Figure 2C). In the EFL2 group, 47 miRNAs were differentially expressed compared with the control group, of which 23 miRNAs were downregulated and 24 mRNAs were upregulated (Figure 2B); 1101 mRNAs were differentially expressed, of which 695 mRNAs were downregulated and 406 mRNAs were upregulated (Figure 2D).

To describe the miRNA-mRNA crosstalk, the top five differentially expressed miRNAs, namely, hsa-miR-6774-3p, hsa-miR-6715a-3p, hsa-miR-4523, hsa-miR-548av-3p, and hsa-miR-616-3p in the EFL1 group, and hsa-miR-5090, hsa-miR-1255b-5p, hsa-miR-4284, hsa-miR-146a-3p, and hsa-miR-4458 in the EFL2 group were selected. As shown in Figure 2E,F, the target genes of miRNA are presented in respect of three databases: miRnada, TargetScan, and RNAhybrid.

The overlaps between the target mRNAs of the top five differentially expressed miRNAs and differentially expressed mRNAs were analyzed. In the EFL1 group, 100 target mRNAs of miRNAs were reported, but no genes were in the overlap. In the EFL2 group, 162 target mRNAs of miRNAs were reported, and 7 of them were in the overlap, namely MDGA1, C3orf18, QRICH2, PROM2, AKR1C1, HPGD, and CCDC148.

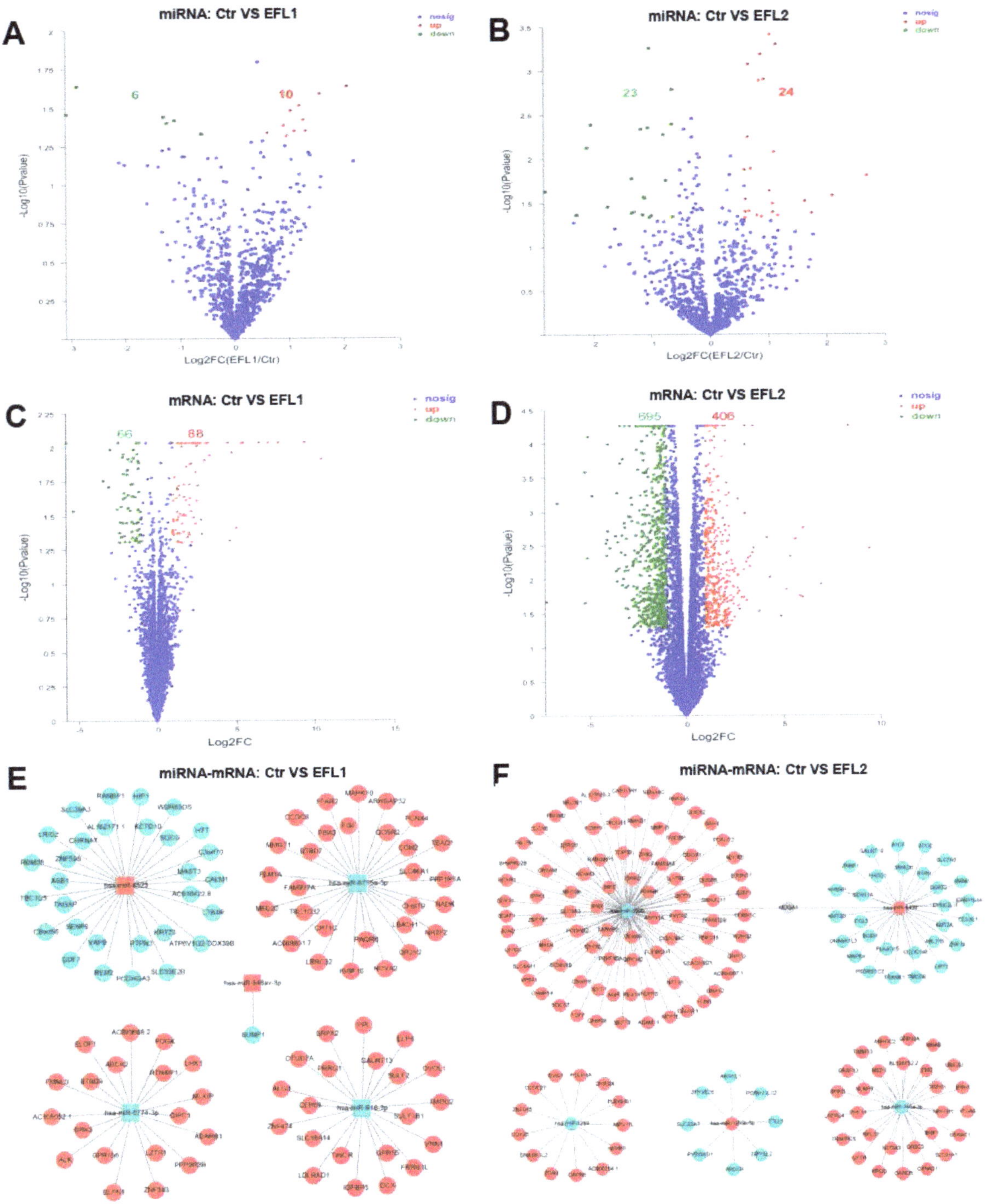


**Figure 2.** The volcano map of differentially expressed mRNA and miRNA between control and EFLs (**A–D**). The crosstalk between top 5 differentially expressed miRNA and corresponding regulated mRNA (**E**,**F**). The square is miRNA, and the circle is mRNA. Red is upregulation, and green is downregulation.

## 2.7. Functional Annotation of Differentially Expressed mRNA in EFL-Treated Cells

The COG functional annotation of differentially expressed mRNA is shown in Figure 3A. Most mRNA functions were enriched in intracellular trafficking, secretion, and vesicular transport, as well as post-translational modification, protein turnover, chaperones, transcription, signal transduction mechanisms, and the cytoskeleton.

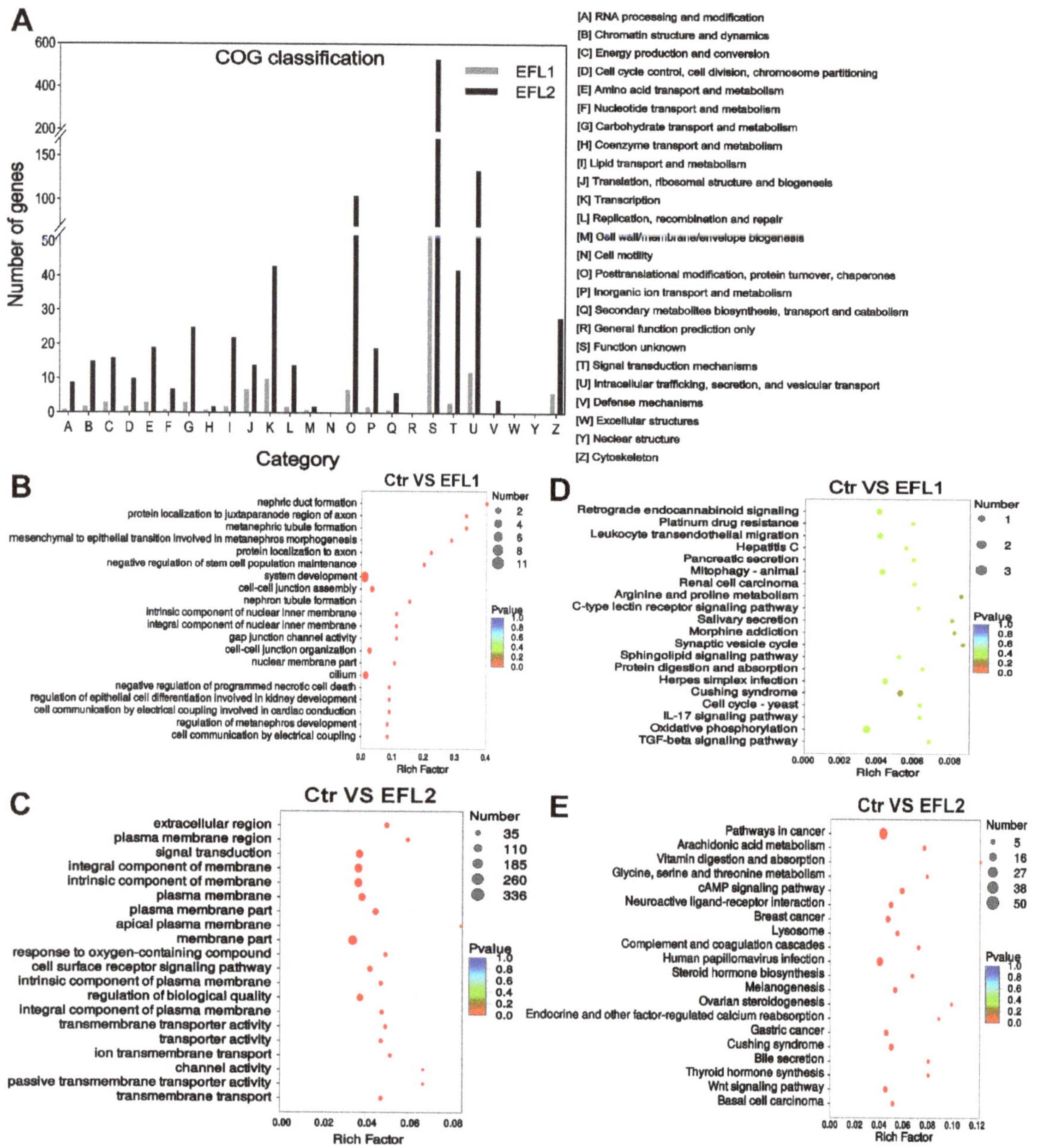


**Figure 3.** (**A**) The COG classification of differentially expressed mRNA in EFL-treated Caco-2 cells. (**B**) GO and (**C**) KEGG annotation of the differentially expressed mRNA in the EFL1 group. (**D**) GO and (**E**) KEGG annotation of the differentially expressed mRNA in the EFL2 group.

The GO enrichment of the differentially expressed mRNA indicated that EFL1 affects the negative regulation of programmed necrotic cell death, cell–cell junction assembly, gap junction channel activity, cell–cell organization, and cell communication by electrical coupling (Figure 3B). EFL2 affects the response to oxygen-containing compounds, the cell surface receptor signaling pathway, transmembrane transporter activity, ion transmembrane transport, channel activity, and passive transmembrane transporter activity (Figure 3D).

The KEGG enrichment of the differentially expressed mRNA indicated that EFL1 mainly affects mitophagy, protein digestion and absorption, the IL-17 signaling pathway, oxidative phosphorylation, and the TGF-beta signaling pathway (Figure 3C). EFL2 affects the arachidonic acid metabolism, cAMP signaling pathway, lysosome, bile secretion, and the Wnt signaling pathway (Figure 3E).

### *2.8. Functional Annotation of the Target Genes of Differentially Expressed miRNA in EFL-Treated Cells*

The COG functional annotation of the target genes of differentially expressed miRNAs is shown in Figure 4A, most functions were enriched in terms of transcription, post-translational modification, protein turnover, chaperones, signal transduction mechanisms, intracellular trafficking, secretion, and vesicular transport.

In the EFL1-treated group, the GO enrichment of the target genes of differentially expressed miRNAs showed enrichment of the term's biological regulation, membrane, protein binding, and intracellular membrane-bounded organelle (Figure 4B). In the EFL2-treated group, the enriched terms were regulation of multicellular organismal development, cell projection, regulation of transcription from the RNA polymerase II promoter, positive regulation of gene expression and membrane region (Figure 4D).

In the EFL1-treated group, the KEGG enrichment of the target genes of differentially expressed miRNA showed enrichment of the terms AMPK signaling pathway, autophagy, and inflammatory mediator regulation of TRP channels (Figure 4C). In the EFL2-treated group, the enriched terms were the MAPK signaling pathway, chemokine signaling pathway, Ras signaling pathway, and cAMP signaling pathway (Figure 4E).

### *2.9. The Protein–Protein Interaction (PPI) of Differentially Expressed mRNA*

The genes network was investigated through analysis of PPIs. After EFL1 treatment for 72 h in Caco-2 cells, the core genes involved in the network were shown to be *RPL7, RPL7A, RPL14, MT-ND5, ATP5G1, BMS1, RPS15, PKM, RPL21, WDR12, SKP1, EPO, SLC12A5, OCLN* and *PAX2* (Figure 5A). These genes regulate peptide chain elongation, translational control, respiratory electron transport, ATP synthesis by chemiosmotic coupling, rRNA processing in the nucleus and cytosol, RNA binding, processing of capped intron-containing pre-mRNA, mitotic G1 phase and G1/S transition, development EPO-induced Jak-STAT pathway, K-Cl cotransporter activity, tight junction paracellular permeability barrier, and transcription factor binding.

In EFL2-treated Caco-2 cells, the core genes involved in the PPI network were shown to be *RPS15, RPL4, RPL7, RPS2, RPL15, RPSAP58, RPS3, RPL13, RPS20, RPS11, ALB, FN1, SERPINE1, IFIT1, IFITM1* and *MT-ATP6* (Figure 5B). These genes were found to be involved in peptide chain elongation, RNA binding and structural constituents of the ribosome, laminin receptor activity, the response to elevated platelet cytosolic $Ca^{2+}$, integrin family cell surface interactions, signaling receptor binding, the interferons-mediated signaling pathway, the innate immune system, respiratory electron transport, ATP synthesis by chemiosmotic coupling, and transcriptional activation of mitochondrial biogenesis.

### *2.10. The Functional Annotation of Differentially Co-Expressed mRNA*

After EFL1 and EFL2 treatment in Caco-2 cells, 57 mRNAs were the differentially co-expressed as the union (Figure 6A), and a cluster analysis heat map was created (Figure 6B). The GO enrichment indicated that the differentially co-expressed mRNA affected plasma lipoprotein particle oxidation, the positive regulation of T cell extravasation, death effector domain binding, and the neuron apoptotic process (Figure 6C). The KEGG enrichment terms included platinum drug resistance, the RIG-I-like receptor signaling pathway, ribosome biogenesis in eukaryotes, IL-17 signaling pathway, apoptosis, the NOD-like receptor signaling pathway, oxidative phosphorylation, cell adhesion molecules, and the cell cycle (Figure 6D).

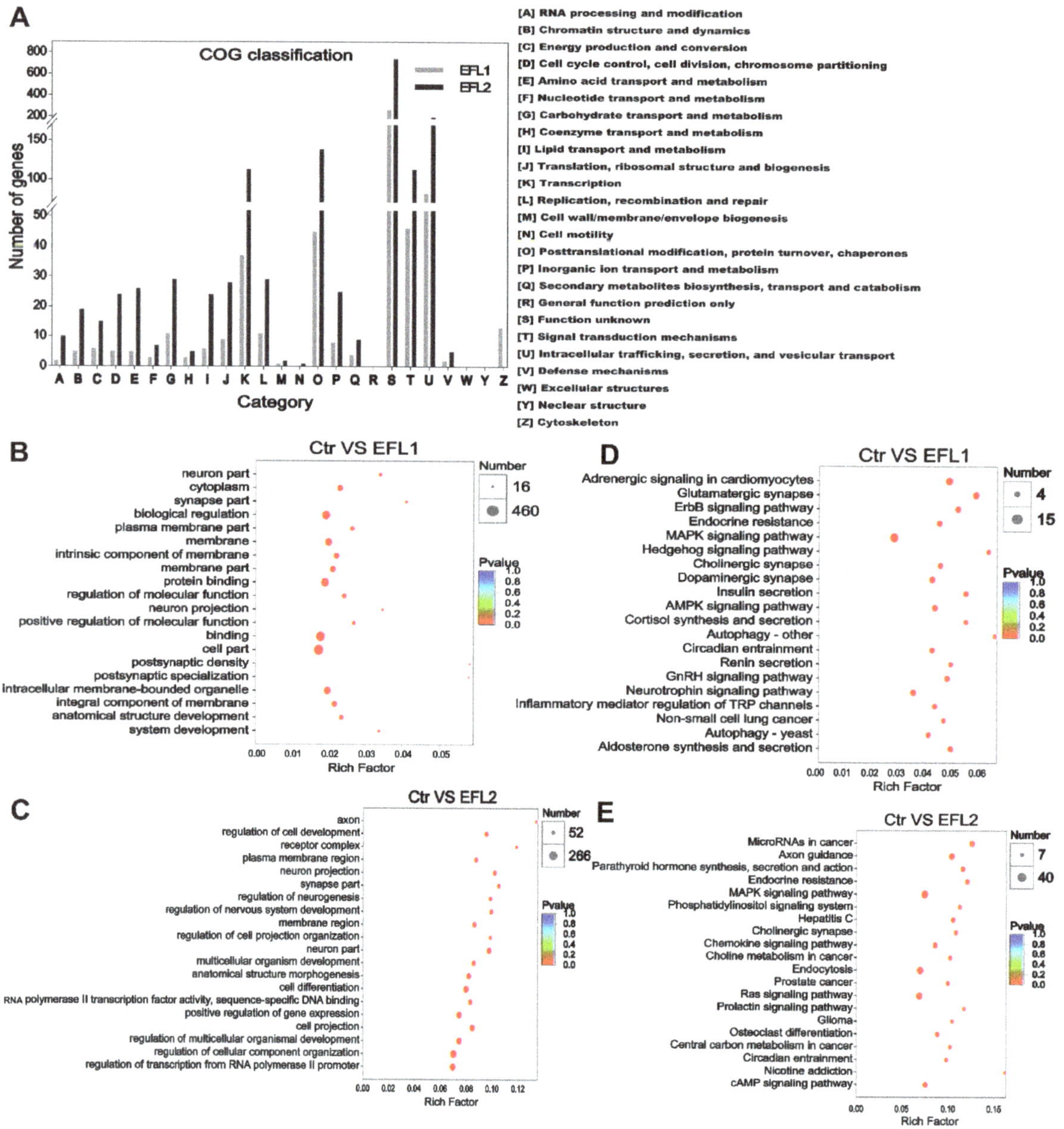


**Figure 4.** (**A**) The COG classification of differentially expressed miRNA in EFL-treated Caco-2 cells. (**B**) GO and (**C**) KEGG annotation of the differentially expressed miRNA in the EFL1 group. (**D**) GO and (**E**) KEGG annotation of the differentially expressed miRNA in the EFL2 group.

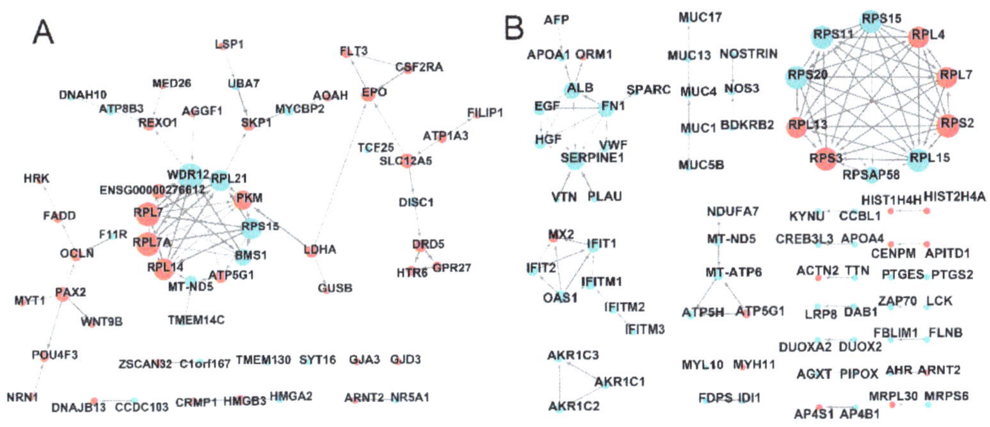


**Figure 5.** The PPI network of differentially expressed mRNA in EFL1 (**A**) and EFL2 (**B**) treated Caco-2 cells.

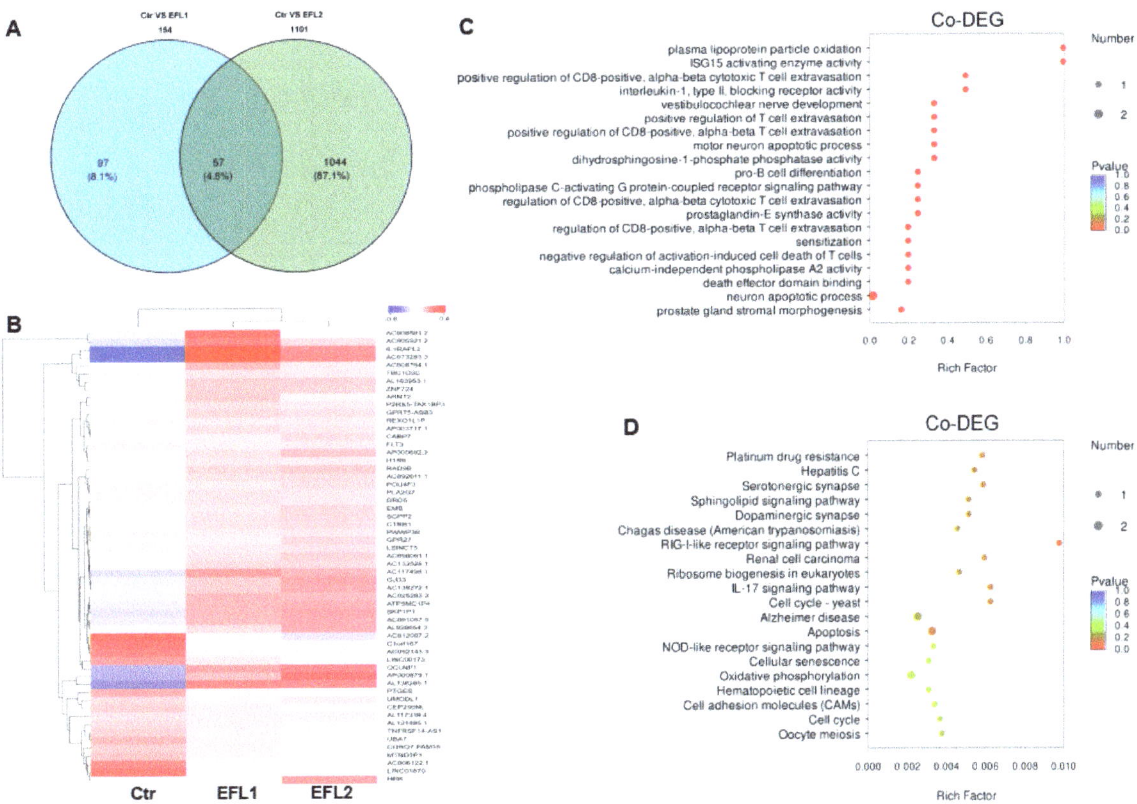


**Figure 6.** (**A**) The Venn diagram of differentially expressed mRNA in each EFL-treated group. The cluster analysis heat map (**B**), GO (**C**) and KEGG (**D**) enrichment of 57 differentially co-expressed mRNAs.

*2.11. Visual Analysis of Metabolic Pathways Involved in Differentially Expressed Genes*

After the treatment of Caco-2 cells for 72 h, EFL1 affected a few metabolic pathways including nucleotide metabolism, the metabolism of cofactors and vitamins, xenobiotic biodegradation and metabolism (glyoxylate and dicarboxylate metabolism, citrate cycle, chlorocyclohexane and chlorobenzene degradation, and oxidative phosphorylation), carbohydrate metabolism (pyruvate metabolism), and amino acid metabolism (phosphonate and phosphinate metabolism, and D-Alanine metabolism) (Figure S3A). EFL2 affected most of the metabolic pathways, including the metabolism of terpenoids and polyketides, xenobiotic biodegradation and metabolism, amino acid metabolism, nucleotide metabolism, the metabolism of cofactors and vitamins, the metabolism of other amino acids, and the biosynthesis of other secondary metabolites, especially glycan biosynthesis and metabolism, lipid metabolism, carbohydrate metabolism, and energy metabolism (Figure S3B).

### *2.12. Relative Expression Levels of Proteins Involves in Oxidative Stress and Mitochondria*

In the PPI network (Figure 5), genes involved in oxidative stress and mitochondrial energy metabolism were presented. In the Western blotting assay, heme oxygenase 1 (HO-1) and heat shock 70 kDa protein (HSP-70) were selected as the biomarkers of oxidative stress, cytochrome c oxidase 5 (COX-5), mitochondrial fission 1 protein (Fis1), mitofusin-1 (Mfn 1), and dynamin-like 120 kDa protein optic atrophy 1 (OPA1) were the biomarkers of mitochondrial energy metabolism. As shown in Figure 7, after treatment of EFLs, HO-1 and HSP-70 were activated, and the COX-5, Fis-1, and Mfn-1 were inhibited, as well as OPA-1 was increased.

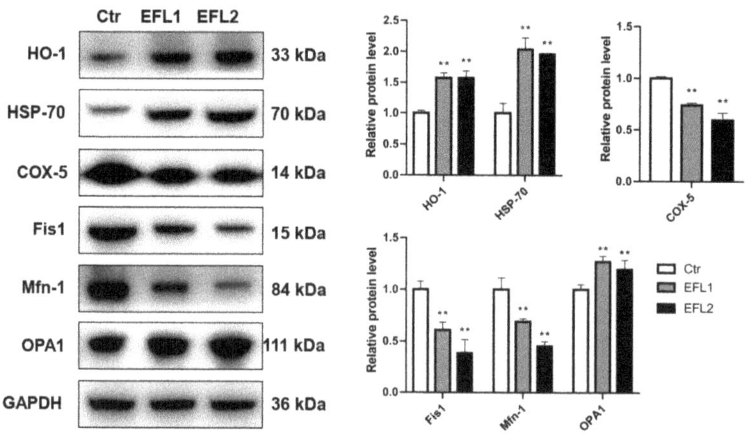


**Figure 7.** The relative expression levels of proteins involved in oxidative stress and mitochondrial energy metabolism. ** $p < 0.01$ compared with vehicle control group.

### *2.13. Gene Structure Change after EFL Treatment in Caco-2 Cells*

In the Caco-2 cells treated with EFLs for 72 h, 2022 and 5385 AS were reported in the EFL1 and EFL2 groups, respectively, and most of them were SEs, with a total of 1499 (74.1%) and 3830 (71.1%), respectively (Table S7).

The Indel numbers of 17 regions were recorded. The totals of 3240.3, 3157.3, and 2988.7 Indels were found in the control, EFL1, and EFL2 groups, respectively, and the number in the EFL2 group was significantly decreased when compared with the control group (Table S8). Most of the Indels were distributed in intronic, UTR3, ncRNA_intronic, intergenic, and ncRNA_exoinc regions, and those with no Indel were distributed in the exonic-splicing and ncRNA_splicing regions.

As shown in Table S9, 564,449, 562,139, and 509,564 SNP sites were recorded in the control, EFL1, and EFL2 groups, respectively. The total numbers of SNP, C/T and G/A transition, and T/A transversion sites in the EFL2 group were significantly decreased when compared with the control group. The SNP sites were distributed in all of the 17 regions, mainly in the intronic, UTR3, intergenic, exonic, and ncRNA_exonic regions (Table S10). No regional distributions were significantly different.

## 3. Materials and Methods

### *3.1. Materials*

EFL1 and EFL2 were purchased from Spring & Autumn Biological Engineering Co., Ltd. (Nanjing, China). The chemical structure and purity of EFL1 were described in our previous study [20]. The chemical structure of EFL2 was confirmed by 400 MHz $^{13}C$ and $^{1}H$ nuclear magnetic resonance (NMR) spectra (Bruker, Rheinstetten, Germany), and the purity of EFL2 was 99.63% as determined by high-performance liquid chromatography (HPLC) (Agilent, Santa Clara, CA, USA) (Figure 8).

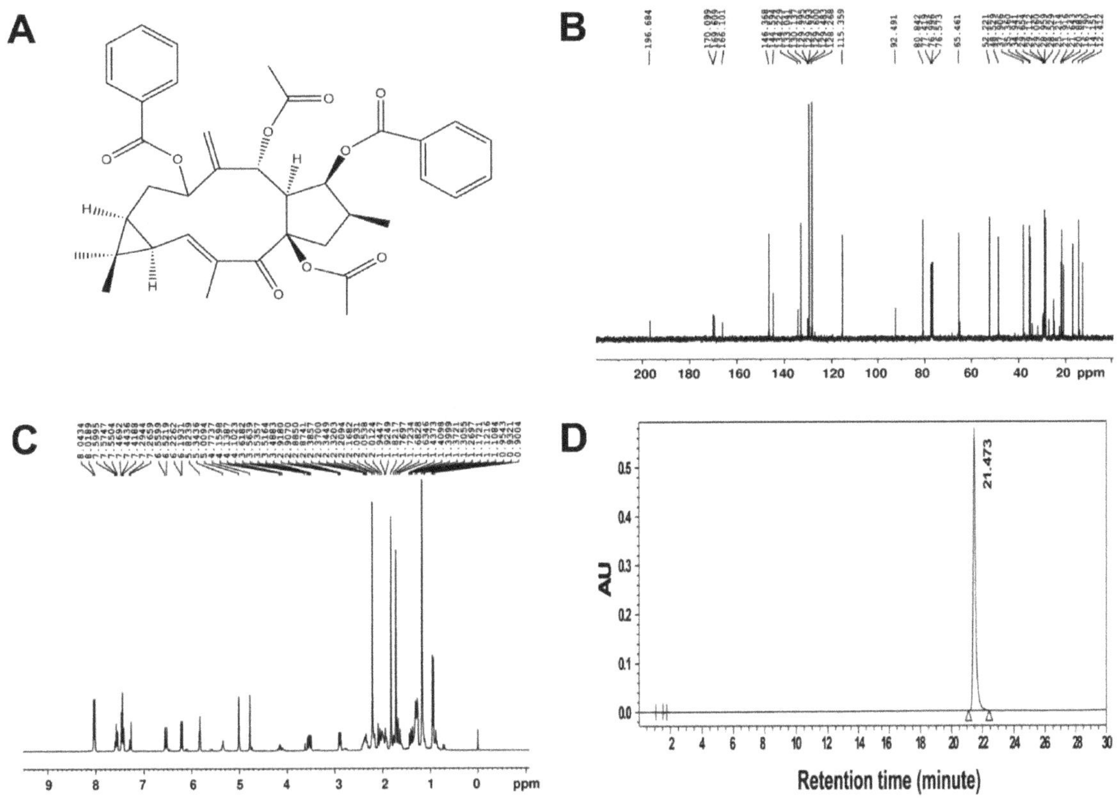


**Figure 8.** (**A**) The chemical structure of EFL2. The nuclear magnetic resonance spectra of EFL2. The left panel was the $^{13}C$ spectra (**B**), and the right panel was the $^{1}H$ spectra (**C**). The purity of EFL2 detected by HPLC (**D**). Chromatographic conditions: Gemini C18 column (250 × 4.6 mm, 4 μm particle size); mobile phase of methanol:water = 65:35; flow rate: 1.0 mL/min; detection wavelength: 250 nm; column temperature: 35 °C; injection volume: 5 μL; injection concentration: 2 mg/mL in methanol. The retention time of EFL1 was 21.473 min, and the area percent of EFL1 was 99.63%.

### *3.2. Cell Culture and Treatment*

Caco-2 cells were obtained from Biotides Biotechnology Co., Ltd. (Beijing, China), and the genetic information was identified by short tandem repeats. The cells were cultured in minimum essential medium (MEM, Gibco, New York, NY, USA) supplemented with 10% fetal bovine serum (Gibco), 100 U/mL penicillin G sodium salt, 100 μg/mL streptomycin sulfate, 1% sodium pyruvate, 1% GlutaMAX and 1% non-essential amino acid. Cells were maintained in an incubator (Thermo Fisher, Langenselbold, Germany) in an atmosphere of 5% $CO_2$ at 37 °C. EFLs were dissolved in dimethyl sulfoxide (DMSO) and then diluted in MEM medium to concentrations of 12.5, 25, 50, 100, and 200 μM. The final working concentration of DMSO in cell culture experiments did not exceed 1%.

### 3.3. Measurement of Cytotoxicity Using the MTT Assay

The viability of Caco-2 cells exposed to EFLs was evaluated using MTT cytotoxicity assay. In brief, $10^4$ cells were seeded in 96-well plates and treated with gradient concentrations of EFLs for 72 h. After incubation, cells received 20 μL 5 mg/mL MTT per well and were maintained at 37 °C for 4 h. After removing the medium, 100 μL of 4% isopropanol acid was added to dissolve the intracellular formazan product. The optical density of formazan was measured at a wavelength of 570 nm using a microplate reader (Omega, Ortenberg, Germany).

### 3.4. RNA Extraction and cDNA Library Construction

Caco-2 cells were seeded in 25 $cm^2$ flasks, and the total RNA was isolated by adding 500 μL TRIzol (Invitrogen, Carlsbad, CA, USA). The RNA quality was determined using 1% agarose gel electrophoresis. RNA samples with a qualified purity of $OD_{260}/OD_{280} \geq 1.8$ and $OD_{260}/OD_{230} \geq 1.5$ were measured using Nanodrop 2000 (Thermo Scientific). The RNA integrity number (RIN) was further assessed using a bioanalyzer 2100 (Agilent).

Poly-(A)-containing mRNA was isolated using oligo (dT) magnetic beads. A fragmentation buffer was added to disrupt mRNA into 300 bp short fragments, and these were then used as templates to synthesize the first-strand cDNA using reverse transcriptase and random hexamers. Second-strand cDNA was synthesized using dNTPs, RNase H, and DNA polymerase I. The double-strand cDNA was subjected to end repair and adapter ligation. Adapter-modified fragments were purified and amplified to create a cDNA library.

The Truseq Small RNA kit (Illumina, San Diego, CA, USA) was used to construct the cDNA library of miRNA. Briefly, 1 μg total RNA was used to add 3′ and 5′ adaptors, for the synthesis of first-strand cDNA by random hexamers. The cDNA was amplified by 12 PCR cycles and purified in 6% Novex TBE PAGE gel, before being quantified by PicoGreen TBS380. After bridge PCR by cBot, a cluster was generated for subsequent sequencing. Assays were performed in triplicate.

### 3.5. Transcriptome Sequencing and Bioinformatics Analysis

Transcriptome sequencing was performed on the second-generation high-throughput sequencing platform of Novaseq 6000 (Illumina). Homo sapiens GRCh38 was applied as the reference genome. The FASTX-Toolkit 0.0.14 was used to calculate the base mass distribution and base error rate distribution. The raw reads were subjected to adapter trimming, and low-quality reads were filtered using SeqPrep and Sickle [21]. The high-quality clean reads were aligned to the human genome using TopHat2 [22]. The transcript assembly was performed in StringTie. The overlapping read pairs were assembled into super reads. Then, the super reads were aligned to the reference genome to construct the graph of the splice site. The paths with high read coverage were retained and assembled to form transcripts. Then, they were annotated in the Clusters of Orthologous Groups (COG), Gene Ontology (GO), Kyoto Encyclopedia of Genes and Genomes (KEGG), Non-Redundant Protein Sequence Database (NR), Swiss-Prot and Pfam databases for mRNA using DIAMOND [23], Blast2GO [24], HMMER [25] and KOBAS2.1 [26] software and servers, and miRBase [27] or miRDeep2 [28] were used for miRNA annotation. The gene expression levels were calculated based on transcripts per million reads (TPM) using RSEM [29]. iPath3.0 (http://pathways.embl.de, accessed on 28 July 2019) was used to visualize and analyze metabolic pathways [30]. It contains three types of pathway maps: the metabolic pathways including 146 KEGG pathways involved in the overall metabolism of biological systems, the regulatory pathways including 22 KEGG regulatory pathways, and the biosynthesis of secondary metabolites including 58 KEGG pathways.

### 3.6. miRNA-mRNA Network

The target gene of miRNA was analyzed by the intersection of three databases: miRnada [31], TargetScan [32], and RNAhybrid [33]. In the paradigm suggesting that miRNAs are restrictive to mRNA, miRNA can couple with mRNA to silence or degrade it. Furthermore, apart from the classic negative regulation, a positive correlation of mRNA and miRNA has also been reported as the complementary mechanism and perturbation of feedback loops. This positive regulation was not included in the present study. Finally, Cytoscape 3.5.1 was used to construct the protein–protein interaction between miRNA and mRNA.

### 3.7. RNA Isolation and Quantitative Real-Time Polymerase Chain Reaction (RT-qPCR)

RT-qPCR was performed to identify the expression of DEGs. Total RNA from Caco-2 cells was isolated using TRIzol reagent (Invitrogen, Carlsbad, CA, USA). To quantify the relative gene expression, the primers were synthesized, as shown in Tables S1 and S2, with β-actin and U6 used as the endogenous controls of mRNA and miRNA, respectively. The universal PCR reverse primer and U6 forward primer were used for the cDNA reverse transcription of miRNA. RT-qPCR was performed using a 20 μL reaction volume. The relative gene expression was calculated using the $2^{-\Delta\Delta Ct}$ method (Schmittgen and Livak, 2008), and the fold change was normalized to that observed in the control group.

### 3.8. mRNA Gene Structure Analysis

The gene structure changes were identified by aligning sequenced sequences with reference genome sequences. Single nucleotide polymorphisms (SNPs) are genetic markers on the genome formed by single nucleotide variations. Another structure change is the insertion and deletion (InDel) of fragments relative to the reference genome, which may contain one or more bases. The SNP types, including base conversion and transversion, mutation frequency and depth, as well as the Indel frequency and depth, were calculated using Samtools software [34]. The gene sequence of eukaryotic cells contains introns and exons. After the gene is transcribed into mRNA precursors, the introns are removed by RNA splicing bodies, while the exons are retained in the mature mRNA. Alternative splicing (AS) means that unspliced RNA can have multiple forms of exon splicing, so that a gene can translate different proteins, thereby improving the complexity and adaptability of the system under various physiological conditions. The AS, including the skipped exon (SE), alternative 5′ splice site (A5SS), alternative 3′ splice site (A3SS), mutually exclusive exon (MXE), and retained intron (RI), were quantified and analyzed in the Junction Count Only mode of rMATS software [35].

### 3.9. Protein Expression Detected by Western Blotting

The cells were cultured in 6-well plates and treated with 200 μM EFL1 or EFL2 for 72 h. Cells were lysed by RIPA and the protein concentrations were quantified by BCA method. Equal amounts of 20 μg proteins were loaded for separation by SDS-page electrophoresis, and then electrophoretically transferred to polyvinylidene difluoride membranes. After blocking in 5% skim milk and successively incubation with primary and secondary antibodies (Proteintech and Abclonal, Wuhan, China), the membranes were exposed to ECL chemiluminescence solution (Beyotime, Shanghai, China) for imaging.

### 3.10. Statistical Analysis

The data are expressed as the mean ± standard deviation. One-way analysis of variance was performed using SPSS software (IBM, Armonk, NY, USA) to compare the differences between groups. Values of * $p < 0.05$ or ** $p < 0.01$ were considered statistically significant.

## 4. Discussion

*Euphorbiaceae* plants are mainly distributed in the tropics and subtropics and include more than 300 genera and more than 8000 species. In China, there are more than 80 plants from the *Euphorbia* genus, and 40 of them are used for traditional medicines, while over 40% of them are toxic [36]. *Euphorbia kansui* S.L.Liou ex S.B.Ho [37], *Euphorbia pekinensis* Rupr. [38], *Euphorbia ebracteolata* Hayata [39], *Euphorbia fischeriana* Steud. [40], *Euphorbia lathyris* L., and *Euphorbia hirta* L. [41], have been reported to be poisonous. There are typical gastrointestinal irritation xenobiotic substances that induce clinical symptoms of acute gastroenteritis, nausea, vomiting, abdominal pain, diarrhea and hematochezia. Although the toxicity effects have been reported in many clinical cases, the toxicity mechanism has still not been fully revealed.

In the 2020 edition of the Chinese Pharmacopoeia, 83 traditional Chinese medicinal materials are considered to be toxic. Depending on the intensity of toxicity, these materials are categorized into three levels, and *Euphorbia* semen belongs to the middle level [42]. The excellent pharmacological efficacy of *Euphorbia* semen makes it difficult to remove from the list of medicinal plants, but its clinical use is severely restricted due to the drug's safety. In a pharmacokinetic study, SD rats and rhesus macaques were used for the intragastric administration (i.g.) and intravenous injection of EFL1, and the $T_{1/2}$ (i.g.) values were 12.6 h and 9.9 h, while the bioavailability values were 2.0 and 4.6, respectively [43]. Although the oral bioavailability of EFL1 was low, toxicological studies has reported that the methanol, ethyl acetate, petroleum ether, and aqueous extract of *Euphorbia* semen cause reduced spontaneous activities, watery stools, wet perioral hair, urinary incontinence, and tics in mice treated with EFL1 (i.g.) once [44,45]. Jaretzky and Köhler used 2.5 mL/100 g seed oil of *Euphorbia* semen for once i.g. dose in rats. Diarrhea was reported as the toxicity endpoint, and this was related to accelerated intestinal rhythm movement [11]. In our previous study, EFL1 impaired the intestinal barrier integrity and accelerated the defecation cycle in the coelomic model organism *Caenorhabditis elegans* [46].

The main toxic component of *Euphorbia* semen is lathyrane, of which EFL1 and EFL2 have high proportions of the mass fraction [47,48]. In the present study, these two toxic compounds were selected to explore the toxicological mechanism of *Euphorbia* semen, and the transcriptome profiles were analyzed based on the Caco-2 intestinal cell line model. After treatment of EFLs for 72 h, the cell viability was inhibited. Through second-generation high-throughput sequencing of mRNA and miRNA, more differentially expressed genes were discovered in the EFL2 group, suggesting that EFL2 is more toxic to Caco-2 cells. Through gene assembly and functional annotation, gene function enrichment was performed. A previous study found that EFL1 inhibited the mRNA expression of aquaporin, a protein facilitating water transport across the plasma membrane, thus contributing to abnormal intestinal permeability [49]. In our study, GO and KEGG enrichment of EFL1 found enrichment of the terms cell–cell junction, transporter activity and transmembrane transport. This factor was also shown to regulate ion and molecule transport and result in osmolality disturbance. The results were consistent with previous reports in the literature. The gene function enrichment in the EFL2 group included the terms cAMP, IL-17, and TGF-beta signaling pathway, suggesting that ATP and energy metabolism dysfunction as well as the inflammatory response occur in intestinal cells. Once the cells are in a state of energy deficiency, the reverse concentration transport will be blocked, resulting in abnormal intestinal osmotic pressure [50]. The inflammation usually damages the integrity of the intestinal barrier, and the tight junctions and adherent junctions between cells break, allowing the intestinal endotoxins produced by the intestinal flora to be transported to the circulatory system [51]. Intestinal inflammation is the leading cause of diarrhea [52].

Next-generation high-throughput sequencing methods have been applied in the study of transcriptomes and have revolutionized the understanding of gene structure changes in AS, Indel and SNP, the most abundant gene polymorphisms in eukaryotic genomes that are usually associated with various diseases [53]. Nearly all transcripts from human protein-coding genes undergo several forms of AS, such as SE, A5SS, A3SS, MXE, and RI. After the gene is transcribed into mRNA precursors, the introns are moved by RNA splicing bodies, while the exons are retained in the mature mRNA. Unspliced RNA has multiple forms of exon splicing, so a gene can translate different proteins at different times and in different environments, thereby increasing the complexity or adaptability of the organism under different physiological conditions [54]. An SNP refers to a genetic marker formed by a single nucleotide variation on the genome. SNPs are numerous in number and rich in polymorphisms. SNPs appear most frequently on CG sequences, and most of them are C/T conversions, since the C in CG is often methylated, and becomes thymine after spontaneous deamination. It is divided into conversion and transversion, according to the characteristics of the SNP [55]. The Indel refers to the difference in the whole genome between two parents, where one parent has a certain number of nucleotide insertions or deletions in the genome relative to the other parent. It involves the insertion or deletion sequence of small fragments in the genome, the lengths of which are between 1 and 50 bp [56]. Indel occurrence is mainly related to genomic features and DNA replication errors. It has an average density of one Indel per 7.2 kb in the human genome [57]. In the present study, after treatment of EFL1 or EFL2 in Caco-2 cells, thousands of AS sites were discovered, and most of them were SEs. This might be the adaptive response for cells. The number of Indels in the EFL2 groups was significantly decreased, suggesting the less genetic length polymorphism. The C/T and G/A transitions, T/A transversion, and total SNP numbers in EFL2-treated Caco-2 cells were decreased, and SNPs occurring in any region of the gene had the potential to affect the protein structure or expression levels of the gene products. The EFLs, especially EFL2, could induce changes in the AS, Indel, and SNP sites in Caco-2 cells and thus contribute to gene expression and affect the biological function.

## 5. Conclusions

In conclusion, as the main diarrheal toxicity components of *Euphorbia* semen, EFL1 and EFL2 caused profiles changes in miRNA and mRNA, and the gene functions were enriched in the terms transmembrane transport, T cell extravasation, the IL-17 signaling pathway, apoptosis, and the cell cycle. The hsa-miR-6774-3p and hsa-miR-5090 were identified as the key miRNAs for the silencing and degradation of target mRNAs after the exposure of EFLs. Regarding the gene structure, EFL2 induced the AS, Indel and SNP components. This study provides gene expression, functional and structural information regarding Caco-2 cells exposed to EFLs. To the main deficit of the present study, all the conclusions were conducted based on bioinformatic analysis, and further molecular biological experiments should be performed to validate whether the DEGs-related functions or pathways were changed in reality.

**Supplementary Materials:** The following supporting information can be downloaded at: https://www.mdpi.com/article/10.3390/molecules27206931/s1, Figure S1: Assembly of mRNA transcripts by StringTie. (A) Transcripts per gene, (B) length distribution of transcripts, and (C) exons per transcript.; Figure S2: The correlation between the fold change of RT-qPCR and mRNA-seq of part of differentially expressed mRNA and miRNA in EFL1 group (A and B) and EFL2 group (C and D).; Figure S3. The visual analysis of metabolic pathways involved in differentially expressed genes after (A) EFL1 or (B) EFL2 treatment in Caco-2 cells.; Table S1. Primer sequences for miRNA of Caco-2 cells; Table S2. Gene descriptions and sequences of qRT-PCR primers for mRNA of Caco-2 cells; Table S3. Summary of reads and mapping rate of mRNA sequences of Caco-2 cells with or without EFLs treatment; Table S4. Summary of reads and mapping rate of miRNA sequences of Caco-2 cells with or without EFLs treatment; Table S5. Functional annotation of transcriptome data in six public databases; Table S6. The miRNA numbers including known and novel miRNAs in each sample; Table S7. The AS number in Caco-2 cells treated with EFLs; Table S8. The Indel regions distribution after EFLs treatment in Caco-2 cells; Table S9. The SNP number in Caco-2 cells treated with EFLs; Table S10. The SNP regions distribution after EFLs treatment in Caco-2 cells.

**Author Contributions:** Conceptualization, A.Z. and Q.W.; data curation, A.Z., L.Z., Y.G., C.L., J.Z. and D.Z.; formal analysis, L.Z., W.B., M.C., Y.W. and S.W.; funding acquisition, A.Z.; supervision, A.Z. and Q.W.; writing—original draft, L.Z.; writing—review and editing, A.Z. All authors have read and agreed to the published version of the manuscript.

**Funding:** This research was funded by the National Natural Science Foundation of China (82104520), Natural Science Foundation of Fujian Province (2021J05045), Fujian Medical University High-level Talent Research Startup Funding Project (XRCZX2020014), and Open Research Fund of Key Laboratory of Gastrointestinal Cancer (Fujian Medical University), Ministry of Education (FMUGIC-202101).

**Institutional Review Board Statement:** Not applicable.

**Informed Consent Statement:** Not applicable.

**Data Availability Statement:** All data are included in this manuscript.

**Acknowledgments:** The authors thank Majorbio Company (Shanghai, China) for the mRNA and miRNA sequencing.

**Conflicts of Interest:** The authors declare no conflict of interest.

**Sample Availability:** Samples of the compounds are not available from the authors.

## Abbreviations

| | |
|---|---|
| 3′-UTR | 3′ untranslated region |
| A3SS | Alternative 3′ splice site |
| A5SS | Alternative 5′ splice site |
| Ago | Agronaut |
| AS | Alternative splicing |
| COG | Clusters of orthologous groups of proteins |
| COX-5 | Cytochrome c oxidase 5 |
| DEG | Differentially expression genes |
| DMSO | Dimethyl sulfoxide |
| EFL1 | *Euphorbia* factor L1 |
| EFL2 | *Euphorbia* factor L2 |
| Fis1 | Mitochondrial fission 1 protein |
| GO | Gene ontology |
| HO-1 | Heme oxygenase 1 |
| HPLC | High performance liquid chromatography |
| HSP-70 | Heat shock 70 kda protein |
| InDel | Insertion and deletion |
| KEGG | Kyoto Encyclopedia of Genes and Genomes |
| Mfn-1 | Mitofusin-1 |
| miRNA | Microrna |
| mRNA | Messenger RNA |
| MXE | Mutually exclusion exon |
| NMR | Nuclear magnetic resonance |
| OPA1 | Optic atrophy 1 |
| RI | Retained intron |
| RIN | RNA integrity number |
| RISC | RNA-induced silencing complex |
| RNA-seq | RNA sequencing |
| RT-qPCR | Quantitative real-time polymerase chain reaction |
| SE | Skipped exon |
| SNP | Single nucleotide polymorphisms |

## References

1. Luo, D.; Callari, R.; Hamberger, B.; Wubshet, S.G.; Nielsen, M.T.; Andersen-Ranberg, J.; Hallström, B.M.; Cozzi, F.; Heider, H.; Lindberg Møller, B.; et al. Oxidation and cyclization of casbene in the biosynthesis of *Euphorbia* factors from mature seeds of *Euphorbia lathyris* L. *Proc. Natl. Acad. Sci. USA* **2016**, *113*, E5082–E5089. [CrossRef]
2. Hong, S.E.; Lee, J.; Seo, D.H.; In Lee, H.; Ri Park, D.; Lee, G.R.; Jo, Y.J.; Kim, N.; Kwon, M.; Shon, H.; et al. *Euphorbia* factor L1 inhibits osteoclastogenesis by regulating cellular redox status and induces Fas-mediated apoptosis in osteoclast. *Free Radic. Biol. Med.* **2017**, *112*, 191–199. [CrossRef] [PubMed]
3. Mesas, C.; Martínez, R.; Doello, K.; Ortiz, R.; López-Jurado, M.; Bermúdez, F.; Quiñonero, F.; Prados, J.; Porres, J.M.; Melguizo, C. In vivo antitumor activity of *Euphorbia lathyris* ethanol extract in colon cancer models. *Biomed. Pharmacother.* **2022**, *149*, 112883. [CrossRef]
4. Appendino, G.; Della Porta, C.; Conseil, G.; Sterner, O.; Mercalli, E.; Dumontet, C.; Di Pietro, A. A new P-glycoprotein inhibitor from the caper spurge (*Euphorbia lathyris*). *J. Nat. Prod.* **2003**, *66*, 140–142. [CrossRef] [PubMed]

5. Zhang, J.Y.; Mi, Y.J.; Chen, S.P.; Wang, F.; Liang, Y.J.; Zheng, L.S.; Shi, C.J.; Tao, L.Y.; Chen, L.M.; Chen, H.B.; et al. *Euphorbia* factor L1 reverses ABCB1-mediated multidrug resistance involving interaction with ABCB1 independent of ABCB1 downregualtion. *J. Cell. Biochem.* **2011**, *112*, 1076–1083. [CrossRef]
6. Zuo, Q.; Mu, H.Y.; Gong, Q.; Ding, X.; Wang, W.; Zhang, H.Y.; Zhao, W.M. Diterpenoids from the seeds of *Euphorbia lathyris* and their effects on microglial nitric oxide production. *Fitoterapia* **2021**, *150*, 104834. [CrossRef] [PubMed]
7. Chinese Pharmacopeia Committee. *Pharmacopoeia of China*; Chinese Medical Press: Beijing, China, 2020.
8. Adolf, W.; Hecker, E. On the active principles of the spurge family. III. skin irritant and cocarcinogenic factors from the caper spurge. *Z. Krebsforsch. Klin. Onkol. Cancer Res. Clin. Oncol.* **1975**, *84*, 325–344. [CrossRef] [PubMed]
9. Li, B.; Liu, S.L.; Zou, C.Z.; Qi, F.Q.; Sun, B.X. Experimental study on acute toxicity of Qianjinzi. *Heilongjiang Med. J.* **2006**, *19*, 96.
10. Zhu, A.; Zhang, T.; Wang, Q. The phytochemistry, pharmacokinetics, pharmacology and toxicity of *Euphorbia* semen. *J. Ethnopharmacol.* **2018**, *227*, 41–55. [CrossRef] [PubMed]
11. Jaretzky, R.; Köhler, W. *Euphorbia lathyris* L. als heilpflanze. pharmakologische prüfungen, insonderheit des fetten samenöls. *Arch. Pharm.* **1943**, *281*, 256–264. [CrossRef]
12. Chen, Z.; Zhou, L.; Ge, Y.; Chen, J.; Du, W.; Xiao, L.; Tong, P.; Huang, J.; Shan, L.; Efferth, T. Fuzi decoction ameliorates pain and cartilage degeneration of osteoarthritic rats through PI3K-Akt signaling pathway and its clinical retrospective evidence. *Phytomedicine* **2022**, *100*, 154071. [CrossRef] [PubMed]
13. Gao, L.; Feng, Y.; Ge, C.; Xu, X.; Wang, S.; Li, X.; Zhang, K.; Wang, C.; Dai, F.; Xie, S. Identification of molecular anti-metastasis mechanisms of lycorine in colorectal cancer by RNA-seq analysis. *Phytomedicine* **2021**, *85*, 153530. [CrossRef]
14. Zhou, Z.Y.; Xiao, Y.; Zhao, W.R.; Zhang, J.; Shi, W.T.; Ma, Z.L.; Ye, Q.; Chen, X.L.; Tang, N.; Tang, J.Y. Pro-angiogenesis effect and transcriptome profile of Shuxinyin formula in zebrafish. *Phytomedicine* **2019**, *65*, 153083. [CrossRef] [PubMed]
15. Burlacu, C.C.; Neag, M.A.; Mitre, A.O.; Sirbu, A.C.; Badulescu, A.V.; Buzoianu, A.D. The role of miRNAs in dexmedetomidine's neuroprotective effects against brain disorders. *Int. J. Mol. Sci.* **2022**, *23*, 5452. [CrossRef] [PubMed]
16. Szymański, M.; Barciszewski, J. Beyond the proteome: Non-coding regulatory RNAs. *Genome Biol.* **2002**, *3*, reviews0005. [PubMed]
17. Esteller, M. Non-coding RNAs in human disease. *Nat. Rev. Genet.* **2011**, *12*, 861–874. [CrossRef] [PubMed]
18. Lytle, J.R.; Yario, T.A.; Steitz, J.A. Target mRNAs are repressed as efficiently by microRNA-binding sites in the 5′ UTR as in the 3′ UTR. *Proc. Natl. Acad. Sci. USA* **2007**, *104*, 9667–9672. [CrossRef]
19. O'Brien, J.; Hayder, H.; Zayed, Y.; Peng, C. Overview of microRNA biogenesis, mechanisms of actions, and circulation. *Front. Endocrinol.* **2018**, *9*, 402. [CrossRef] [PubMed]
20. Zhu, A.; Sun, Y.; Zhong, Q.; Yang, J.; Zhang, T.; Zhao, J.; Wang, Q. Effect of *Euphorbia* factor L1 on oxidative stress, apoptosis, and autophagy in human gastric epithelial cells. *Phytomedicine* **2019**, *64*, 152929. [CrossRef] [PubMed]
21. Kilicoglu, H.; Rosemblat, G.; Fiszman, M.; Shin, D. Broad-coverage biomedical relation extraction with SemRep. *BMC Bioinform.* **2020**, *21*, 188. [CrossRef]
22. Kim, D.; Pertea, G.; Trapnell, C.; Pimentel, H.; Kelley, R.; Salzberg, S.L. TopHat2: Accurate alignment of transcriptomes in the presence of insertions, deletions and gene fusions. *Genome Biol.* **2013**, *14*, R36. [CrossRef]
23. Buchfink, B.; Xie, C.; Huson, D.H. Fast and sensitive protein alignment using DIAMOND. *Nat. Methods* **2015**, *12*, 59–60. [CrossRef] [PubMed]
24. Conesa, A.; Götz, S.; García-Gómez, J.M.; Terol, J.; Talón, M.; Robles, M. Blast2GO: A universal tool for annotation, visualization and analysis in functional genomics research. *Bioinformatics* **2005**, *21*, 3674–3676. [CrossRef] [PubMed]
25. Finn, R.D.; Clements, J.; Eddy, S.R. HMMER web server: Interactive sequence similarity searching. *Nucleic Acids Res.* **2011**, *39*, W29–W37. [CrossRef] [PubMed]
26. Xie, C.; Mao, X.; Huang, J.; Ding, Y.; Wu, J.; Dong, S.; Kong, L.; Gao, G.; Li, C.Y.; Wei, L. KOBAS 2.0: A web server for annotation and identification of enriched pathways and diseases. *Nucleic Acids Res.* **2011**, *39*, W316–W322. [CrossRef] [PubMed]
27. Kozomara, A.; Birgaoanu, M.; Griffiths-Jones, S. miRBase: From microRNA sequences to function. *Nucleic Acids Res.* **2019**, *47*, D155–D162. [CrossRef]
28. Friedländer, M.R.; Mackowiak, S.D.; Li, N.; Chen, W.; Rajewsky, N. miRDeep2 accurately identifies known and hundreds of novel microRNA genes in seven animal clades. *Nucleic Acids Res.* **2012**, *40*, 37–52. [CrossRef]
29. Li, B.; Dewey, C.N. RSEM: Accurate transcript quantification from RNA-Seq data with or without a reference genome. *BMC Bioinform.* **2011**, *12*, 323. [CrossRef]
30. Darzi, Y.; Letunic, I.; Bork, P.; Yamada, T. iPath3.0: Interactive pathways explorer v3. *Nucleic Acids Res.* **2018**, *46*, W510–W513. [CrossRef] [PubMed]
31. Li, X.N.; Wang, Z.J.; Ye, C.X.; Zhao, B.C.; Li, Z.L.; Yang, Y. RNA sequencing reveals the expression profiles of circRNA and indicates that circDDX17 acts as a tumor suppressor in colorectal cancer. *J. Exp. Clin. Cancer Res.* **2018**, *37*, 325. [CrossRef]
32. McGeary, S.E.; Lin, K.S.; Shi, C.Y.; Pham, T.M.; Bisaria, N.; Kelley, G.M.; Bartel, D.P. The biochemical basis of microRNA targeting efficacy. *Science* **2019**, *366*, eaav1741. [CrossRef]
33. Krüger, J.; Rehmsmeier, M. RNAhybrid: microRNA target prediction easy, fast and flexible. *Nucleic Acids Res.* **2006**, *34*, W451–W454. [CrossRef] [PubMed]
34. Li, H.; Handsaker, B.; Wysoker, A.; Fennell, T.; Ruan, J.; Homer, N.; Marth, G.; Abecasis, G.; Durbin, R. The Sequence Alignment/Map format and SAMtools. *Bioinformatics* **2009**, *25*, 2078–2079. [CrossRef] [PubMed]

35. Shen, S.; Park, J.W.; Lu, Z.X.; Lin, L.; Henry, M.D.; Wu, Y.N.; Zhou, Q.; Xing, Y. rMATS: Robust and flexible detection of differential alternative splicing from replicate RNA-Seq data. *Proc. Natl. Acad. Sci. USA* **2014**, *111*, E5593–E5601. [CrossRef] [PubMed]
36. Baole, C.L.; Na, S.S. Research progress of toxicity of medicinal plants of *Euphorbia* L. and attenuation by processing. *World Sci. Technol. Mod. Tradit. Chin. Med. Materia Med.* **2018**, *20*, 585–591.
37. Lou, J.W.; Cao, L.L.; Zhang, Q.; Jiang, D.J.; Yao, W.F.; Bao, B.H.; Cao, Y.D.; Tang, Y.P.; Zhang, L.; Wang, K.; et al. The toxicity and efficacy evaluation of different fractions of Kansui fry-baked with vinegar on Walker-256 tumor-bearing malignant ascites effusion rats and normal rats. *J. Ethnopharmacol.* **2018**, *219*, 257–268. [CrossRef]
38. Cao, Y.; Cheng, F.; Yao, W.; Bao, B.; Zhang, K.; Zhang, L.; Ding, A. Toxicity of Pekinenin C from *Euphorbia pekinensis* Radix on rat small intestinal crypt epithelial cell and its apoptotic mechanism. *Int. J. Mol. Sci.* **2016**, *17*, 850. [CrossRef]
39. Yang, T.; He, J.; Yan, Y.; Lian, W.W.; Xia, C.Y.; Xu, J.K.; Zhang, W.K. *Euphorbia ebracteolata* Hayata (*Euphorbiaceae*): A systematic review of its traditional uses, botany, phytochemistry, pharmacology, toxicology, and quality control. *Phytochemistry* **2021**, *186*, 112736. [CrossRef]
40. Li, Y.N.; He, J.; Zhang, J.; Shi, Y.X.; Guo, L.B.; Peng, Z.C.; Yang, T.; Ding, K.; Zhang, W.K.; Xu, J.K. Existing knowledge on *Euphorbia fischeriana* Steud. (*Euphorbiaceae*): Traditional uses, clinical applications, phytochemistry, pharmacology and toxicology. *J. Ethnopharmacol.* **2021**, *275*, 114095. [CrossRef]
41. Yuet Ping, K.; Darah, I.; Chen, Y.; Sreeramanan, S.; Sasidharan, S. Acute and subchronic toxicity study of *Euphorbia hirta* L. methanol extract in rats. *BioMed Res. Int.* **2013**, *2013*, 182064. [CrossRef]
42. Gao, Y.F.; Wang, R.Y.; Yi, L.; Ouyang, R.; Zhang, Z.G. Statistical analysis of toxic herbs and points for attention in Chinese Pharmacopoeia (2020 edition, part I). *China Pharm.* **2021**, *24*, 1133–1137.
43. Fu, H.Y. *Study on the Pharmacokinetics of Euphorbiasteroid*; Guangzhou University of Chinese Medicine: Guangzhou, China, 2013.
44. Huang, X.T.; Huang, G.Y.; Xue, C.K.; Kong, C.X.; He, X.B. Experimental study on the anti-tumor effect of methanolic extractive from *Euphorbia lathyris* L. in vitro and in vivo. *Cancer Res. Prev. Treat.* **2004**, *31*, 556–558.
45. Liang, Y.J.; Zheng, F.L.; Tang, D.X.; Luo, Y.H.; Tan, Z.H. Study on the potency and toxicity of different extracts from *Euphorbia lathyris* in mice. *West China J. Pharm. Sci.* **2011**, *26*, 27–29.
46. Zhu, A.; Ji, Z.; Zhao, J.; Zhang, W.; Sun, Y.; Zhang, T.; Gao, S.; Li, G.; Wang, Q. Effect of *Euphorbia* factor L1 on intestinal barrier impairment and defecation dysfunction in *Caenorhabditis elegans*. *Phytomedicine* **2019**, *65*, 153102. [CrossRef] [PubMed]
47. Fan, L.; Zhu, H.; Tao, W.; Liu, L.; Shan, X.; Zhao, M.; Sun, D. *Euphorbia* factor L2 inhibits TGF-β-induced cell growth and migration of hepatocellular carcinoma through AKT/STAT3. *Phytomedicine* **2019**, *62*, 152931. [CrossRef] [PubMed]
48. Park, S.J.; Park, J.H.; Han, A.; Davaatseren, M.; Kim, H.J.; Kim, M.S.; Hur, H.J.; Sung, M.J.; Hwang, J.T.; Yang, H.J.; et al. Euphorbiasteroid, a component of *Euphorbia lathyris* L.; inhibits adipogenesis of 3T3-L1 cells via activation of AMP-activated protein kinase. *Cell Biochem. Funct.* **2015**, *33*, 220–225. [CrossRef] [PubMed]
49. Wang, Y.Z.; Zhang, X.T.; Li, S.J.; Zhang, Y.; Li, F.Y.; Zhang, C.N.; Wang, Q.; Li, W.H.; Luo, S.X. Expression of AQP2, AQP4 and AQP 8 in mouse intestine induced by unprocessed and processed *Euphorbia lathyris*. *Pak. J. Pharm. Sci.* **2018**, *31*, 1229–1235.
50. Li, Y.; Wang, S.; He, G.; Wu, H.; Pan, F.; Jiang, Z. Facilitated transport of small molecules and ions for energy-efficient membranes. *Chem. Soc. Rev.* **2015**, *44*, 103–118. [CrossRef]
51. Luissint, A.C.; Parkos, C.A.; Nusrat, A. Inflammation and the intestinal barrier: Leukocyte-epithelial cell interactions, cell junction remodeling, and mucosal repair. *Gastroenterology* **2016**, *151*, 616–632. [CrossRef]
52. Ko, G.; Jiang, Z.D.; Okhuysen, P.C.; DuPont, H.L. Fecal cytokines and markers of intestinal inflammation in international travelers with diarrhea due to noroviruses. *J. Med. Virol.* **2006**, *78*, 825–828. [CrossRef]
53. Wang, L.; Zheng, G.; Yuan, Y.; Wang, Z.; Liu, C.; Zhang, H.; Lian, L. Exploration of alternative splicing (AS) events in MDV-infected chicken spleens. *Genes* **2021**, *12*, 1857. [CrossRef] [PubMed]
54. Seo, J.H.; Dhungana, S.K.; Kang, B.K.; Baek, I.Y.; Sung, J.S.; Ko, J.Y.; Jung, C.S.; Kim, K.S.; Jun, T.H. Development and validation of SNP and InDel markers for pod-shattering tolerance in soybean. *Int. J. Mol. Sci.* **2022**, *23*, 2382. [CrossRef] [PubMed]
55. Song, L.; Wu, J.Q.; Dong, C.M.; Park, R.F. Integrated analysis of gene expression, SNP, InDel, and CNV identifies candidate avirulence genes in Australian isolates of the wheat leaf rust pathogen *Puccinia triticina*. *Genes* **2020**, *11*, 1107. [CrossRef] [PubMed]
56. Li, D.; Zeng, R.; Li, Y.; Zhao, M.; Chao, J.; Li, Y.; Wang, K.; Zhu, L.; Tian, W.M.; Liang, C. Gene expression analysis and SNP/InDel discovery to investigate yield heterosis of two rubber tree F1 hybrids. *Sci. Rep.* **2016**, *6*, 24984. [CrossRef]
57. Mills, R.E.; Luttig, C.T.; Larkins, C.E.; Beauchamp, A.; Tsui, C.; Pittard, W.S.; Devine, S.E. An initial map of insertion and deletion (INDEL) variation in the human genome. *Genome Res.* **2006**, *16*, 1182–1190. [CrossRef]

*molecules*

*Article*

# Improving the Transduction Efficiency and Antitumor Effect of Conditionally Replicative Adenovirus by Application of 6-Cyclohexyl Methyl-β-D-maltoside

**Wenjing Lu, Yaping Fang, Xue Meng, Xiaoli Wang, Wenbo Liu, Mengdong Liu and Ping Zhang ***

Beijing International Science and Technology Cooperation Base of Antivirus Drug, Faculty of Environment and Life, Beijing University of Technology, Beijing 100124, China
* Correspondence: zplife@bjut.edu.cn

**Abstract:** As a tumor-targeting oncolytic adenovirus (Ad), conditionally replicating adenovirus (CRAd) can access the cell interior by binding to coxsackievirus-Ad receptors (CARs) and specifically replicate and destroy cancer cells without lethal effects on normal cells. The transduction efficiency of CRAd is highly dependent on the number of CARs on the cell membrane. However, not all tumor cells highly express CARs; therefore, improving the transduction efficiency of CRAd is beneficial for improving its antitumor effect. In this study, 6-cyclohexyl methyl-β-D-maltoside (6-β-D), as maltoside transfection agent, showed several advantages, including high transfection efficiency, low toxicity, and potential for intensive use and easy operation. With pretreatment of cancer cells with low concentration of 6-β-D (≤5 μg/mL), the transduction efficiency of "model" Ad (eGFP-Ad) was improved 18-fold compared to eGFP-Ad alone. 6-β-D improved the antitumor effect of CRAd while being safe for normal cells, in which treatment with 6-β-D helped the lethal effects of CRAd at a multiplicity-of-infection ratio of 10 (MOI 10) achieve the oncolytic outcomes of MOI 50. This means that if CRAd is combined with 6-β-D, the amount of CRAd used in clinical practice could be greatly reduced without diminishing its curative effect or exposing patients to the potential side effects of high-titer CRAd. Finally, the underlying mechanism of antitumor effect of CRAd + 6-β-D was primarily investigated, and we found that 6-β-D increased the virus's replication in cancer cells at the early stage of infection and activated the apoptosis signaling pathway at the late stage of the cell cycle. This research will provide an effective technical reference for further improving Ad-mediated cancer gene therapy in clinical practice.

**Keywords:** 6-cyclohexyl methyl-β-D-maltoside; transduction efficiency; conditionally replicative adenovirus; lethal effect; gene therapy

**Citation:** Lu, W.; Fang, Y.; Meng, X.; Wang, X.; Liu, W.; Liu, M.; Zhang, P. Improving the Transduction Efficiency and Antitumor Effect of Conditionally Replicative Adenovirus by Application of 6-Cyclohexyl Methyl-β-D-maltoside. *Molecules* **2023**, *28*, 528. https://doi.org/10.3390/molecules28020528

Academic Editors: Guohui Sun and Chongwen Wang

Received: 17 November 2022
Revised: 7 December 2022
Accepted: 7 December 2022
Published: 5 January 2023



## 1. Introduction

Conditionally replicating adenovirus (CRAd) is a tumor-targeting oncolytic adenovirus used in cancer gene therapy [1–4]. As a carrier of tumor-specific survivin promoter, CRAd can selectively replicate in cancer cells under the effect of survivin and lyse those cells without lethal effect on normal cells [5–9]. The adenoviruses mainly attach to coxsackie/Ad receptors (CARs) on cell surfaces to access into cells. However, CARs are lacking on the membrane of some kinds of tumor tissues and cells, which leads to a lower Ad transduction efficiency [2,10,11]. Therefore, it is important to efficiently transfer exogenous genes into cancer cells to improve therapeutic outcomes in cancer gene therapy [12–14].

Transfection agents with high efficiency and low toxicity have been considered a promising approach to overcoming this serious obstacle. Currently, commonly used transfection reagents include calcium phosphate, nanomaterials, liposomes, and non-liposomal agents [3,15–19]. The glycoside agent 6-cyclohexyl methyl-β-D-maltoside (6-β-D) is a bipolar molecule with a hydrophilic polar group and a hydrophobic nonpolar group. It has been reported that 6-β-D can insert into the cell membrane, break down membrane lipids, and release membrane proteins without disturbing the structural stability of the membrane proteins [20–22], thereby altering cell membrane permeability and mediating exogenous substances into cells. Theoretically, as a non-liposomal agent, 6-β-D shows several advantages in gene- and drug-delivery systems, including high transduction efficiency, low toxicity, intensive applications, and easy operation. However, up to now, research on the use of 6-β-D as a new gene delivery carrier in Ad-mediated gene therapies has been limited.

In this study, the appropriate doses of 6-β-D with high transduction efficiency and low toxicity were determined. Based on the safe dose, we measured the lethal effects of CRAd with and without 6-β-D on different human cancer cells. In addition, the antitumor mechanism of CRAd was explored. The results showed that 6-β-D at very low concentration (several ppm) could significantly improve the transduction efficiency of Ad into cells. CRAd had a broad-spectrum killing effect on different types of cancer cells, while safe to human normal cells. The lethal effect of CRAd showed positive relationships with multiplicity-of-infection ratio (MOI) and treatment time. CRAd replication significantly increased in 6-β-D-pretreated human cancer cells. When combined with 6-β-D, CRAd at MOI 10 achieved the killing effect of MOI 50 on cancer cells. It is suggested that the viral titer could be greatly reduced in clinical practice, which could free patients from the potential adverse effects of high-titer adenoviruses. This study will provide technical support and a theoretical basis for improving the clinical application of Ad in cancer gene therapy.

## 2. Results and Discussion

### *2.1. Conditional Replication Adenoviruses Can Kill Cancer Cells Specifically*

As shown in Figure 1a, CRAds enter into cells by recognizing the glycoprotein CAR receptors on the cell surface. CRAd contains a specific survivin promoter before the E1A gene. It has been widely shown that the survivin gene is highly and specifically expresses in many human tumors, but absent in normal adult differentiated cells [6–9]. The incorporation of a survivin promoter leads to CRAd's specific replication in human cancer cells, and eventually kills them to release offspring viruses to further infect neighboring cancer cells. Therefore, the Ad vector with a survivin promoter has potential to target human cancer cells.

CRAds had a broad-spectrum killing effect on different types of cancer cells, while being safe for human normal cells. There is a positive antitumor correlation with the multiplicity of infection or duration of treatment (Figure 1b,c). However, different types of cancer cells showed different sensitivities to CRAds. Specifically, the most efficient oncolytic effects occurred on esophageal cancer cells EC109 and breast cancer cells MCF-7. As for the human normal skin fibroblast cells, the survival rate of BJ-1 cells was 85.43 ± 3.56% at MOI 10, indicating that CRAd was safe for human normal cells.

The CRAds were more effective in the treatment of esophageal and breast cancers, but less so in ovarian cancer, because only 60% of SKOV-3 cells were killed even as the multiplicity of infection reached 50. This may be due to the membrane of ovarian cancer cells lacking CAR receptors [2,10,11], thus hindering adenoviruses entering the SKOV-3 cells through the CAR receptors. Moreover, survivin gene expression was lower in ovarian cancer cells SKOV-3 and normal BJ-1 cells than that in breast cancer cells MCF-7 and esophageal cancer cells EC109 (Figure 1d,e). These above results collectively explain the varying oncolytic effects of CRAds on different cancer cells.

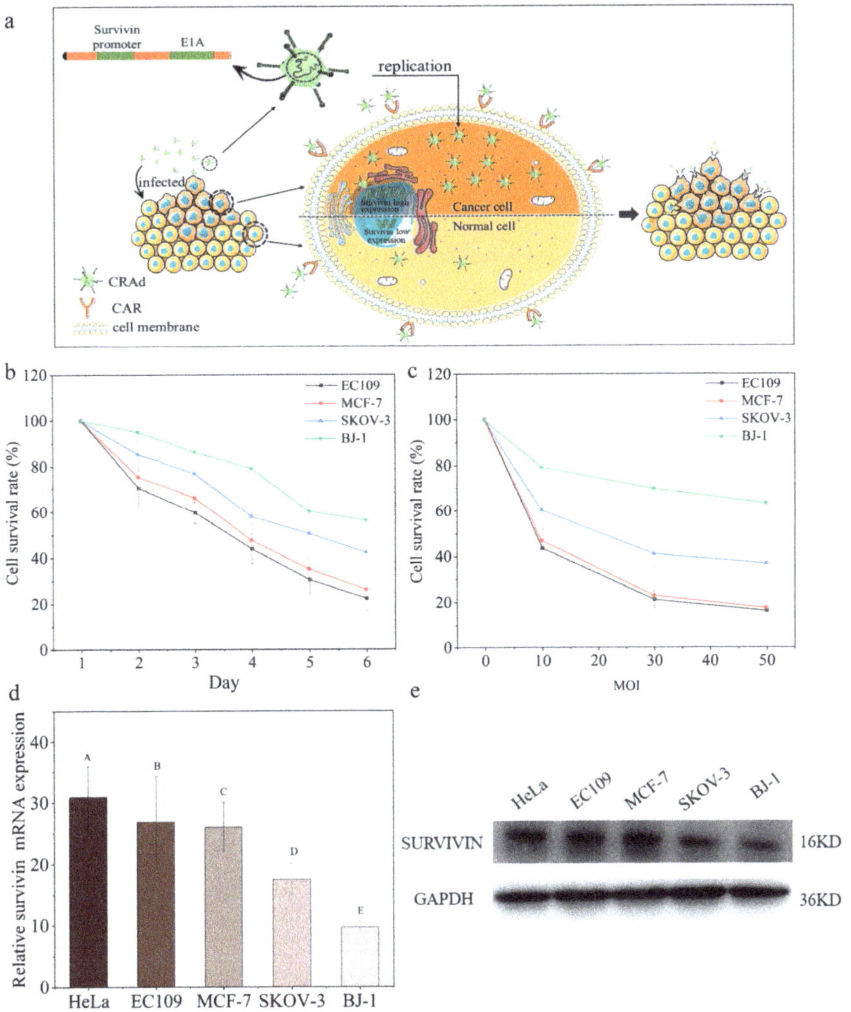


**Figure 1.** CRAd can replicate and package the offspring adenoviruses specifically in cancer cells. (**a**) Proposed antitumor mechanism of CRAd; (**b**) Antitumor effects of CRAd (MOI 10) on different cells for different durations (in days); (**c**) Antitumor effects of CRAd at varying MOI ratios on the 4th day; (**d**) Survivin mRNA expression in various cell lines; (**e**) Survivin protein in various cells, as detected by Western blot. Quantitative analysis of survivin protein could been seen in Supplementary Materials. Note: The different letters indicate that differences reached extremely significant levels ($p < 0.001$).

### *2.2. High Transduction Efficiency and Low Cytotoxicity of 6-β-D*

The killing effect of CRAd on cancer cells can be improved by increasing the viral titer in vitro. However, this is not desirable in vivo because of the potential side effects in practice. It is important to allow more CRAds to invade cancer cells and replicate and package more offspring viruses to achieve better curative effects. Therefore, using transfection reagents to improve the transduction efficiency of CRAd was considered.

In this study, a glycoside material, 6-cyclohexyl methyl-β-D-maltoside (6-β-D), was selected and used to explore its transduction efficiency. The 6-β-D is a bipolar molecule with a hydrophilic polar group and a hydrophobic non-polar group that can help foreign substances invade target cells by changing the permeability of cell membranes [20–22]. As shown in Figure 2a, the cell viability showed no difference between 6-β-D-treated (12 μg/mL) and untreated cells, indicating 6-β-D at this concentration exhibited an unobvious inhibitory effect on cell growth ($p > 0.05$).

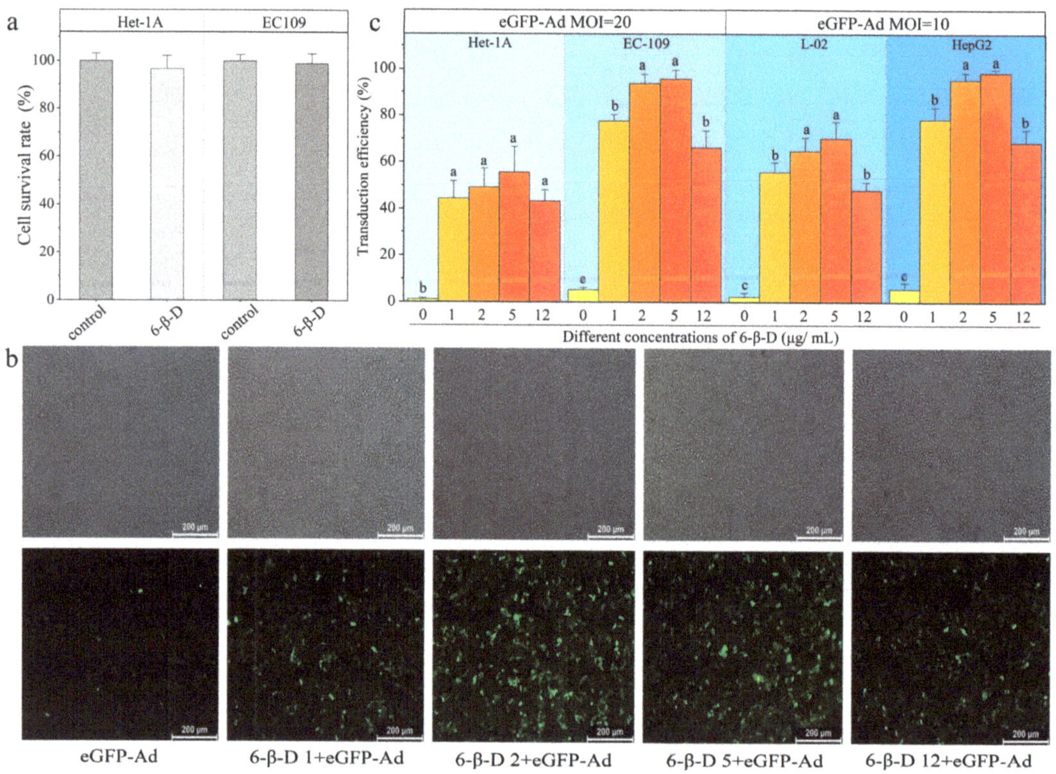


**Figure 2.** When eGFP was used as a reporter gene, 6-β-D showed low cytotoxicity and high transduction efficiency in different human cell lines. (**a**) Cytotoxicity of 6-β-D on Het-1A and EC109 cells; (**b**) 6-β-D promoted eGFP expression in EC109 cells; (**c**) Transduction efficiency of eGFP-Ad was affected by different concentrations of 6-β-D. Note: The same letter means no significant difference ($p > 0.05$), while different letters indicate that differences reached significant levels ($p < 0.05$).

We adopted the Ad with enhanced GFP expression (eGFP-Ad) as a "reporter" to rapidly and quantitatively evaluate the transduction effect of 6-β-D. By directly observing the amount of lightened $GFP^+$ cells, we found that the transduction rate of Ad alone was only 5%, while the 6-β-D treatment showed remarkably more GFP expression (Figure 2b). It indicated that more eGFP-Ad invaded cells with application of 6-β-D. There was a correlation between the 6-β-D concentration and the transduction efficiency. The most efficient concentration of 6-β-D was 2~5 μg/mL in vitro cellular system, at which >95% of EC109 and HepG2 cells were successfully transfected by eGFP-Ad (Figure 2b,c). The transduction efficiency was improved 18-fold with application of 6-β-D (Figure 2c).

In addition, comparing the transduction rates between cancer cells and normal cells, eGFP-Ad was more efficient in infecting cancer cells than normal cells under the action of 6-β-D. In Figure 2c, the transduction rates of cancer cells HepG2 and EC109 were 98.37 ± 1.56% and 95. 67 ± 4.04%, respectively, while they were 70.27 ± 7.16% and 55.60 ± 11.02% in normal cells L02 and Het-1A, respectively. Moreover, the 6-β-D was safe for cultured cells at a concentration of below 5 μg/mL. This meant that it was unnecessary to remove the 6-β-D solution after pretreating cells at low concentrations (<5 μg/mL). Thus, the experimental steps can be simplified, making the operation more convenient. Collectively, the results proved that 6-β-D showed more advantages as a transfection reagent, including high efficiency, low toxicity, low cost, and easy operation.

### 2.3. Application of 6-β-D Improved the Oncolytic Effect of CRAd

Compared with normal cells BJ-1, CRAd had a significant killing effect on cancer cells ($p < 0.01$, Figure 1b), and the killing effect was positively correlated with multiplicity of infection (Figure 1c). As shown in Figure 3, when CRAd of MOI 10 infected cancer cells, on the fourth day, the survival rates of EC109 and MCF-7 cells were 43.49 ± 3.26% and 47.04 ± 2.53%, respectively. As MOI reached 50, cell survival rate decreased to 15% (Figure 3a). After 6-β-D pretreatment, CRAd at MOI 10 could further improve the killing effect on cancer cells. The lethal rates were as high as 85.33 ± 1.64% and 84.68 ± 2.74% in EC109 and MCF-7 cells, respectively, which were significantly different from that of CRAd alone ($p < 0.01$, Figure 3b).

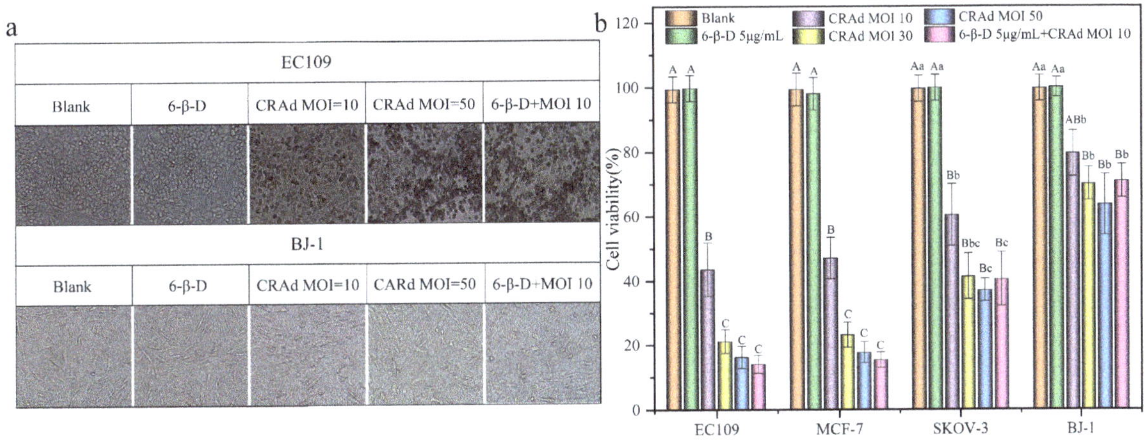


**Figure 3.** Application of 6-β-D improved the oncolytic effect of CRAd. (**a**) Comparison of cytopathic effects on esophageal cancer cells EC109 vs. human normal skin fibroblast cells BJ-1 after infection with different titers of CRAd (×200); (**b**) The synergistic killing effect of CRAd (MOI 10) + 6-β-D achieved outcomes of higher MOI 50 on various cancer cell lines. Note: Different letters indicate significant difference in pairwise comparison, while the same letter means no significant difference. The difference was significant in lowercase letters ($p < 0.05$) and extremely significant in uppercase letters ($p < 0.01$).

The oncolytic effect of CRAd (MOI 10) + 6-β-D was similar to that of CRAd with high viral titer (MOI 50) ($p > 0.05$, Figure 3). In these treated groups, almost all the cancer cells were killed and floated, and the few remaining cells were weak and round in shape (Figure 3a). This meant the viral titer could be reduced by combining 6-β-D with CRAd in clinical practice. It can be used as an auxiliary reagent to obtain higher gene-transduction efficiency, thus reducing the amount of CRAd in Ad-based gene therapy.

However, for normal cells BJ-1, most of them still adhered to the culture flask, and the cellular morphology was intact without obvious damage (Figure 3a). This result further proved that CRAd selectively killed cancer cells and had a lower attacking effect on normal cells.

### 2.4. Combining 6-β-D with CRAd Increased Virus Replication in Cancer Cells

E1A is an early-expressed gene related to adenovirus replication [5,23,24]. In this experiment, we detected the expression of E1A to evaluate the amount of CRAds entering cells after drug treatment. First, cells were treated with 6-β-D for about 20 min and then infected with CRAd of MOI 10. After 4 h, the cells were flushed with PBS three times to remove the superfluous viruses and replaced with fresh medium to cultivate for another 24 h. Thereafter, the DNA copy number and mRNA expression of E1A were detected.

As shown in Figure 4a,b, E1A gene expression in 6-β-D + CRAd-treated cells increased significantly over that in 6-β-D-untreated cells; especially in EC109 and MCF-7 cells, the DNA copy number increased nearly four-fold (Figure 4a). This indicates that CRAd invasion of cancer cells was improved by 6-β-D treatment. Meanwhile, CRAd-E1A mRNA expression was significantly higher in EC109 and MCF-7 cells than in BJ-1 cells (Figure 4b). The higher expression of E1A was consistent with the killing effect of CRAd.

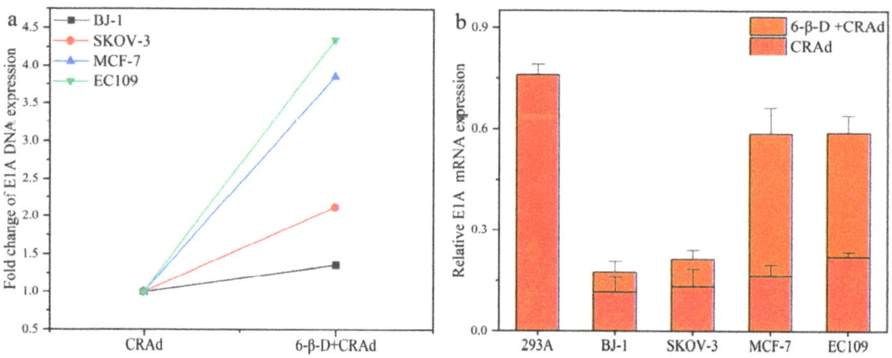


**Figure 4.** Combination of 6-β-D with CRAd increased viral replication in various cancer cell lines. (**a**) 6-β-D promoted CRAds invasion into cancer cells; (**b**) As detected by Real time-PCR, 6-β-D increased CRAd–E1A mRNA expression in cancer cells.

Furthermore, the $TCID_{50}$ method was applied to detect CRAd replication and amplification in EC109 and SKOV-3 cells with and without 6-β-D. It was reported that most adenoviruses could enter cells within 3 or 4 h of infection. During this period, the intracellular viruses had not yet packaged the offspring viruses. Therefore, the viral titer detected at 3 h reflected the amount of CRAds entering cells. After 96 h of infection, CRAds replicated numerous progeny viruses. The results collected at this stage reflected the replication ability of CRAds. Therefore, the viral titers were analyzed after 3 h and 96 h infection.

As shown in Table 1, 6-β-D treatment facilitated more CRAds entering cancer cells at the initial stage of infection. The viral titer had increased 1.5-fold and 1.58-fold in SKOV-3 and EC109 cells, respectively, while 96 h later, the CRAds proliferated more rapidly in EC109 cells than in SKOV-3 cells. The 6-β-D-treated group was 4.9-fold of the untreated one. This explains why the killing effect of CRAds on EC109 cells was better than that on SKOV-3 cells.

**Table 1.** Viral titer of CRAd detected by $TCID_{50}$.

| Time | Treatment | Fold (Titer/PFU/mL) | |
|---|---|---|---|
| | | SKOV-3 | EC109 |
| **3 h** | CRAd | 1 ($10^{5.3}$) | 1 ($10^{5.5}$) |
| | 6-β-D + CRAd | 1.50 ($10^{5.5}$) | 1.58 ($10^{5.7}$) |
| **96 h** | CRAd | 1 ($10^{9.5}$) | 1 ($10^{10.4}$) |
| | 6-β-D + CRAd | 1.86 ($10^{10}$) | 4.90 ($10^{11.2}$) |

In brief, the above results indicated that intracellular CRAds can exert a strong cytopathic effect on cancer cells with high E1A and survivin expression. The more oncolytic

CRAds entered cells, the more CRAds replicated in cancer cells, and thus more significant killing effects were exhibited on cancer cells.

According to the flow cytometry (FCM) results, the proportions of apoptotic and necrotic cells in CRAd-treated groups were significantly higher than those in 6-β-D-treated and untreated groups (Figure 5a–e). Moreover, there was no significant difference between 6-β-D-treated and untreated groups, indicating the safety of 6-β-D to human cells (Figure 5a–e). In Figure 5c,d, the cellular apoptotic rate increased from 73.6% to 86.9%, indicating that 6-β-D promoted the killing effect of CRAd at cell-safe doses.

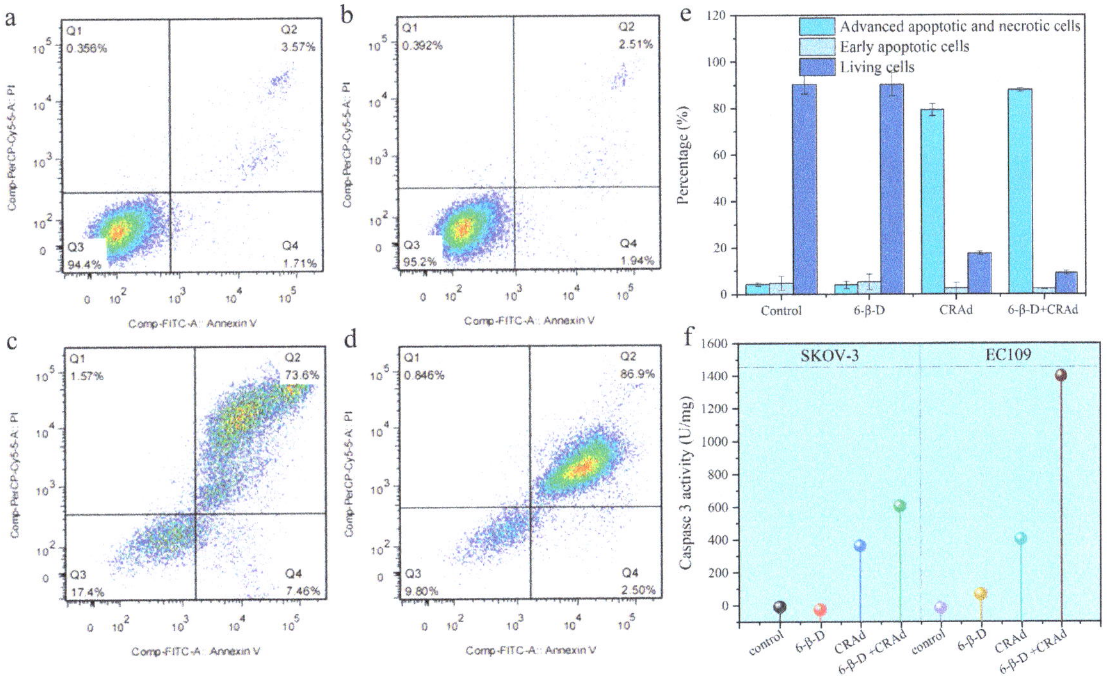


**Figure 5.** Effects of CRAd on Caspase-3 activity and cellular apoptosis. CRAd-induced cellular apoptosis detected by flow cytometry. (**a**) Control: EC109 cells; (**b**) 6-β-D-treated EC109 cells; (**c**) CRAd-treated EC109 cells; (**d**) 6-β-D- and CRAd–treated EC109 cells; (**e**) Quantitative analysis of cellular apoptosis; (**f**) CRAd stimulated Caspase-3 activity in EC109 and SKOV-3 cells.

Caspase-3 is an initial protein of the apoptosis pathway. In Figure 5f, Caspase-3 activity in both CRAd-treated EC109 and SKOV-3 cells was significantly higher than that of untreated cells. This result suggested that CRAd treatment stimulated Caspase-3 activity and initiated the cellular apoptosis pathway. Furthermore, Caspase-3 activity in EC109 cells was generally greater than that in SKOV-3 cells under 6-β-D + CRAd treatment (Figure 5f), verifying the better killing effect of CRAd on esophageal cancer cells. These data corresponded to the previous results.

Collectively, we speculate that CRAd may exert its lethal effect on cancer cells through two different mechanisms. First, at the early stage of infection, it mainly does so via specific replication in cancer cells to produce more progeny adenoviruses to cause cytolysis. Second, CRAd replication stimulates Caspase-3 activity, which triggers the apoptosis signaling pathway at the late stage of the cell cycle.

## 3. Materials and Methods

### 3.1. Cell Lines and Drugs

The human cancer cells (including esophageal cancer cell EC109, liver cancer cell HepG2, breast cancer cell MCF-7, ovarian cancer cell SKOV-3, and cervical cancer cell HeLa), human normal esophageal cell (Het-1A), and human normal skin fibroblast (BJ-1) cell lines were obtained from the National Collection of Authenticated Cell Cultures (Shanghai, China). The human normal liver cell line (L-02) was collected from the China Center for Type Culture Collection (Wuhan, China). 6-β-D was purchased from Sigma-Aldrich (St. Louis, MO, USA). The CCK-8 kit and bicinchoninic acid (BCA) kit were acquired from Beyotime Institute of Biotechnology (Shanghai, China). An AceQ quantitative polymerase chain reaction (qPCR) kit was obtained from Vazyme Biotech Co., Ltd. (Nanjing, China). TRIzol LS Regent, which was used for extraction of total ribonucleic acid (RNA), was purchased from Tiangen Biotech Co., Ltd. (Beijing, China). CRAd and eGFP-Ad were constructed and preserved in our laboratory in accordance with previously published methods [5]. Briefly, CRAd with strong cancer-dependent proliferation was created by introducing the cancer-specific survivin promoter to regulate adenoviral E1A gene expression, and the E1B gene was deleted from the Ad.

### 3.2. Cytotoxicity Assay

The cellular toxicity of 6-β-D was detected using the CCK-8 method [25,26]. First, EC109 and Het-1A cells were seeded in a 96-well plate at a density of $8 \times 10^3$/well and cultured at 37 °C with 5% $CO_2$ for 24 h. Then, each well was incubated with 100 μL 6-β-D (max. 12 μg/mL) for about 20 min. The regent was discarded and replaced with fresh Dulbecco's Modified Eagle's medium (DMEM), and cells were cultivated for another 48 h. Finally, the cytotoxicity test was conducted according to the CCK-8 protocol. Optical density (OD) was measured at 450 nm using a Multiskan Sky microplate spectrophotometer (Thermo Fisher Scientific, Waltham, MA, USA). The cell viability was calculated according to Equation (1), as follows. The experiment was repeated 3 times.

$$\text{Cell survival rate (\%)} = (OD_{\text{treatment}} - OD_{\text{blank}})/(OD_{\text{control}} - OD_{\text{blank}}) \times 100\% \quad (1)$$

### 3.3. Transduction Efficiency

We adopted a "model" Ad (eGFP-Ad) to evaluate how well 6-β-D improved the transduction efficiency. The concentrations of 6-β-D solution were 1, 2, 5, and 12 μg/mL. Before virus infection, 100 μL of 6-β-D solution at different concentrations was separately added to each well. After 20 min, HepG2 and L-02 cells were transducted with MOI 10 and EC109 and Het-1A cells with MOI 20 of eGFP-Ad for 4 h. Then, the drug solution was replaced with 500 μL fresh DMEM medium, and cells were cultivated for another 48 h. Next, the GFP expression was observed under a fluorescence microscope, counting $GFP^+$-cells and comparing counts between the 6-β-D-treated and -untreated groups. The transduction efficiency was calculated as the ratio of the percentage of GFP-expressing cells to that of whole cells under the observation field. For each sample, five observation fields were randomly selected for data collection, and each experiment was repeated 3 times.

### 3.4. Survival Inhibition Effect Assay

According to the selected dosage of 6-β-D used in the transduction efficiency assay, the synergistic effect of 6-β-D and CRAd in killing cancerous cells was analyzed using the CCK-8 method. Three cancer cell lines (EC109, MCF-7, and SKOV-3 cells) and one human normal cell line (BJ-1) were selected and seeded in a 96-well plate at a density of $8 \times 10^3$/well and cultured for 24 h. For each cell line, four treatment groups were established: the CRAd (MOI 10, 30, 50)-treated group, 6-β-D (5 μg/mL)-treated group, 6-β-D (5 μg/mL) + CRAd-treated (MOI 10) group, and blank control group. After treatment for 4 days, cell survival rate was calculated according to Equation (1).

### 3.5. RNA Preparation and RT-PCR

Total RNA was prepared using TRIzol LS Regents (DP430, Tiangen, Beijing, China). Reverse transcription was performed using Hiscipt®IIQ Select RT SuperMix for qPCR (+gDNA wiper) (Vazyme Biotech Co., Ltd., Nanjing, China). The E1A and survivin gene expressions were quantitatively analyzed using SYBR Green qRT-PCR premix (TaKaRa, Kyoto, Japan) ($2^{-\Delta\Delta Ct}$ method) on a fluorescent quantitative PCR system (MX3000P, Stratagene, La Jolla, CA, USA). PCR amplification used an AceQ qPCR kit (Q111-02/03, Vazyme Biotech Co., Ltd., Nanjing, China) and started with an initial denaturation step at 95 °C for 5 min; followed by 40 cycles of 95 °C for 30 s, 58 °C for 30 s, and 72 °C for 30 s; and a final elongation step at 72 °C for 3 min. Primer sequences for E1A were F: 5′-TCCTCACCCTCTTCATCCTC-3′, R: 5′-GAACCACCTACCCTTCACGA-3′. Primer sequences for survivin were F: 5′-CAGCCCTTTCTCAAGGACCAC-3′, R: 5′-TTTCTCCGCAGTTTCCTCAAA-3′. Glyceraldehyde 3-phosphate dehydrogenase (GAPDH) was an internal reference gene and its primer sequences were F: 5′-TTCCGTGTTCCTACCCCCAA-3′, R: 5′-AGCCCAAGATGCCCTTCAG-3′. These gene primers were designed using Premier software version 5.0 and synthesized by Sangon Biotech (Shanghai, China).

### 3.6. Viral Titer Assay

The viral titer was tested using the median tissue culture infectious dose ($TCID_{50}$) assay, which was calculated according to Karber's formula.

### 3.7. Flow Cytometry (FCM)

FCM determination was performed on 6-β-D-treated (5 µg/mL), CRAd (MOI 50)-treated, 6-β-D (5 µg/mL) + CRAd (MOI 10)-treated, and untreated cells. After treatment for 72 h, cells were harvested and stained with Annexin-V-Fluorescein isothiocyanate (FITC) and propidium iodide (PI). Then, the cells were analyzed on a FACScan flow cytometer equipped with CellQuest software version 5.1 (BD Biosciences, Franklin Lakes, NJ, USA).

### 3.8. Caspase-3 Activity Analysis

Caspase-3 activity was analyzed in four treatment groups: 6-β-D (5 µg/mL) only, CRAd (MOI 10) only, 6-β-D (5 µg/mL) + CRAd (MOI 10), and blank control. For this assay, a Caspase-3 Activity Assay Kit (C1116, Beyotime Biotechnology, Shanghai, China) was used according to the manufacturer's protocol.

### 3.9. Statistical Analysis

Results were expressed as mean ± standard deviation (SD) for continuous variables. One-way analysis of variance (ANOVA) of Bonferroni's post hoc test was used to compare differences among various groups using GraphPad Prism software version 8.0 (GraphPad Software, Inc., San Diego, CA, USA). $p < 0.05$ was considered statistically significant, and $p < 0.01$ was extremely significant.

## 4. Conclusions

Conditionally replicating adenoviral vectors, carrying a specific survivin gene promoter, tend to propagate in cancer cells and eventually lyse cancer cells. In this study, 6-β-D as a new transduction agent showed several advantages, including high transduction efficiency, low toxicity, intensive application, cost effectiveness, and easy operation. With application of 6-β-D, the lethal effect of CRAd at MOI 10 improved significantly, achieving the oncolytic outcomes associated with the use of a higher titer (MOI 50). This meant that by using 6-β-D + CRAd, clinicians could greatly reduce the amount of CRAd without reducing its curative effect, thereby sparing patients from most of CRAd's potential side-effects caused by higher viral titer. Finally, the mechanism underlying the antitumor effect of CRAd was primarily investigated. This study will increase the feasibility of reducing viral titer to improve antitumor effects through the use of a safe transfection agent in adenovirus-based gene therapy. These insights may prove to be a timely opportunity for the application of CRAd in clinical treatment of cancers.

**Supplementary Materials:** The following supporting information can be downloaded at: https://www.mdpi.com/article/10.3390/molecules28020528/s1, Figure S1: Quantitative analysis of survivin protein (Figure 1e); Figure S2: Survivin protein in various cells (HeLa, EC109, MCF-7, SKOV-3, BJ-1), as detected by Western blot (Figure 1e); Figure S3: GAPDH protein in various cells (HeLa, EC109, MCF-7, SKOV-3, BJ-1), as detected by Western blot (Figure 1e).

**Author Contributions:** Conceptualization, W.L. (Wenjing Lu) and P.Z.; methodology, W.L. (Wen-jing Lu), X.M. and X.W.; software, W.L. (Wenjing Lu) and Y.F.; validation, X.W., W.L. (Wenbo Liu), M.L. and P.Z.; formal analysis, W.L. (Wenjing Lu) and P.Z.; investigation, W.L. (Wenjing Lu), W.L. (Wenbo Liu) and M.L.; resources, X.M. and P.Z.; data curation, W.L. (Wenjing Lu); writing—original draft preparation, W.L. (Wenjing Lu) and Y.F.; writing—review and editing, W.L. (Wenjing Lu) and P.Z.; visualization, W.L. (Wenjing Lu) and Y.F.; supervision, P.Z.; project administration, W.L. (Wenjing Lu) and P.Z. All authors have read and agreed to the published version of the manuscript.

**Funding:** This research was funded by the Scientific Research Program of Beijing Education Commission, P. R. China, grant number KM201810005031.

**Institutional Review Board Statement:** Not applicable.

**Informed Consent Statement:** Not applicable.

**Data Availability Statement:** Not applicable.

**Conflicts of Interest:** The authors declare no conflict of interest.

## References

1. Bulcha, J.T.; Wang, Y.; Ma, H.; Tai, P.W.L.; Gao, G. Viral vector platforms within the gene therapy landscape. *Signal Transduct. Target. Ther.* **2021**, *6*, 53. [CrossRef] [PubMed]
2. Gonzalez-Pastor, R.; Goedegebuure, P.S.; Curiel, D.T. Understanding and addressing barriers to successful adenovirus-based virotherapy for ovarian cancer. *Cancer Gene Ther.* **2021**, *28*, 375–389. [CrossRef] [PubMed]
3. Sayed, N.; Allawadhi, P.; Khurana, A.; Singh, V.; Navik, U.; Pasumarthi, S.K.; Khurana, I.; Banothu, A.K.; Weiskirchen, R.; Bharani, K.K. Gene therapy: Comprehensive overview and therapeutic applications. *Life Sci.* **2022**, *294*, 120375. [CrossRef] [PubMed]
4. Watanabe, M.; Nishikawaji, Y.; Kawakami, H.; Kosai, K.-I. Adenovirus Biology, Recombinant Adenovirus, and Adenovirus Usage in Gene Therapy. *Viruses* **2021**, *13*, 2502. [CrossRef]
5. Muhammad, T.; Sakhawat, A.; Khan, A.A.; Ma, L.; Gjerset, R.A.; Huang, Y. Mesenchymal stem cell-mediated delivery of therapeutic adenoviral vectors to prostate cancer. *Stem Cell Res. Ther.* **2019**, *10*, 190. [CrossRef]
6. Sasso, E.; Froechlich, G.; Cotugno, G.; D'Alise, A.M.; Gentile, C.; Bignone, V.; De Lucia, M.; Petrovic, B.; Campadelli-Fiume, G.; Scarselli, E.; et al. Replicative conditioning of Herpes simplex type 1 virus by Survivin promoter, combined to ERBB2 retargeting, improves tumour cell-restricted oncolysis. *Sci. Rep.* **2020**, *10*, 4307. [CrossRef]
7. Chen, J.-S.; Liu, J.-C.; Shen, L.; Rau, K.-M.; Kuo, H.-P.; Li, Y.M.; Shi, D.; Lee, Y.-C.; Chang, K.-J.; Hung, M.-C. Cancer-specific activation of the survivin promoter and its potential use in gene therapy. *Cancer Gene Ther.* **2004**, *11*, 740–747. [CrossRef]
8. Ryan, B.M.; O'Donovan, N.; Duffy, M.J. Survivin: A new target for anti-cancer therapy. *Cancer Treat. Rev.* **2009**, *35*, 553–562. [CrossRef]
9. Mu, X.; Wang, X.; Wei, Y.; Wen, C.; Zhang, Q.; Xu, C.; Liu, C.; Zhang, C.; Meng, F.; Zhao, N.; et al. ApoE-modified liposomes mediate the antitumour effect of survivin promoter-driven HSVtk in hepatocellular carcinoma. *Cancer Gene Ther.* **2020**, *27*, 754–767. [CrossRef]
10. You, Z.; Fischer, D.C.; Tong, X.; Hasenburg, A.; Aguilar-Cordova, E.; Kieback, D.G. Coxsackievirus–adenovirus receptor expression in ovarian cancer cell lines is associated with increased adenovirus transduction efficiency and transgene expression. *Cancer Gene Ther.* **2001**, *8*, 168–175. [CrossRef]
11. Kim, J.-S.; Lee, S.-H.; Cho, Y.-S.; Choi, J.-J.; Kim, Y.H.; Lee, J.-H. Enhancement of the Adenoviral Sensitivity of Human Ovarian Cancer Cells by Transient Expression of Coxsackievirus and Adenovirus Receptor (CAR). *Gynecol. Oncol.* **2002**, *85*, 260–265. [CrossRef] [PubMed]
12. Matsunaga, W.; Hamada, K.; Tagawa, M.; Morinaga, T.; Gotoh, A. Cancer Cell-specific Transfection of hCas9 Gene Using Ad5F35 Vector. *Anticancer Res.* **2021**, *41*, 3731–3740. [CrossRef] [PubMed]
13. Cui, Y.; Li, Y.; Li, S.; Li, W.; Zhu, Y.; Wang, J.; Liu, X.; Yue, Y.; Jin, N.; Li, X. Anti-tumor effect of a dual cancer-specific recombinant adenovirus on ovarian cancer cells. *Exp. Cell Res.* **2020**, *396*, 112185. [CrossRef]
14. Tian, Y.; Yao, W.; He, D.; Xu, Y.; Li, Y.; Zhu, Y.; Fang, J.; Bai, B.; Li, X.; Sun, L.; et al. A dual cancer-specific recombinant adenovirus suppresses the growth of liver cancer cells in vivo and in vitro. *Anti-Cancer Drugs* **2020**, *31*, 110–122. [CrossRef]

15. Xiang, C.; Tenkumo, T.; Ogawa, T.; Kanda, Y.; Nakamura, K.; Shirato, M.; Sokolova, V.; Epple, M.; Kamano, Y.; Egusa, H.; et al. Gene transfection achieved by utilizing antibacterial calcium phosphate nanoparticles for enhanced regenerative therapy. *Acta Biomater.* **2021**, *119*, 375–389. [CrossRef]
16. Sokolova, V.; Epple, M. Biological and Medical Applications of Calcium Phosphate Nanoparticles. *Chemistry* **2021**, *27*, 7471–7488. [CrossRef] [PubMed]
17. Kumar, A.R.K.; Shou, Y.; Chan, B.; Logan, K.; Tay, A. Materials for Improving Immune Cell Transfection. *Adv. Mater.* **2021**, *33*, e2007421. [CrossRef] [PubMed]
18. Kaczmarek, J.C.; Patel, A.K.; Rhym, L.H.; Palmiero, U.C.; Bhat, B.; Heartlein, M.W.; DeRosa, F.; Anderson, D.G. Systemic delivery of mRNA and DNA to the lung using polymer-lipid nanoparticles. *Biomaterials* **2021**, *275*, 120966. [CrossRef]
19. Ravula, V.; Lo, Y.-L.; Wang, L.-F.; Patri, S.V. Gemini Lipopeptide Bearing an Ultrashort Peptide for Enhanced Transfection Efficiency and Cancer-Cell-Specific Cytotoxicity. *ACS Omega* **2021**, *6*, 22955–22968. [CrossRef]
20. Stetsenko, A.; Guskov, A. An Overview of the Top Ten Detergents Used for Membrane Protein Crystallization. *Crystals* **2017**, *7*, 197. [CrossRef]
21. Wehbie, M.; Onyia, K.K.; Mahler, F.; Le Roy, A.; Deletraz, A.; Bouchemal, I.; Vargas, C.; Babalola, J.O.; Breyton, C.; Ebel, C.; et al. Maltose-Based Fluorinated Surfactants for Membrane-Protein Extraction and Stabilization. *Langmuir* **2021**, *37*, 2111–2122. [CrossRef] [PubMed]
22. Xia, H.; Liu, L.; Reinhart, C.; Michel, H. Heterologous expression of human Neuromedin U receptor 1 and its subsequent solubilization and purification. *Biochim. Biophys. Acta* **2008**, *1778*, 2203–2209. [CrossRef] [PubMed]
23. Sakurai, F.; Nishimae, F.; Takayama, K.; Mizuguchi, H. Optimization of an E1A Gene Expression Cassette in an Oncolytic Adenovirus for Efficient Tumor Cell Killing Activity. *Anticancer Res.* **2021**, *41*, 773–782. [CrossRef] [PubMed]
24. Brachtlova, T.; Van Beusechem, V.W. Unleashing the Full Potential of Oncolytic Adenoviruses against Cancer by Applying RNA Interference: The Force Awakens. *Cells* **2018**, *7*, 228. [CrossRef]
25. Zhou, C.; Liu, C.; Liu, W.; Chen, W.; Yin, Y.; Li, C.-W.; Hsu, J.L.; Sun, J.; Zhou, Q.; Li, H.; et al. SLFN11 inhibits hepatocellular carcinoma tumorigenesis and metastasis by targeting RPS4X via mTOR pathway. *Theranostics* **2020**, *10*, 4627–4643. [CrossRef]
26. He, R.; Cui, M.; Lin, H.; Zhao, L.; Wang, J.; Chen, S.; Shao, Z. Melatonin resists oxidative stress-induced apoptosis in nucleus pulposus cells. *Life Sci.* **2018**, *199*, 122–130. [CrossRef]

molecules

MDPI

*Article*

# AHNAK Contributes to Hepatocellular Carcinoma Growth by Interacting with IGF-1R

**Kang Li [1], Ke Song [2], Yuli Hou [3], Yuan Tian [4], Huijuan Wang [4], Libo Sun [5], Ang Li [1] and Yonghong Zhang [1,*]**

1 Biomedical Information Center, Beijing You'An Hospital, Capital Medical University, Beijing 100069, China
2 Department of Histology and Embryology, School of Basic Medical Sciences, Capital Medical University, Beijing 100069, China
3 Department of Clinical Laboratory, Xuanwu Hospital, National Clinical Research Center for Geriatric Diseases, Capital Medical University, Beijing 100053, China
4 Beijing Institute of Hepatology, Beijing You'An Hospital, Capital Medical University, Beijing 100069, China
5 Department of Hepatobiliary Surgery, Beijing You'An Hospital, Capital Medical University, Beijing 100069, China
* Correspondence: zhangyh@ccmu.edu.cn

**Abstract:** Neuroblast differentiation-associated protein AHNAK, a large structural scaffold protein, remains mysterious in biological processes. AHNAK plays a suppressive or progressive role in different types of cancers. To investigate the role of the AHNAK in hepatocellular carcinoma (HCC), cell viability assays were performed to determine the cell proliferation of the stable AHNAK-knockdown HepG2 cell line; co-immunoprecipitation (Co-IP) and liquid chromatography coupled with tandem mass spectrometry (LC-MS/MS) were performed on HCC and matched paracancerous (MPC) tissues. The Metascape platform was used for enrichment analyses; the "ComplexHeatmap" package was applied for cluster analyses and visualization. Co-IP, Western botting and immunofluorescence double staining were performed to assess the interactions between AHNAK and insulin-like growth factor 1 receptor (IGF-1R). AHNAK silencing reduced the viability of HepG2 cells; the interactome in HCC and MPC tissues enriched 204 pathways and processes, which partially reflected the signature of HCC field cancerization. AHNAK could co-localize and interact with IGF-1R. These results suggested that the AHNAK complex contributes to HCC growth, potentially by interacting with IGF-1R.

**Keywords:** neuroblast differentiation-associated protein AHNAK; hepatocellular carcinoma (HCC); insulin-like growth factor 1 receptor (IGF-1R); interactome; field cancerization

**Citation:** Li, K.; Song, K.; Hou, Y.; Tian, Y.; Wang, H.; Sun, L.; Li, A.; Zhang, Y. AHNAK Contributes to Hepatocellular Carcinoma Growth by Interacting with IGF-1R. *Molecules* **2022**, *27*, 8680. https://doi.org/10.3390/molecules27248680

Academic Editor: Filippo Minutolo

Received: 7 November 2022
Accepted: 30 November 2022
Published: 8 December 2022



## 1. Introduction

Hepatocellular carcinoma (HCC) is one of the most common and lethal tumors worldwide and the most common type of primary liver cancer, accounting for 75–85% of cases [1]. Despite several signaling pathways having been identified to be related with the progression of HCC, the molecular mechanism of HCC development remains complicated. Adjacent tumor tissues are not completely normal tissues [2], which is known as field cancerization [3,4]. Although histologically normal, they share some common molecular alterations with tumors. Field cancerization has been shown to affect circRNA expression profiles in gastric cancer [3]. This understanding might also provide new enlightenment for the study of the occurrence and development of HCC.

Neuroblast differentiation-associated protein AHNAK, also known as desmoyokin [5], is a large structural scaffold protein (molecular mass >620 kDa) that remains mysterious and plays diverse roles in the biological processes of various cancers [6]. As reported in the literature, AHNAK suppressed the proliferation and invasion of triple-negative breast cancer via different signaling pathways [7]. AHNAK also suppressed the progress of ovarian cancer by impairing the canonical Wnt/β-catenin pathway [8]. Similar results occurred in colorectal cancer [9] and melanoma [10]. However, a high level of AHNAK was associated with poor outcome in pancreatic ductal adenocarcinoma (PDAC) [11] and laryngeal carcinoma [12]. It could enhance the metastasis of lung cancer via epithelial–mesenchymal transition [13] and promote migration and invasion in mesothelioma [14]. In liver cancer, our previous research found that the mRNA expression of AHNAK was elevated in liver cancer tissues and the methylation level of AHNAK decreased from liver disease to HCC [15]. However, the molecular mechanism via which AHNAK is involved in HCC progression is unclear.

Therefore, this study utilized co-immunoprecipitation (Co-IP) and liquid chromatography coupled with tandem mass spectrometry (LC-MS/MS) to analyze the AHNAK interactome in liver cancer and matched paracancerous (MPC) tissues, aiming to investigate the AHNAK involving signaling pathways. Our findings indicate that AHNAK could serve as a potential candidate targeting the IGF-1R signaling pathway to promote the development of HCC.

## 2. Results

### 2.1. Specific Knockout of AHNAK Protein Can Inhibit the Proliferation of HepG2 Cells

To determine the effect of AHNAK on HepG2 cell proliferation, modified lentiviral AHNAK vectors were used to knock out the expression of AHNAK in HepG2 cells. The expression of AHNAK was markedly diminished with the application of sgAHNAK-1, sgAHNAK-2 and sgAHNAK-3 lentiviruses (Figure 1A). Due to the higher efficiency, the sgAHNAK-1 lentivirus was utilized in our subsequent experiments. Cell counting kit-8 (CCK8) assays demonstrated that AHNAK silencing led to reduced viability of HepG2 cells (Figure 1B).

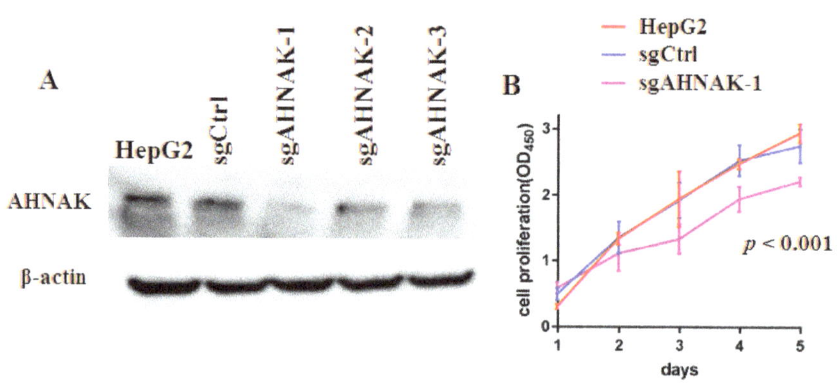


**Figure 1.** AHNAK expression was essential for HepG2 cell growth: (**A**) Results of Western blotting analysis of AHNAK expression in sgCtrl, sgAHNAK-1, sgAHNAK-2 and sgAHNAK-3 HepG2 cells, and HepG2 cells. (**B**) Cell counting kit-8 assays were used to determine cell viability in sgCtrl and sgAHNAK-1 HepG2 cells, and HepG2 cells (n = 3).

### 2.2. Expression Analysis of AHNAK Protein in Liver Cancer Tissues

In previous studies [15], we found high-level mRNA expression of AHNAK in HCC tissue, which was consistent with the results of The Cancer Genome Atlas (TCGA) and the Genotype-Tissue Expression (GTEx) database (Figure 2A). Here, we analyzed the expression of AHNAK protein in HCC and MPC tissues. We found that the AHNAK protein expression level in tumor tissues was higher than that in MPC tissues, and the results were confirmed with Western blotting (Figure 2B) and immunohistochemistry (Figure 2C–E). These results showed that AHNAK overexpressed both in HCC and MPC tissues compared with donor liver tissues. The presence of field cancerization has been confirmed in HCC [16,17]. The cancer-promoting AHNAK expressed in HCC field cancerization might be a new opportunity worthy of study to explore the signature of HCC field cancerization and to comprehend the important role that AHNAK plays in HCC development and progression.

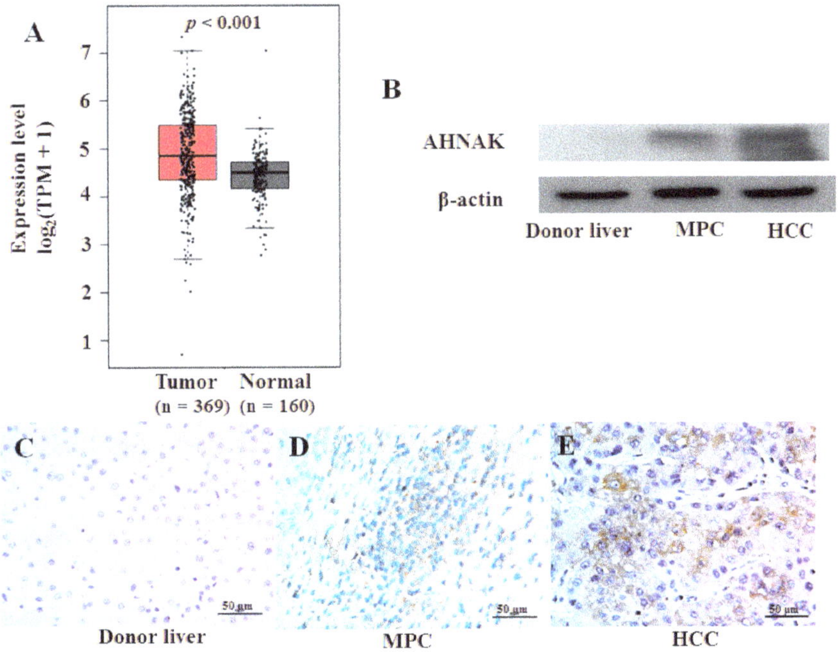


**Figure 2.** AHNAK was overexpressed in HCC tissues: (**A**) Up-regulated mRNA expression of AHNAK in HCC tissues samples from TCGA database (369 tumor and 50 normal samples) and GTEx database (110 normal samples); $p < 0.001$. (**B**) Results of Western blotting analysis of protein expression of AHNAK in donor liver, tumor and MPC tissues. Results of immunohistochemical analysis of protein expression of AHNAK in donor liver (**C**), MPC tissues (**D**) and HCC tissues (**E**). Scale bar, 50 μm. TCGA, The Cancer Genome Atlas; GTEx, Genotype-Tissue Expression; MPC, matched paracancerous.

### 2.3. *Identification of AHNAK as a Candidate Involved in IGF-1 Signaling Transduction*

The HCC field cancerization concept proposes that the paracancerous tissues of HCC patients are at a high risk of HCC [18]. To determine the role of AHNAK in the field cancerization of HCC, we analyzed the AHNAK interactome in HCC and MPC tissues. Three HCC tissues and their MPC tissues were tested using endogenous Co-IP and LC-MS/MS. To reduce false positives, Co-IP was performed on samples using antibody isotype IgG as control. Therefore, a total of 12 samples were tested using LC-MS/MS, and their total ion current chromatograms are shown in Supplementary Materials: Figure S1. By comparing the results of anti-IgG and anti-AHNAK, the proteins that could not be detected in the anti-IgG samples but could be detected in the anti-AHNAK ones were selected as the total interaction proteins of AHNAK in the samples. The total and specific non-redundant proteins interacting with AHNAK in HCC and MPC tissues of three HCC patents are shown in Figure 3A,B. The protein functional annotation and enrichment of signaling pathways were analyzed with the Reactome Gene Sets in Metascape platform. The enriched pathways using proteins interacting with AHNAK both in HCC and MPC tissues from Pat.1, Pat.2 and Pat.3 were 241, 324 and 280, respectively. The intersection of three Venn diagrams identified 204 enriched pathways shared by all three HCC patents ($p < 0.001$) (Figure 3C). The AHNAK protein involving 204 enriched pathways could be mainly divided into 10 categories: cellular signal transduction, cell cycle, cellular response to stress, degradation of proteins, metabolism of biological macromolecules, citric acid cycle, apoptosis, transport of proteins and small molecules, transcription and translation, and immune response. As can be seen in the heat map, the 204 enrichment pathways involving AHNAK protein were different comparing HCC with MPC tissues, which suggested that AHNAK might perform different functions in MPC and HCC tissues (Figure 3D).

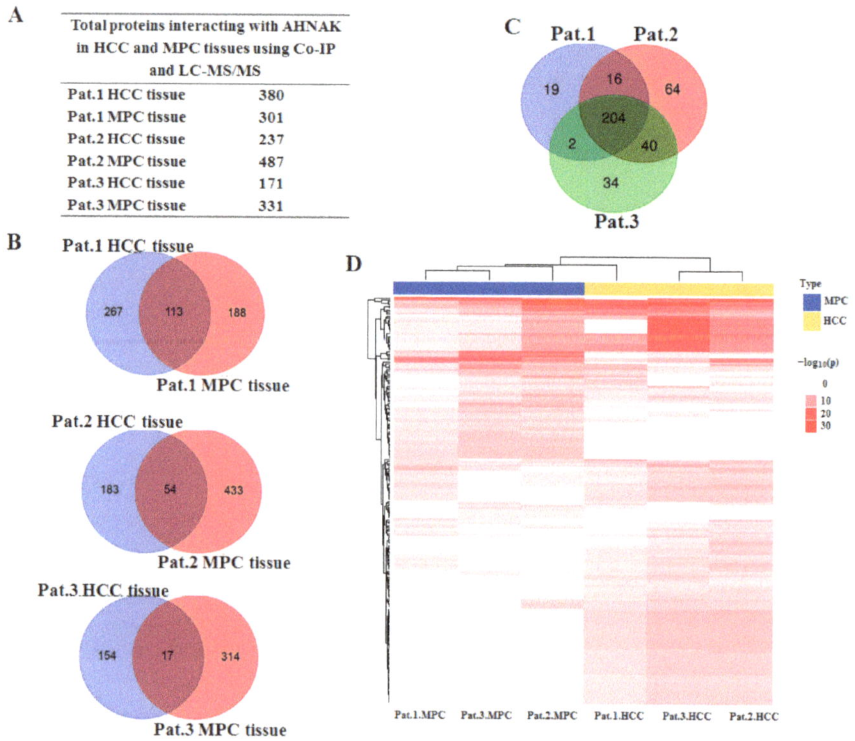


**Figure 3.** AHNAK involving pathways identified using LC-MS/MS: (**A**) Total proteins interacting with AHNAK in HCC and MPC tissues using Co-IP and LC-MS/MS. (**B**) Venn diagram graphic showing common and specific protein interaction with AHNAK between HCC and MPC tissues from Pat.1, Pat.2 and Pat.3. (**C**) Venn diagram graphics showing 204 enriched pathways shared by all three HCC patients ($p < 0.001$). (**D**) Heatmap showing the 204 enrichment pathways in HCC and MPC tissues. The discrete red scale represents statistical significance, and gray indicates a lack of significance. LC-MS/MS, liquid chromatography coupled with tandem mass spectrometry; Co-IP, co-immunoprecipitation; MPC, matched paracancerous.

Of the 10 categories of enriched pathways and processes involving AHNAK protein, 4 are shown in Figure 4, and 6 are shown in supplementary Figure S2. The differences in the enriched statistical significance of apoptosis, cellular response to stress, metabolism of biomacromolecules, and transcription and translation (Figure 4A) were not obvious in MPC and HCC tissues, which suggests that HCC field cancerization might first occur in these pathways and processes. While most cell cycle pathways (Figure 4C), part of cellular signal transduction pathways (Figure 4D), the degradation of proteins, the transport of proteins and the immune response showed statistical significance in HCC tissues, it was not the same case in MPC tissues. In other words, these pathways were enriched in HCC tissues, while they were little or no enriched in MPC tissues. The results showed that these pathways involving AHNAK were activated in HCC but not in MPC tissues. In MPC tissues, citric acid cycle (Figure 4B) and respiratory chain processes were still in dominant positions among the energy generation sources. Taken together, these results indicated differences of AHNAK involved enrichment pathways and processes between HCC and MPC tissues and partially reflected the signature of HCC field cancerization.

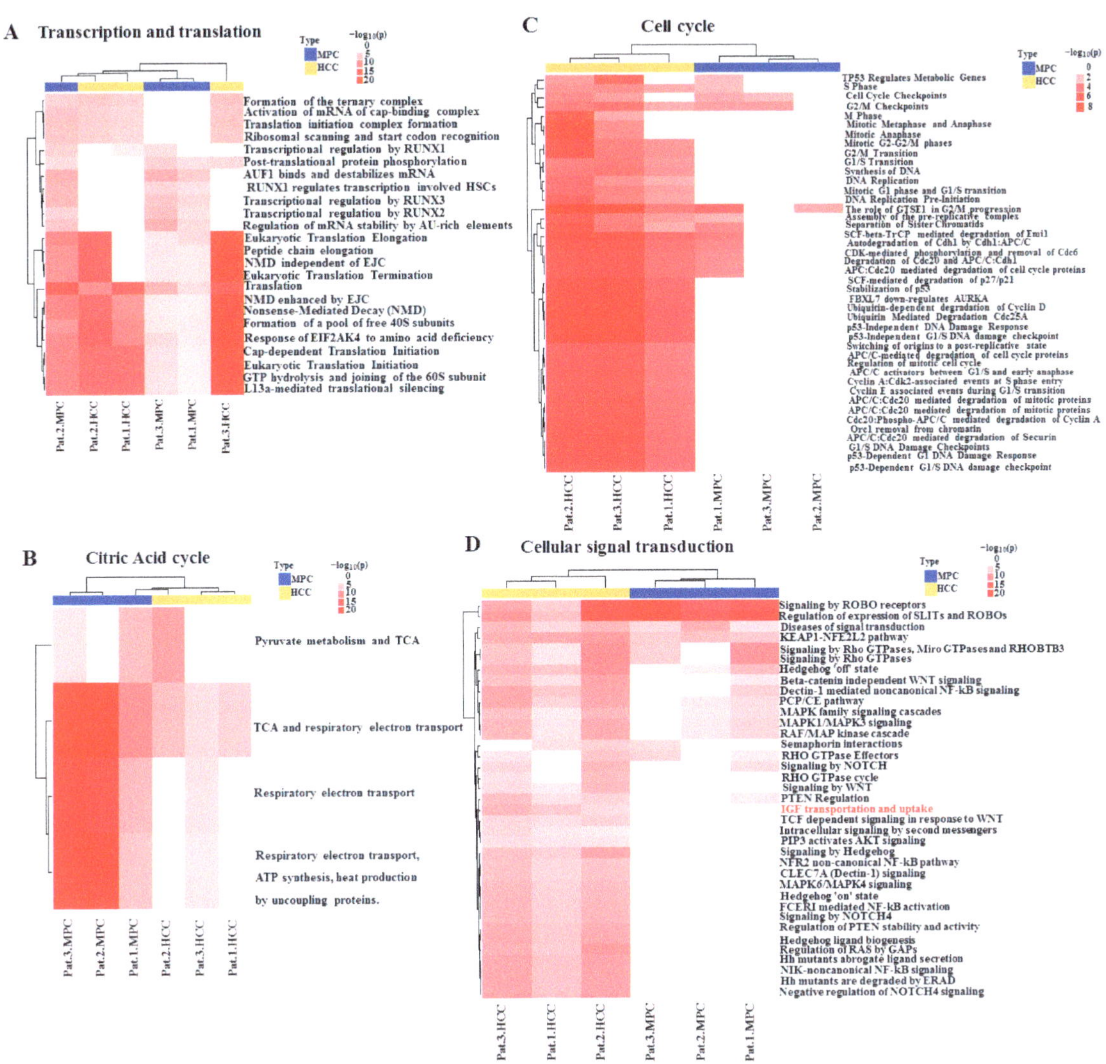


**Figure 4.** Heatmap showing enrichment pathways involving AHNAK protein in HCC and MPC tissues: (**A**) Transcription and translation. (**B**) Citric acid cycle. (**C**) Cell cycle; (**D**) Cellular signal transduction. Red text indicating enriched pathway: "IGF uptake and transportation". MPC, matched paracancerous.

It is known that giant AHNAK is located on the cell membrane and in the cytoplasm and it possesses an N-terminal PDZ-like region (postsynaptic density protein-95, disc large and zonula occludens-1) [19], which could interact with the C-terminal tail (serine-threonine-cysteine) of insulin-like growth factor 1 receptor (IGF-1R) [20]. IGF-1 is a ligand of IGF-1R, which is phosphorylated and activated when binding to the factor. According to the results of Co-IP, LC-MS/MS and previous studies [19,20], we hypothesize that AHNAK might bind to IGF-1R and subsequently has an effect on the transportation and uptake of IGF-1 or the activation of the receptor.

### 2.4. Identification of AHNAK as a Novel IGF-1R-Interacting Protein

To test the hypothesis that AHNAK protein might interact with IGF-1R, serum-starved HepG2 cells were examined. HepG2 cells were starved overnight and were then stimulated with IGF-1 (50 ng/mL) for a time interval, as shown in Figure 5A, prior to harvesting. The Western blotting analysis of whole-cell extracts demonstrated that treatment with IGF-1 led to the potentiating of AHNAK expression. To further investigate the role of AHNAK in IGF-1 signaling transduction, immunofluorescence double staining and Co-IP were used to assess the interactions between AHNAK and IGF-1R. As shown in Figure 5B, HepG2 cells were starved overnight and were then stimulated with IGF-1 (50 ng/mL) for 1 h. Confocal microscopy analyses revealed that IGF-1 could induce AHNAK (green puncta) expression on the cell membrane and in the cytoplasm of serum-starved HepG2 cells; meanwhile, the result demonstrated the co-localization of AHNAK with phosphorylated IGF-1R (red puncta). Based on protein sequences, an online tool, PPA-Pred2, was used to make a prediction on the protein binding affinity between AHNAK and IGF-1R. The predicted value of Delta G (binding free energy) indicates spontaneous binding between two proteins, and the higher the absolute value of Delta G is, the more stable the binding affinity is. The value of Delta G was predicted to be −37.66 kcal/mol between AHNAK and IGF-1R using the "Miscellaneous" category. Then, the interaction was confirmed with Co-IP experiments. HepG2 cells were starved overnight and then stimulated with IGF-1 (50 ng/mL) for 1 h. Immunoprecipitation followed by Western blotting was performed to detect the interactions between AHNAK and IGF-1R. Our data showed that AHNAK was pulled down by anti-IGF-1R; meanwhile, IGF-1R was also pulled down by anti-AHNAK. Then, they were detected using the corresponding antibody (Figure 5C). These data verified the direct physical binding of AHNAK and IGF-1R in vitro, which suggested that the AHNAK complex promoted HCC growth, potentially by interacting with IGF-1R.

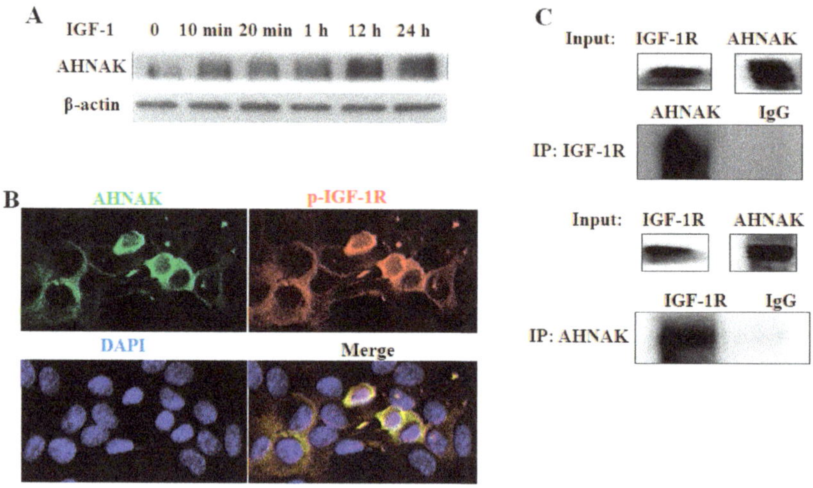


**Figure 5.** Identification of AHNAK as a novel IGF-1R-interacting protein: (**A**) Western blotting experiments showed that IGF-1 induced AHNAK expression. (**B**) Confocal microscopy scans of immunofluorescence double staining showed that AHNAK (green) co-localized with IGF-1R (red) in serum-starvation-treated HepG2 cells. (**C**) Co-immunoprecipitation and Western blotting experiments showed interactions between AHNAK and IGF-1R.

## 3. Discussion

AHNAK has been previously identified as a structural scaffold protein [5], implicated in a range of cancer-related pathways and processed as an ambiguous factor [7,8,11,14,17], which might depend on the type of tumor cells [21]. AHNAK has been described as a nucleoprotein that is significantly suppressed in neuroblastoma cell lines [22]. Immunoreaction to AHNAK has mainly been observed in the cytoplasm of normal cells compared with melanoma [10] and bladder urothelial carcinoma cells [22], whereas high levels of AHNAK have been shown to predict a poor outcome in PDAC [11], HCC [17] and laryngeal carcinoma [12]. AHNAK has also been reported to enhance metastasis in lung cancer [13] and mesothelioma [14]. Based on these observations, AHNAK might be a multifunctional protein and play variable roles in different types of cancers.

To investigate what interacted with AHNAK and its role in HCC, we performed Co-IP and LC-MS/MS using anti-AHNAK in HCC and MPC tissues, due to AHNAK being also expressed in MPC tissues. It has been shown that MPC tissues are not completely normal tissues, which is known as field cancerization and was first studied by Slaughter and Southwick [23]. It has been proposed that tumors originate from a field of mutation cells, such as aberrations of genetic, epigenetic or biochemical nature, which creates a permissive environment for malignant evolution, which could occur with or without morphological changes [24]. The molecular and cellular mechanisms underpinning the etiology of field cancerization remain largely unknown. In this study, AHNAK immunoreaction was observed both in MPC and HCC tissues, but not in donor liver. That proved to be an opportunity to explore the relationship between HCC field cancerization and the AHNAK interactome. The results showed that a total of 204 enriched pathways and processes involved by AHNAK in each HCC patient. Some pathways and processes, including transcription and translation, apoptosis, transport of proteins and small molecules, cellular response to stress and metabolism of biomacromolecules represented similar enrichment statistical significance both in HCC and MPC tissues, which suggests that AHNAK is involved in them both in MPC and HCC tissues. HCC field cancerization might first occur in these pathways and processes. The citric acid cycle and respiratory chain processes were still in dominant positions among the energy generation sources in MPC tissues. Generally, most types of cancer cells produce energy predominantly through the Warburg effect, which is a less efficient process of "aerobic glycolysis" consisting of a high level of glucose uptake followed by lactic acid fermentation taking place in the cytosol, not the mitochondria, and the preferential production of lactate even in the presence of abundant oxygen [25].

About half of the 204 pathways and processes mainly focusing on cell cycle, cellular signal transduction, degradation of proteins and immune response represented enrichment statistical significance in HCC tissues but a lack of significance in MPC tissues. AHNAK might perform more functions in HCC than in MPC tissues. In tumor cells, AHNAK mRNA was found to be significantly enriched in the G0 and G1 phases and substantially reduced in S/G2 of the cell cycle [26]. AHNAK serves as a G1-enriched interactor of p53-binding protein 1 (53BP1) to regulate the tumor cell cycle [26]. The AHNAK-53BP1 complex has been shown to suppress p53 target gene networks in multiple cancer types; meanwhile, AHNAK also directly interacts with p53 and inhibits p53-mediated target gene expression [27]. The concealed role of AHNAK in curbing the spontaneous activation of p53 response to fine-tune the levels of "G1-S checkpoint" has been identified. These results together indicate that high-level expression of AHNAK is relevant to cancer cell proliferation, in particular, that caused by a p53 function defect [26]. These results are consistent with our observation of AHNAK-enriched G0/G1- or p53-related pathways and processes in HCC tissues (Figure 4C).

IGF/IGF-1R signaling plays a crucial role in tumorigenesis, proliferation and metastasis through the regulation of multiple downstream signaling pathways, including PI3K/AKT and MAPK/ERK pathways [28]. In a previous study, following the analysis results from TCGA database, AHNAK and IGF-1 were selected as independent risk factors associated with the prognosis of patients with bladder urothelial carcinoma [29]. The activation of IGF-1R signaling was associated with poor prognosis in HBV-related HCC [30]. In this study, IGF-1 could intrude the overexpression of AHNAK in starvation-treated HepG2 cells. Additionally, the N-terminal PDZ-like region possessed by AHNAK could interact with the C-terminal tail (serine-threonine-cysteine) of IGF-1R [19,20]. We applied immunofluorescence double staining and Co-IP to confirm the hypothesis of the interaction between AHNAK and IGF-1R. In the future, we need to study whether AHNAK plays a key role in IGF-1R phosphorylation and IGF/IGF-1R signaling pathway activation.

## 4. Materials and Methods

### 4.1. Cell Culture

Hepatoma cell line HepG2 cells were maintained in Dulbecco's Minimal Essential Medium (DMEM) supplemented with 10% fetal bovine serum, 50 units/milliliter of penicillin and 50 microgram/milliliter streptomycin sulfate. Then, they were incubated at 37 °C in a humidified atmosphere of 5% $CO_2$. Prior to the addition of IGF-1 (Cell Signaling Technology, Danvers, MA, USA), cells were maintained overnight in serum-free medium.

### 4.2. Tissue Samples

In total, four HCC patients who were admitted at Beijing You'An Hospital, Capital Medical University, from 1 June 2013 to 30 December 2016 were recruited. Liver donor tissues, tumor tissues and MPC tissues located at least 2 cm from the tumor were obtained at the time of segmental surgical resection or liver transplantation. The diagnosis of HCC was confirmed with a histopathological examination, and medical records were reviewed for clinicopathological characteristics, according to the following 2012 EASL clinical practice guidelines: Management of chronic hepatitis B virus infection and Barcelona clinic liver cancer staging system. Exclusion criteria were as follows: less than 18 years old; co-infection with human immunodeficiency virus, hepatitis C virus or hepatitis D virus; coexistence of liver injury caused by drug intake, alcohol consumption or autoimmune hepatitis; pregnancy; lactation. This study was approved by the medical research ethics committee of Beijing You'An Hospital, Capital Medical University, and adhered to the 1975 Declaration of Helsinki.

### 4.3. CoIP for Nano-LC-MS/MS Analysis

Tissue lysate was centrifuged at 14,000× *g* for 15 min at 4 °C, and protein A agarose beads (50%) were added to the supernatant; then, it was shaken at 4 °C for 10 min on horizontal ice to remove non-specific foreign proteins and reduce the background. From the samples, we removed protein A beads after centrifuging at 14,000× *g* for 15 min at 4 °C. We slowly shook the mixture of antibody and tissue lysate at 4 °C overnight. We added 100 μL of protein A agarose beads to capture the antibody and its bound proteins, and we slowly shook the antigen–antibody mixture at 4 °C overnight. We then centrifuged the mixture at 14,000× *g* for 5 s, collected the agarose bead antibody complex, removed the supernatant, and washed with cooled phosphate-buffered saline (PBS) buffer for three times.

### 4.4. Trypsin Digestion

The samples were treated with 5 mM dithiothreitol (dissolved in 25 mM ammonium bicarbonate) at room temperature for 40 min for disulfide bond reduction. The samples were treated with 15 mM iodoacetamide (dissolved in 25 mM ammonium bicarbonate) and kept away from light at room temperature for 40 min. Trypsin was added in a ratio of 1:50 for protease digestion at 37 °C overnight. After the desalination process, the peptide sample eluted with the highly hydrophobic eluent was vacuumized for standby.

### 4.5. Liquid Chromatography-Tandem Mass Analysis

The extracted peptide sample was redissolved with 0.1% acetic acid and centrifuged at 13,000× *g* for 10 min for mass spectrometry analyses. Peptides were separated using liquid chromatography using an Acclaim Pep Map 100 column and an EASY-Spray column on an EASY-NLC 1000 system (ThermoFisher Scientific, Waltham, MA, USA). The loading sample volume was 3 μL. The gradient was generated using mobile phase A (0.1% formic acid in water) and mobile phase B (0.1% formic acid in acetonitrile). Mass spectrometry was performed using an Obitrap Fusion Lumos mass spectrometer (ThermoFisher Scientific). The spray voltages were set at 2.2 kV, and the heated capillary temperature was 270 °C. The parameters of the MS/MS scan were as follows: resolution, 60,000 at 400 $m/z$; maximum isolation time, 30 ms; normalized collision energy, 40%. Data-dependent MS/MS: up to top five most intense peptide ions from the preview scan in Obitrap.

### 4.6. Protein Identification and Annotation

The raw MS files were analyzed and searched against the Uniprot-Homo Sapiens protein sequence database using Maxquant [31]. The parameters were set as follows: the protein modifications were carbamidomethylation (C) (fixed) and oxidation (M) (variable); the enzyme specificity was set to trpsin; the maximum missed cleavages were set to 2; the precursor ion mass tolerance was set to 3 ppm; and MS/MS tolerance was 0.01 Da. Only highly confidently identified peptides were chosen.

The protein functional annotation and identification of signaling pathways of potential targets were performed using Gene Ontology (GO) and Kyoto Encyclopedia of Genes and Genomes (KEGG) enrichment analyses using the Metascape platform. The "ComplexHeatmap" software package was used in R (Version 3.6.3) for cluster analyses and visualization. The online prediction software Protein-Protein Affinity Predictor (PPA-Pred2) is available at https://www.iitm.ac.in/bioinfo/PPA_Pred/ (accessed on 19 May 2022).

### 4.7. Immunoprecipitation (IP) and Western Blotting Analysis

Cells were lysed, and proteins were extracted and quantified. The cell lysates were incubated with specific antibodies or the anti-IgG control for 3–4 h at 4 °C with rotation. Then, samples were incubated with protein A/G agarose beads (Santa Cruz Biotechnology, Santa Cruz, CA, USA) overnight at 4 °C with rotation. The beads were pelleted using centrifugation at 1000× $g$ for 30 s at 4 °C; then, they were boiled for 10 min at 95 °C and resolved with 10% SDS-polyacrylamide gel electrophoresis. The separated proteins were transferred to PVDF membranes (Millipore, Bedford, MA, USA), and the membranes were blocked in 5% skim milk and incubated overnight at 4 °C with anti-AHNAK or anti-IGF-1R at a dilution of 1:200. After washing with TBST, the membranes were incubated with a horseradish peroxidase-conjugated secondary antibody for 1 h at room temperature. Finally, the membranes were incubated with enhanced chemiluminescent reagents (Millipore) and exposed to autoradiography film in the dark.

### 4.8. Immunohistochemistry Array

The slides were dewaxed, rehydrated, and blocked with 3% hydrogen peroxide, and antigen retrieval was performed with citrate buffer (Dako target retrieval solution; citrate buffer at pH 6.0). Slides were incubated with AHNAK antibody at a dilution of 1:50 (Santa Cruz Biotechnology) overnight at 4 °C. Then, the slides were incubated with secondary antibody and daminobenzidine after washing with phosphate-buffered saline.

### 4.9. SgRNA Plasmid Constructs and Generation of a Stable AHNAK-Knockout HepG2 Cell Line

AHNAK was knocked out using the CRISPR/cas9 method. In brief, the main steps were as follows: First, four sgRNAs were designed: sgCtrl (5′-CCTCGTTCACCGCCGTCG CG-3′), sgAHNAK-1 (5′-TCTTCGTGGTGTAGATGCGC-3′), sgAHNAK-2 (5′-CCATCTTCC GACTTCAGCCG-3′) and sgAHNAK-3 (5′-CTGAAAGGCCCTAACGTAAA-3′). The primers were inserted into the sgRNA backbone to construct the lentivirus plasmid vectors. Lentiviral production was performed by transfecting the lentivirus vectors into HEK 293H cells and collecting the resulting supernatant after 48 h. Titers were determined by detecting the WPRE sequence that was located on the lentiviral vector and could be integrated into the cell genome with a target gene with a qRT-PCR assay. HepG2 cells were infected with sgCtrl, and AHNAK sgRNA-1, -2 and -3 lentiviruses and were screened using basticidin resistance.

### 4.10. Cell Viability Assays

CCK8 assays were performed to determine cell proliferation. An equal number of the indicated cells in 200 μL of culture medium was seeded into 96-well plates for 24 h. At the indicated time, 20 μL of CCK-8 working solution (Boster, Wuhan, China) was added into each well and incubated with culture medium at 37 °C for 2 h. Cell viability was determined using an absorbance value of OD450 nm on a microplate reader.

### 4.11. Confocal Immunofluorescence

Briefly, cells on chamber slides were fixed with 4% paraformaldehyde and permeabilized with 0.2% TritonX-100. The samples were probed with specific antibodies against AHNAK (Santa Cruz) and p-IGF-1R (Cell Signaling Technology) at 4 °C for 12 h and then incubated with FITC- and TRITC-labeled secondary antibodies (1:200 dilution) for 1 h. After each step, cells were washed two times using PBS. The cell nuclei were stained with DAPI for 3 min, and a Leica DM14000B confocal microscope was used to capture images.

## 5. Conclusions

In conclusion, our results partially reflected the signature of HCC field cancerization and showed that AHNAK promoted HCC growth, potentially by interacting with IGF-1R. We performed Co-IP and LC-MS/MS on HCC and MPC tissues to explore the relationship between HCC field cancerization and the AHNAK interactome. The results of immunofluorescence double staining and Co-IP confirmed the interaction between AHNAK and IGF-1R. The role of AHNAK in IGF-1R phosphorylation and IGF/IGF-1R signaling pathway activation is an important research direction.

**Supplementary Materials:** The following supporting information can be downloaded at: https://www.mdpi.com/article/10.3390/molecules27248680/s1, Figure S1: Total ion current chromatograms of 12 samples tested with LC-MS/MS, Figure S2: Six categories of enriched pathways and processes involving AHNAK protein.

**Author Contributions:** Conceptualization, K.L. and Y.Z.; Data curation, K.L., K.S. and Y.H.; Funding acquisition, K.L., L.S. and Y.Z.; Methodology, Y.T., H.W., L.S. and A.L.; Writing—original draft, K.L. All authors have read and agreed to the published version of the manuscript.

**Funding:** This research was funded by Beijing Natural Science Foundation (7191004), Capital's Funds of Health Improvement and Research (CFH2020-1-2182), Beijing You'An Hospital (BJYAHKF2021003, YNKTQN2021007), Beijing Key Laboratory (BZ0373), "Sail Plan" from Beijing Hospital Authority (XMLX201711), Beijing Municipal Science & Technology Commission (Z171100001017078) and National Key Research and Development Project (2020YFE0202400).

**Institutional Review Board Statement:** The study was conducted in accordance with the Declaration of Helsinki and was approved by the ethics committee of Beijing You'An Hospital, Capital Medical University (Jingyoukelunzi [2017] 21# (in Chinese); date of approval: 28 August 2017).

**Informed Consent Statement:** Informed consent was obtained from all subjects involved in the study.

**Data Availability Statement:** Not applicable.

**Conflicts of Interest:** The authors declare no conflict of interest.

## References

1. Siegel, R.L.; Miller, K.D.; Fuchs, H.E.; Jemal, A. Cancer statistics, 2022. *CA Cancer J. Clin.* **2022**, *72*, 7–33. [CrossRef] [PubMed]
2. Aran, D.; Camarda, R.; Odegaard, J.; Paik, H.; Oskotsky, B.; Krings, G.; Goga, A.; Sirota, M.; Butte, A.J. Comprehensive analysis of normal adjacent to tumor transcriptomes. *Nat. Commun.* **2017**, *8*, 017–01027. [CrossRef] [PubMed]
3. Vidal, A.F.; Ribeiro-Dos-Santos, A.M.; Vinasco-Sandoval, T.; Magalhães, L.; Pinto, P.; Anaissi, A.K.M.; Demachki, S.; de Assumpção, P.P.; Dos Santos, S.E.B.; Ribeiro-Dos-Santos, Â. The comprehensive expression analysis of circular RNAs in gastric cancer and its association with field cancerization. *Sci. Rep.* **2017**, *7*, 15061. [CrossRef] [PubMed]

4. Magalhães, L.; Ribeiro-dos-Santos, A.M.; Cruz, R.L.; Nakamura, K.D.d.M.; Brianese, R.; Burbano, R.; Ferreira, S.P.; Oliveira, E.L.F.d.; Anaissi, A.K.M.; Nahúm, M.C.d.S.; et al. Triple-Negative Breast Cancer circRNAome Reveals Hsa_circ_0072309 as a Potential Risk Biomarker. *Cancers* **2022**, *14*, 3280. [CrossRef]
5. Amagai, M. A mystery of AHNAK/desmoyokin still goes on. *J. Investig. Dermatol.* **2004**, *123*, xiv–xv. [CrossRef]
6. Davis, T.A.; Loos, B.; Engelbrecht, A.M. Corrigendum to AHNAK: The giant jack of all trades. *Cell Signal.* **2014**, *26*, 2683–2693, Erratum in *Cell Signal.* **2015**, *27*, 187–188. [CrossRef]
7. Chen, B.; Wang, J.; Dai, D.; Zhou, Q.; Guo, X.; Tian, Z.; Huang, X.; Yang, L.; Tang, H.; Xie, X. AHNAK suppresses tumour proliferation and invasion by targeting multiple pathways in triple-negative breast cancer. *J. Exp. Clin. Cancer Res.* **2017**, *36*, 0522. [CrossRef]
8. Cai, Y.; Hu, Y.; Yu, F.; Tong, W.; Wang, S.; Sheng, S.; Zhu, J. AHNAK suppresses ovarian cancer progression through the Wnt/β-catenin signaling pathway. *Aging* **2021**, *13*, 23579–23587. [CrossRef]
9. Cho, W.C.; Jang, J.E.; Kim, K.H.; Yoo, B.C.; Ku, J.L. SORBS1 serves a metastatic role via suppression of AHNAK in colorectal cancer cell lines. *Int. J. Oncol.* **2020**, *56*, 1140–1151. [CrossRef]
10. Sheppard, H.M.; Feisst, V.; Chen, J.; Print, C.; Dunbar, P.R. AHNAK is downregulated in melanoma, predicts poor outcome, and may be required for the expression of functional cadherin-1. *Melanoma Res.* **2016**, *26*, 108–116. [CrossRef]
11. Zhang, Z.; Liu, X.; Huang, R.; Liang, Z.; Liu, T. Upregulation of nucleoprotein AHNAK is associated with poor outcome of pancreatic ductal adenocarcinoma prognosis via mediating epithelial-mesenchymal transition. *J. Cancer* **2019**, *10*, 3860–3870. [CrossRef] [PubMed]
12. Dumitru, C.A.; Bankfalvi, A.; Gu, X.; Zeidler, R.; Brandau, S.; Lang, S. AHNAK and inflammatory markers predict poor survival in laryngeal carcinoma. *PLoS ONE* **2013**, *8*, e56420. [CrossRef] [PubMed]
13. Sohn, M.; Shin, S.; Yoo, J.-Y.; Goh, Y.; Lee, I.H.; Bae, Y.S. Ahnak promotes tumor metastasis through transforming growth factor-β-mediated epithelial-mesenchymal transition. *Sci. Rep.* **2018**, *8*, 14379. [CrossRef] [PubMed]
14. Sudo, H.; Tsuji, A.B.; Sugyo, A.; Abe, M.; Hino, O.; Saga, T. AHNAK is highly expressed and plays a key role in cell migration and invasion in mesothelioma. *Int. J. Oncol.* **2014**, *44*, 530–538. [CrossRef] [PubMed]
15. Sun, L.; Li, K.; Liu, G.; Xu, Y.; Zhang, A.; Lin, D.; Zhang, H.; Zhao, X.; Jin, B.; Li, N.; et al. Distinctive pattern of AHNAK methylation level in peripheral blood mononuclear cells and the association with HBV-related liver diseases. *Cancer Med.* **2018**, *7*, 5178–5186. [CrossRef] [PubMed]
16. Ding, X.; He, M.; Chan, A.W.H.; Song, Q.X.; Sze, S.C.; Chen, H.; Man, M.K.H.; Man, K.; Chan, S.L.; Lai, P.B.S.; et al. Genomic and Epigenomic Features of Primary and Recurrent Hepatocellular Carcinomas. *Gastroenterology* **2019**, *157*, 1630–1645.e6. [CrossRef]
17. Shimizu, M.; Shirakami, Y.; Imai, K.; Takai, K.; Moriwaki, H. Acyclic retinoid in chemoprevention of hepatocellular carcinoma: Targeting phosphorylated retinoid X receptor-α for prevention of liver carcinogenesis. *J. Carcinog.* **2012**, *11*, 1477–3163. [CrossRef] [PubMed]
18. Qin, X.-Y.; Suzuki, H.; Honda, M.; Okada, H.; Kaneko, S.; Inoue, I.; Ebisui, E.; Hashimoto, K.; Carninci, P.; Kanki, K.; et al. Prevention of hepatocellular carcinoma by targeting MYCN-positive liver cancer stem cells with acyclic retinoid. *Proc. Natl. Acad. Sci. USA* **2018**, *115*, 4969–4974. [CrossRef]
19. Han, H.; Kursula, P. Periaxin and AHNAK Nucleoprotein 2 Form Intertwined Homodimers through Domain Swapping. *J. Biol. Chem.* **2014**, *289*, 14121–14131. [CrossRef]
20. Ligensa, T.; Krauss, S.; Demuth, D.; Schumacher, R.; Camonis, J.; Jaques, G.; Weidner, K.M. A PDZ Domain Protein Interacts with the C-terminal Tail of the Insulin-like Growth Factor-1 Receptor but Not with the Insulin Receptor. *J. Biol. Chem.* **2001**, *276*, 33419–33427. [CrossRef]
21. Lee, H.; Kim, K.; Woo, J.; Park, J.; Kim, H.; Lee, K.E.; Kim, Y.; Moon, K.C.; Kim, J.Y.; Park, I.A.; et al. Quantitative Proteomic Analysis Identifies AHNAK (Neuroblast Differentiation-associated Protein AHNAK) as a Novel Candidate Biomarker for Bladder Urothelial Carcinoma Diagnosis by Liquid-based Cytology. *Mol. Cell. Proteoms.* **2018**, *17*, 1788–1802. [CrossRef]
22. Shtivelman, E.; Cohen, F.E.; Bishop, J.M. A human gene (AHNAK) encoding an unusually large protein with a 1.2-microns polyionic rod structure. *Proc. Natl. Acad. Sci. USA* **1992**, *89*, 5472–5476. [CrossRef] [PubMed]
23. Slaughter, D.P.; Southwick, H.W.; Smejkal, W. Field cancerization in oral stratified squamous epithelium; clinical implications of multicentric origin. *Cancer* **1953**, *6*, 963–968. [CrossRef] [PubMed]
24. Gadaleta, E.; Thorn, G.J.; Ross-Adams, H.; Jones, L.J.; Chelala, C. Field cancerization in breast cancer. *J. Pathol.* **2022**, *257*, 561–574. [CrossRef] [PubMed]
25. Chen, Z.; Lu, W.; Garcia-Prieto, C.; Huang, P. The Warburg effect and its cancer therapeutic implications. *J. Bioenerg. Biomembr.* **2007**, *39*, 267–274. [CrossRef]
26. Ghodke, I.; Remisova, M.; Furst, A.; Kilic, S.; Reina-San-Martin, B.; Poetsch, A.R.; Altmeyer, M.; Soutoglou, E. AHNAK controls 53BP1-mediated p53 response by restraining 53BP1 oligomerization and phase separation. *Mol. Cell* **2021**, *81*, 2596–2610.e7. [CrossRef] [PubMed]
27. Gu, J.; Mao, W.; Ren, W.; Xu, F.; Zhu, Q.; Lu, C.; Lin, Z.; Zhang, Z.; Chu, Y.; Liu, R.; et al. Ubiquitin-protein ligase E3C maintains non-small-cell lung cancer stemness by targeting AHNAK-p53 complex. *Cancer Lett.* **2019**, *443*, 125–134. [CrossRef]
28. Cao, Z.; Liu, L.Z.; Dixon, D.A.; Zheng, J.Z.; Chandran, B.; Jiang, B.H. Insulin-like growth factor-I induces cyclooxygenase-2 expression via PI3K, MAPK and PKC signaling pathways in human ovarian cancer cells. *Cell. Signal.* **2007**, *19*, 1542–1553. [CrossRef]

29. Na, L.; Bai, Y.; Sun, Y.; Wang, Z.; Wang, W.; Yuan, L.; Zhao, C. Identification of 9-Core Immune-Related Genes in Bladder Urothelial Carcinoma Prognosis. *Front. Oncol.* **2020**, *10*, 1142. [CrossRef]
30. Chang, T.S.; Wu, Y.C.; Chi, C.C.; Su, W.C.; Chang, P.J.; Lee, K.F.; Tung, T.H.; Wang, J.; Liu, J.J.; Tung, S.Y.; et al. Activation of IL6/IGFIR confers poor prognosis of HBV-related hepatocellular carcinoma through induction of OCT4/NANOG expression. *Clin. Cancer Res.* **2015**, *21*, 201–210. [CrossRef]
31. Tyanova, S.; Temu, T.; Cox, J. The MaxQuant computational platform for mass spectrometry-based shotgun proteomics. *Nat. Protoc.* **2016**, *11*, 2301–2319. [CrossRef] [PubMed]

*molecules*

MDPI

*Article*

# 8–17 DNAzyme Silencing Gene Expression in Cells via Cleavage and Antisense

**Zhongchun Zhou [1], Wen Sun [1] and Zhen Huang [1,2,*]**

[1] Key Laboratory of Bio-Resource and Eco-Environment of Ministry of Education, College of Life Sciences, Sichuan University, Chengdu 610064, China
[2] SeNA Research Institute & Szostak-CDHT Large Nucleic Acids Institute, Chengdu 610041, China
* Correspondence: huang@senaresearch.org

**Abstract:** Gene silencing is an important biological strategy for studying gene functions, exploring disease mechanisms and developing therapeutics. 8–17 DNAzyme is of great potential for gene silencing, due to its higher RNA-cleaving activity. However, it is not generally used in practice, due to its divalent cation dependence and poor understanding of its cellular mechanisms. To address these issues, we have explored its activity in vitro and in cells and found that it can cleave RNA substrates under the simulated physiological conditions, and its gene-silencing activity is additionally enhanced by its RNase H compatibility, offering both cleavage and antisense activities in cells. Further, chemical modifications can facilitate its stability, substrate binding affinity and gene-silencing activity. Our research results suggest that this DNAzyme can demonstrate high levels of activities for both actions in cells, making it a useful tool for exploring biomedical applications.

**Keywords:** 8–17 DNAzyme; gene silencing; catalysis; antisense; chemical modifications

**Citation:** Zhou, Z.; Sun, W.; Huang, Z. 8–17 DNAzyme Silencing Gene Expression in Cells via Cleavage and Antisense. *Molecules* **2023**, *28*, 286. https://doi.org/10.3390/molecules28010286

Academic Editors: Chongwen Wang and Guohui Sun

Received: 21 December 2022
Revised: 28 December 2022
Accepted: 28 December 2022
Published: 29 December 2022



## 1. Introduction

Gene silencing takes place naturally in cells by suppressing the expression of a gene at transcriptional or translational levels [1]. Following the discovery and demonstration of RNA interference (RNAi) [2], many astounding advances have been accomplished through gene silencing, in studying gene functions [3], inter-genic interactions and molecular mechanisms of diseases. Most importantly, it is a promising strategy to tackle various diseases, as many diseases are caused by the wrong expression of genes.

There are many gene-silencing methodologies based on nucleic acid sequences, such as antisense oligonucleotides (ASOs), small interfering RNAs (siRNAs), microRNAs (miRNAs), ribozymes, DNAzymes and others [4,5], making it possible to treat any difficult diseases. Among them, ASOs and siRNAs have been approvedfor use as therapeutics in disease treatments [6]. However, they still have shortcomings, for example, siRNAs are of relatively poor stability, immunogenicity and "off-target" effects in vivo, and ASOs have relatively low efficacy and high "off-target" effects when used at high concentrations in practices [7]. Although both RNA-cleaving ribozymes and DNAzymes can specifically cleave RNA substrates, DNAzymes are simpler and more stable than ribozymes. Further, DNAzymes have many advantages, including stability, small molecular size, high selectivity, designability, modifiability and availability [8], making them potential gene-silencing tools.

Two chiefly studied RNA-cleaving DNAzymes are 10–23 DNAzyme (Dz10–23) and 8–17 DNAzyme (Dz8–17), due to their RNA-cleaving activities and substrate selection in vitro and in cells [9]. Each consists of a catalytic loop (13–15 nt) and two substrate-binding arms of variable lengths and sequences, recognizing RNA substrates through the Watson–Crick base pairing [10]. While Dz10–23 is successfully used for gene silencing [11,12], it cannot be applicable to every target gene due to its cleavage site choice and its accessibility to RNA targets. 8–17 DNAzyme, with the determined structure [10] for rational design, is presumably of greater potentials for gene silencing. Unfortunately, research on Dz8–17 is still not sufficient for silencing gene expression efficiently in cells. The lower utility of Dz8–17 is attributed to the poor understanding of its cellular mechanisms and divalent cation dependence [10,13]. Since the physiological ionic strengths of free $Mg^{2+}$ (approximately 0.5–1.0 mM) [14] are reasonably high, likely sufficient to support DNAzyme catalysis, other reasons (such as the DNAzyme stability, nonspecific binding, and target accessibility) might play major roles in causing its poor cellular activities.

To face these problems, we decided to explore the in vitro cleavage of Dz8–17 under simulated physiological conditions, Dz8–17 modifications, and target accessibility. DNAzymes without any modifications can be quickly cleaved by endonucleases and exonucleases in cells and serum. Therefore, in order to make DNAzymes more active in cells, molecular modifications are necessary. Chemical modification is currently a smart strategy to maintain the intracellular activity of DNAzymes. Most commonly, phosphorothioate (PS) linkages are used. PS modification was first used to protect ASOs from the degradation of serum exonuclease and endonuclease [15], and later it was also used to protect DNAzymes [16]. Compared with unmodified DNAzymes, PS-modified DNAzymes have reduced catalytic activity, but their ability to inhibit gene expression in cultured cells is much stronger than that of their unmodified form [17], as it is more resistant to cellular nuclease degradation. The second most popular is 2′-O-Me (OMe), a natural modification widely used for ASOs and aptamers [18]; it can also protect DNAzymes from nuclease degradation [19]. The third most popular is locked nucleic acids (LNAs), a remarkable duplex stabilizer, which has been successfully applied to the modifications of siRNAs [20] and ASOs [21]. Although chemical modifications can enhance the gene-silencing activity of Dz10–23 in cells [16,19], the intracellular activity of chemically modified Dz8–17 remains unknown.

Further, we investigated its cellular activity using the mRNA of EGFP as the target, which is broadly used in gene-silencing studies [12,22]. For identifying the accessible target mRNA sites in cells, we designed a set of Dz8–17 enzymes and examined their activities on cleaving the mRNA fragment in vitro. Interestingly, we found that some Dz8–17 enzymes identified in vitro also worked best in cells, indicating the consistency between in vitro and cellular activities. In addition, for the activity consideration, we studied this DNAzyme and performed in vitro and cellular investigations in order to provide insights into the mechanisms and cellular activities of Dz8–17. Finally, we successfully identified the chemically modified Dz8–17 with both cleavage and antisense activities on EGFP mRNA in cells (Figure 1).

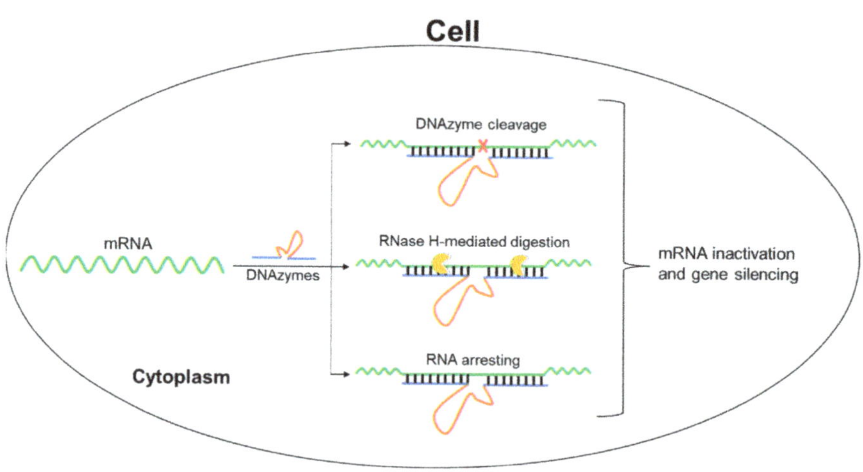


**Figure 1.** 8–17 DNAzyme exerts high gene-silencing effect via catalytic cleavage and antisense activities.

## 2. Results and Discussion

### 2.1. The Design of DNAzymes Targeting EGFP mRNA

According to the EGFP mRNA sequence (Figure S1) and cleavage site of Dz8–17 (AG or GG) and the basis of bioinformatic analysis, we designed and synthesized a set of Dz8–17 DNAzymes (12 DNAzymes, Dz01–Dz12, Table 1) targeting the EGFP-coding region.

**Table 1.** The sequences of twelve Dz8–17 enzymes targeting EGFP mRNA.

| Name | Sequence (5′-3′) [1] | Target Region | Tm (°C) |
|---|---|---|---|
| Dz01 | GGGCAGCTTGTCCGAGCCGGTCGAAGGTGGTGCAGA | 143–165 | 72.5 |
| Dz02 | GCACTGCACGTCCGAGCCGGTCGAAGTAGGTCAGGG | 191–213 | 70.0 |
| Dz03 | GGCTGAAGCATCCGAGCCGGTCGAAGCACGCCGTAG | 198–220 | 70.8 |
| Dz04 | GGGGTAGCGGTCCGAGCCGGTCGAAGAAGCATGCA | 206–228 | 70.1 |
| Dz05 | AGAAGTCGTGTCCGAGCCGGTCGAAGCTTCATGTGG | 231–253 | 65.6 |
| Dz06 | GACGTTGTGGTCCGAGCCGGTCGAAGTTGTAGTTGT | 431–453 | 63.7 |
| Dz07 | CGTTCTTCTGTCCGAGCCGGTCGAATGTCGGCCATG | 459–481 | 66.5 |
| Dz08 | TGCCGTTCTTTCCGAGCCGGTCGAAGCTTGTCGGCC | 462–484 | 70.9 |
| Dz09 | TGCAGATGAATCCGAGCCGGTCGAATCAGGGTCAGC | 126–148 | 65.5 |
| Dz10 | TGAAGTTCACTCCGAGCCGGTCGAATGATGCCGTTC | 477–499 | 63.8 |
| Dz11 | GAGCTGCACGTCCGAGCCGGTCGAAGCCGTCCTCGA | 515–537 | 73.7 |
| Dz12 | ACGCTGCCGTTCCGAGCCGGTCGAATCGATGTTGTG | 508–530 | 68.8 |

[1] Red font, catalytic core sequence of Dz8–17.

### 2.2. Identification of the DNAzyme with Gene-Silencing Activity in Cells

First, we transcribed the full-length mRNA of EGFP (882 nt, Figure S2) as the long substrate for these twelve DNAzymes. In order to better investigate the cleavage of Dz8–17 in cells, we simulated physiological conditions for performing in vitro cleavage experiments. The in vitro cleavage studies indicated that eleven of the twelve Dz8–17 enzymes had various degrees of cleavage activity (Figure 2a). We found that Dz04 and Dz08 had relatively higher activities (Figure 2a). By using short RNA substrates (23 nt, Table S1), we confirmed the catalytic activities of Dz04 and Dz08 (data not shown).

Later, we tested the eleven DNAzymes for gene-silencing activities in cells. To resist nuclease degradation, two PS linkages were introduced to each end of their binding arms. Then, they were co-transfected into 293T cells with EGFP expression plasmids (transient transfection). By measuring the intensity of EGFP fluorescence after 48 h, we found that Dz04 had the strongest activity in gene silencing (Figure 2b, Table 2). We performed a secondary structure prediction of mRNA using RNAfold, and found that the target region (206–228) of Dz04 formed a big open window, which can ease the target binding for Dz04. Although some Dz8–17 enzymes such as Dz06 and Dz08 had relatively higher cleavage activities in vitro, they had poor gene-silencing effects due to their poor target accessibility in cells.

To confirm the substrate-specific recognition of DNAzyme Dz04, we incorporated two mismatches into each arm individually or both (Table S2). We found that after incorporating two mismatches on each arm (Figure 3a), the silencing activity was reduced (compared with the wildtype Dz04), while it was most reduced when there were four mismatches on both arms. Interestingly, the two mismatches on the left arm caused a stronger reduction in activity than those on the right arm.

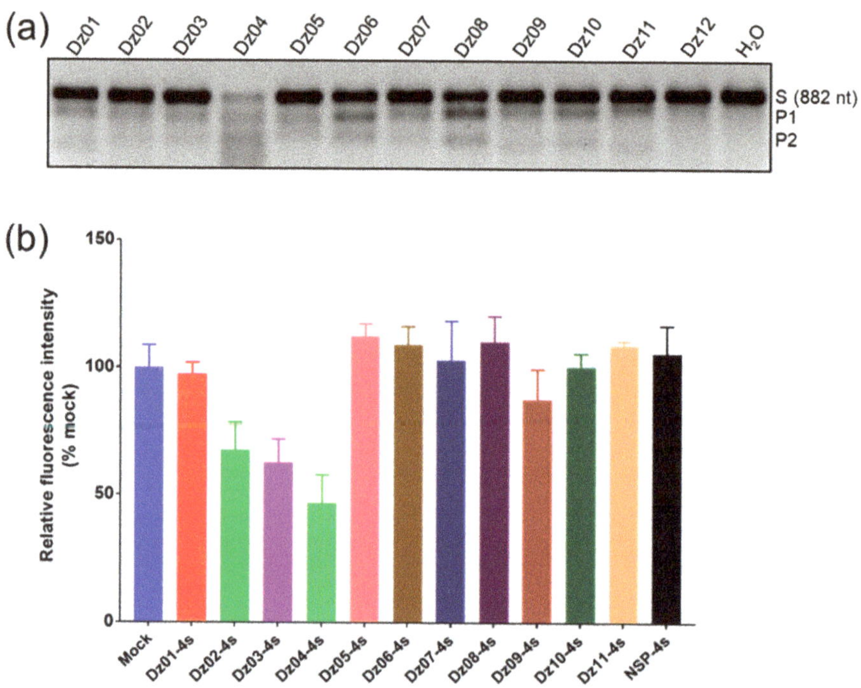


**Figure 2.** Finding the DNAzyme with gene-silencing activity in cells. (**a**) In vitro cleavage study of twelve Dz8–17 DNAzymes via agarose gel analysis. All reactions were carried out under simulated physiological conditions (150 mM KCl, 0.5 mM $MgCl_2$, pH 7.5) at 37 °C for 2 h. s: substrate; P1: product 1; P2: product 2. (**b**) EGFP gene-silencing study with the DNAzymes in cells. The fluorescent data were collected 48 h after transfection and reported as mean value ± SD of three repeats. The transfection concentrations of the DNAzymes were 400 nM, and Mock was a negative control using ultrapure water instead of a DNAzyme. NSP: negative DNAzyme control with nonspecific arm sequences. Each DNAzyme and control were labeled with different colors.

**Table 2.** The inhibition ratios of 11 DNAzymes targeting EGFP mRNA and NSP control.

| DNAzymes | Inhibition Ratio (%) | DNAzymes | Inhibition Ratio (%) |
|---|---|---|---|
| Dz01 | <5 | Dz07 | 0 |
| Dz02 | 32 | Dz08 | 0 |
| Dz03 | 37 | Dz09 | 12 |
| Dz04 | 53 | Dz10 | 0 |
| Dz05 | 0 | Dz11 | 0 |
| Dz06 | 0 | NSP | 0 |

Further, to confirm the activity–concentration relationship, we investigated various DNAzyme concentrations in the cellular experiments. The silencing activity was concentration-dependent (Figure 3b). These results collectively indicated that Dz04 indeed targeted EGFP mRNA, causing cellular gene silencing. Our results suggest that the physiological ionic strengths may be sufficient for DNAzyme catalysis in cells. In addition, the in vitro catalytic experiments can guide efficient DNAzyme identification in cells.

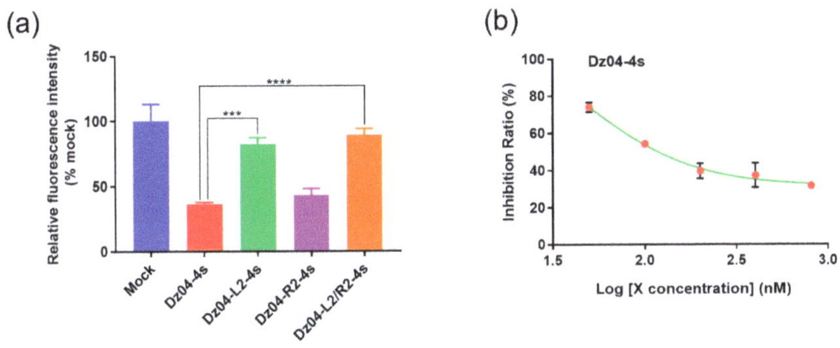


**Figure 3.** Identification the DNAzyme (Dz04) gene-silencing activity in cells. (**a**) The mismatched arm sequences of Dz04 compromised the EGFP gene silencing, and the data significance was determined by a Student's *t*-test (*** $p < 0.001$, **** $p < 0.0001$); Each DNAzyme and mock control were labeled with different colors. (**b**) The gene-silencing activities of Dz04 at different concentrations.

### *2.3. The DNAzyme with Both Cleavage and Antisense Activities in Cells*

Our experiments above indicated that DNAzyme Dz04 is fairly active in vitro and in cells. To clarify its cellular mechanisms, we mutagenized the DNAzyme by changing $T_1$ to $C_1$ (Dz4M, Figure 4a,b), inactivating its cleavage activity without perturbing the overall structure (Figure 4c). Dz4M can bind to its target through hybridization as well as Dz04 (Figure 4b). We co-transfected PS-modified Dz04 and Dz4M (with two PS linkages at both 5′ and 3′ ends, Table S2) individually into 293T cells with EGFP expression plasmids. Twenty-four hours later, after transfection, we quantified the RNA expression levels by RT-PCR. We found that compared with the mock control, Dz04 offered a 70% reduction in gene silencing, while catalytically inactive Dz4M merely offered a 50% reduction (Figures 4d and S3). This difference (20%, $p < 0.05$) indicated that Dz04 was catalytically active in cells. In addition, the cellular activity with this single-base mutation suggested that Dz04 had antisense activity as well. Our results confirmed our hypothesis that the physiological ionic strengths are sufficient for DNAzyme catalysis in cells. However, the antisense nucleic acid effect was dominant in the gene-silencing activity of Dz8–17, which is consistent with the 10–23 DNAzyme [23,24].

### *2.4. In Vitro Studies of the Stability, Catalysis and RNase H Compatibility of the DNAzymes Modified on the Arms*

To increase the biological stability of the DNAzyme, we introduced modifications at its ends OMe and LNA, in addition to the PS linkages, to increase duplex binding and reduce degradation by intracellular nucleases [11]. However, on the basis of our investigations above, we need to balance chemical modifications with the stability, DNAzyme cleavage activity and antisense function (such as RNase H compatibility or RNase H-caused RNA digestion).

To address these issues collectively, we designed and synthesized the DNAzyme with PS, OMe and LNA modifications on the arms (Figure 5a). First, we examined the stability of these modified enzymes (Table S3) in fetal bovine serum (FBS). We found that LNA was the most stable and PS and OMe were less stable, while non-modified Dz04 was almost completely degraded by FBS within 24 h (Figure 5b).

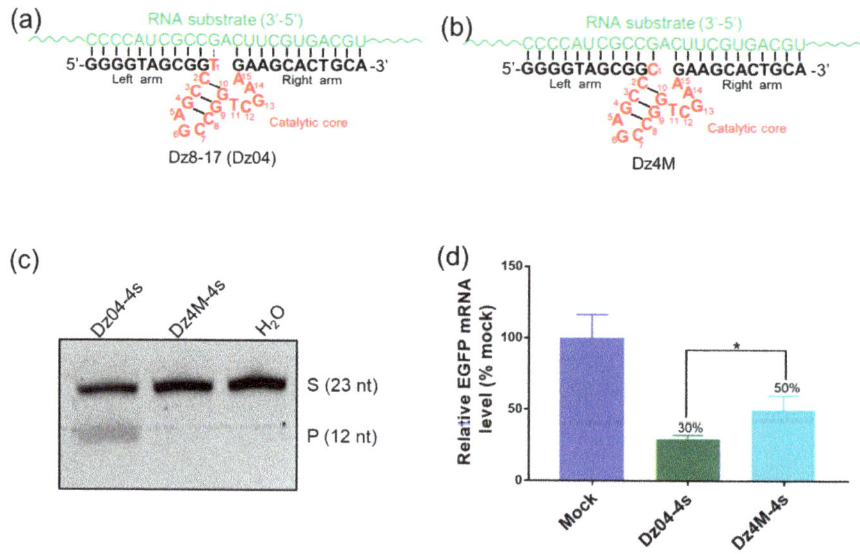


**Figure 4.** The DNAzyme (Dz04) with both cleavage and antisense activities in cells. (**a**,**b**) are secondary structures of Dz04, Dz4M and Dz04-4s. (**c**) The in vitro cleavage activities of Dz04-4s and Dz4M-4s under simulated physiological conditions. All reactions carried out in 1.0 μM Dz8–17, 1.0 μM substrates, 150 mM KCl and 0.5 mM $MgCl_2$ at 37 °C for 2 h. (**d**) The cellular mRNA degradation by Dz04-4s and Dz4M-4s was analyzed by RT-qPCR, 24 h after the DNAzyme transfection. Stu-dent's *t*-tests were used to assess significance of the data according to mean values and SDs of three repeats (* $p < 0.05$). Dz04, Dz4M and mock control were labeled with different colors.

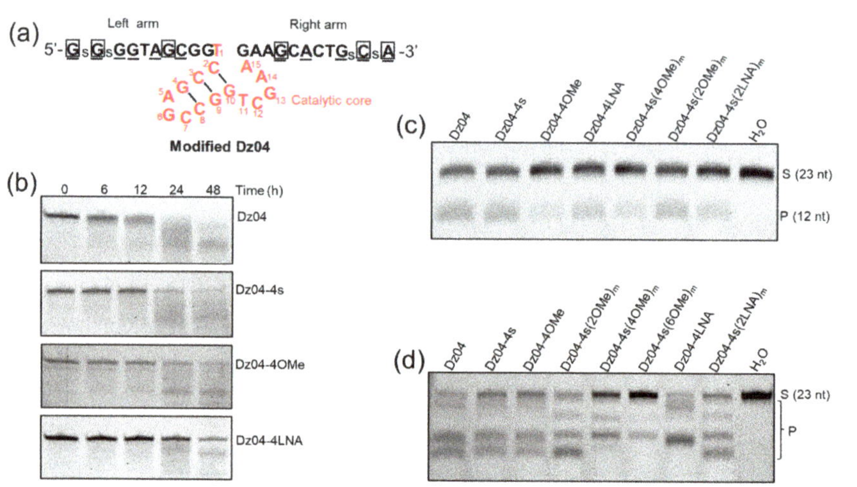


**Figure 5.** In vitro studies of stability, catalysis and RNase H compatibility of the DNAzymes with arm modifications. (**a**) Phosphorothioate linkage (PS), 2′-O-methyl ribonucleotide (OMe), 2′-O and 4′-C LNA (LNA) are represented by "s", "underlined" and "box" labels, respectively. (**b**) Stability study of the modified DNAzymes under FBS serum treatment. (**c**) In vitro cleavage study of the modified DNAzymes under physiological conditions. S: substrate; P: product. (**d**) Modified DNAzyme compatibility with RNase H.

Second, we performed the in vitro cleavage of the modified Dz04 on the short RNA substrate (Table S3) and we found that the chemical modifications did not alter their activity significantly, excluding Dz04-4Ome and Dz04-4s(4Ome)m (Figure 5c). Third, to investigate the impacts of the modifications on the antisense function (such as RNase H compatibility), we performed in vitro experiments on RNase H-directed hydrolysis of RNA. The results showed that the OMe and LNA in the middle of each arm reduced the RNase H compatibility (Figure 5d), while the PS, OMe and LNA modifications at both ends did not.

Our results indicated that most of these modifications do not significantly reduce the cleavage activity of Dz8–17, and the two modifications (such as LNA) at each end are sufficient for the stability requirements. Since OMe and LNA in the middle of the arms are incompatible with RNase H-mediated RNA degradation [25], OMe and LNA at the end of each arm allowed for RNase H compatibility (antisense-mediated cleaving activity).

### 2.5. The Modified DNAzyme with High Activities in Cells

To find out the optimal number of modified nucleotides on each end, we designed Dz04 containing one, two, three and four PS linkages at both ends (Table S4). Our results showed that the modified DNAzymes had higher activity than the unmodified one, while the higher number of PS did not significantly increase the cellular activity (Figure 6a), indicating that two modified nucleotides at each end (for PS linkages) were sufficient.

To further explore the gene-silencing activities of DNAzymes with different chemical modifications, we transfected cells with DNAzymes containing the PS, OMe or LNA modifications with EGFP-expression plasmids, and analyzed the fluorescence intensity after 48 h. We discovered that at the DNAzyme concentration of 100 nM, the gene-silencing efficiencies of DNAzymes with terminal LNA, PS and OMe (Figure 6b,c; Table S3) were 82, 60 and 36%, respectively. Further, we found that the LNA-modified cleaving DNAzyme (Dz04-4LNA) offered 82% inhibition, while the mutant (Dz4M-4LNA) without cleaving activity offered 75% inhibition. In addition, OMe or LNA combined with the terminal PS did not further increase the activity. Moreover, the introduction of OMe and LNA in the middle of the arms largely reduced the cellular activities of DNAzymes, consistent with previous in vitro studies.

In order to eliminate the possibility of non-specific gene silencing by the chemical modifications, we designed scrambled controls with the same modifications (Dz4S, the scrambled sequences of the corresponding Dz04, Table S3). As expected, we found that none of the scrambled controls had gene-silencing activities (Figure 6b,c).

Since LNA modifications are incompatible with RNase H-mediated RNA degradation [25], we have concluded that in cells, DNAzymes modified with two LNAs at each end have higher stability, catalytic activity, RNA substrate affinity (antisense-mediated arresting activity) and RNase H compatibility (antisense-mediated cleaving activity), thereby offering fine gene-silencing tools.

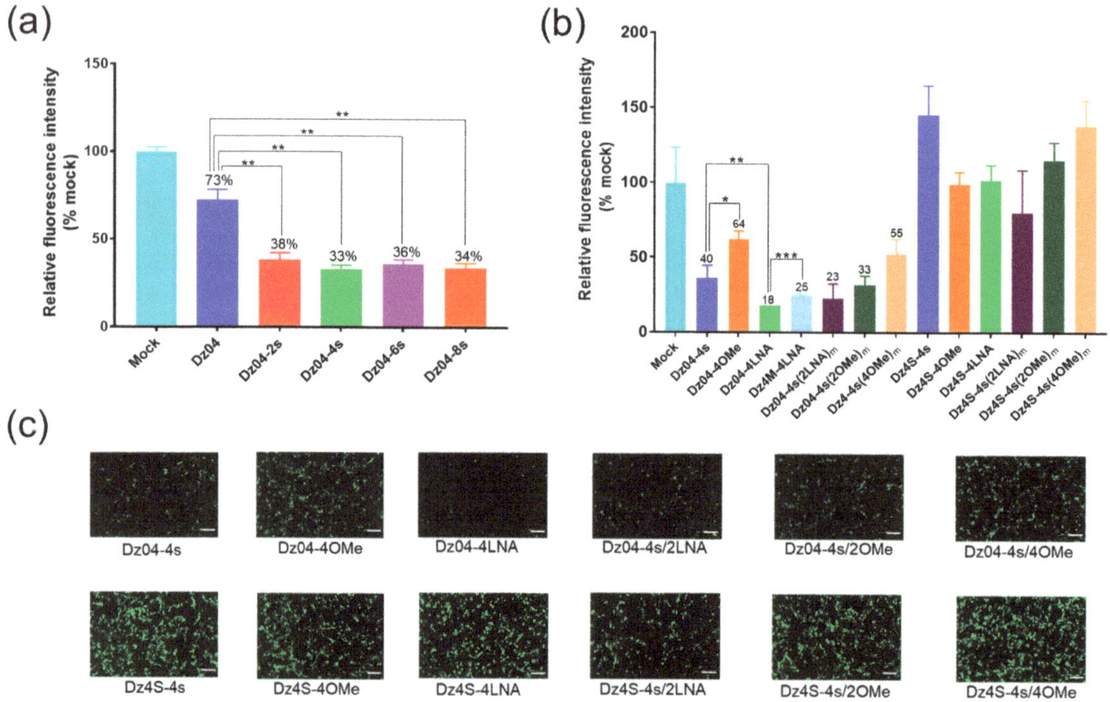


**Figure 6.** The modified DNAzyme (Dz04) with high gene-silencing effect in cells. (**a**) Effect of Dz04 PS modifications on gene silencing. Statistical significance was determined by a Student's *t*-test of three repeats (** $p < 0.01$). (**b**) Gene silencing of Dz04 with different chemical modifications. The fluorescent data significance was determined by a Student's *t*-test (* $p < 0.05$, ** $p < 0.01$, *** $p < 0.001$). Dz04-4s, Dz04-4OMe and Dz04-4LNA contained two PS, OMe and LNA modifications at both 5′ and 3′ ends of Dz04, respectively. Dz04-4s(2LNA)m contained one LNA in the middle of each Dz04-4s arm. Dz04-4s(2OMe)m and Dz04-4s(4OMe)m contained one and two OMe modifications in the middle of each Dz04-4s arm, respectively; Dz4M-4LNA contained two LNAs at both 5′ and 3′ ends of Dz4M (Figure 4b). Dz4S-4s, Dz4S-4OMe and Dz4S-4LNA contained two PS, OMe and LNA modifications at both 5′ and 3′ ends of Dz4S, respectively. Dz4S-4s(2LNA)m contained one LNA in the middle of each Dz4S-4s arm. Dz4S-4s(2OMe)mand Dz4S-4s(4OMe)m contained one and two OMe modifications in the middle of each Dz4S-4s arm, respectively. Their detailed sequences are shown in Table S3. (**c**) Cellular fluorescent images of EGFP gene silencing with the modified DNAzymes in Figure 6b. Scale bar: 500 μm.

## 3. Materials and Methods

### *3.1. Materials*

293T cell lines and pEGFP-C1 plasmids were derived from our laboratory. Growth media (DMEM) was supplemented with 10% FBS and 1% antibiotic–antimycotic. 293T cell lines were maintained up to 30 passages in a 37 °C/5% $CO_2$ incubator. All oligonucleotides were synthesized by Sangon (Shanghai, China). Trypsin EDTA and penicillin streptomycin were purchased from Biosharp (Hefei, China). DMEM basic media and fetal bovine serum and Opti-MEM were purchased from Gibco. Lipofectamine 3000 was purchased from Invitrogen. A CCK-8 kit was purchased from AbMole. A HiScribe™ T7 in vitro Transcription Kit and RNase H were purchased from NEB. A Total RNA Isolation Kit, HiScript III RT SuperMix for qPCR, and ChamQ Universal SYBR qPCR Master Mix were purchased from Vazyme (Nanjing, China).

### 3.2. Methods

#### 3.2.1. Bioinformatics Tools for Designing 8–17 DNAzymes

We used OLIGO Primer Analysis Software (Oligo 7) to design DNAzymes semi-automatically (the oligo length and the Tm value were set at 23 nt and 65–80 °C, respectively). The secondary structures of DNAzymes were analyzed by Predict a Bimolecular Secondary Structure Web Server (https://rna.urmc.rochester.edu/RNAstructureWeb (accessed on 9 September 2021)) "URL". The binding specificities of designed DNAzymes were analyzed by the basic local alignment search tool (BLAST) from NCBI (www.ncbi.nlm.nih.gov/BLAST (accessed on 10 September 2021)) "URL" to avoid binding to the host transcriptome non-specifically.

#### 3.2.2. In Vitro Substrate Cleavage

For the long RNA substrates (882 nt), the reaction components were as follows: 5 μL 2× Reaction buffer (100 mM HEPSE, pH 7.5, 300 mM $K^+$, 1 mM $Mg^{2+}$), 1 μL RNA substrates (300 ng/μL), 1 μL DNAzymes (10 μM), 3 μL DEPC $ddH_2O$. For short RNA substrates (23 nt), the reaction components were as follows: 5 μL 2× Reaction buffer, 1 μL RNA substrates (2 μM), 1 μL DNAzymes (1 μM), 3 μL DEPC $ddH_2O$. The above reaction components were mixed thoroughly and then incubated at 37 °C for 2 h. After the reaction, the cleavage products were analyzed by agarose gel electrophoresis (882 nt) or PAGE (23 nt).

#### 3.2.3. Co-Transfection of 293T Cells with DNAzymes and pEGFP-C1

293T cells were plated in a 96-well plate ($2 \times 10^4$ cells/well, four replicates/sample) or a 12-well plate ($2 \times 10^5$ cells/well) the day before co-transfection, and then were co-transfected with 100 ng pEGFP along with different DNAzymes (final concentration of 100 nM or 400 nM) using Lipofectamine 3000 according to the manual's instructions. After 48 h of incubation, the cells expressing EGFP were scanned for fluorescence intensity, measured with a microplate reader (Varioskan LUX, Thermofisher, Beijing, China) or visualized with fluorescence microscopy (Nikon, Beijing, China).

#### 3.2.4. RNA Quantification (qRT-PCR)

Twenty-four hours after transfection, RNA was extracted from the transfected cells using a FastPure Cell/Tissue Total RNA Isolation Kit according to the protocol. HiScript III RT SuperMix for qPCR was applied for genomic DNA removal and cDNA synthesis. Real-time PCR reactions were carried out in LightCycler® 96 (Roche, Shanghai, China) using ChamQ Universal SYBR qPCR Master Mix according to the protocol. The expression of EGFP gene was normalized to the GAPDH, and the primers are listed as follows: EGFP forward primer: 5′-CGGCAAGCTGACCCTGAA-3′, EGFP reverse primer: 5′-GACGTAGCCTTCGGGCA-3′; GAPDH forward primer: 5′-ACCCACTCCTCCACCTTTGA-3′, GAPDH reverse primer: 5′-CTGTTGCTGTAGCCAAATTCGT-3′.

#### 3.2.5. Stability Tests of Modified DNAzymes in FBS

The reaction components were assembled as follows: 1.0 μL DNAzymes (10 μM), 0.5 μL FBS and 8.5 μL ultrapure water. The reaction mixes were incubated at 37 °C for 0 h, 6 h, 12 h, 24 h or 48 h. Reactions were stopped by addition of 10 μL denaturing gel-loading buffer and denatured at 95 °C for 2 min. Finally, 5 μL denatured products were placed on a 16% PAGE denaturing gel for electrophoresis and analysis.

#### 3.2.6. RNase H-Induced Hydrolysis of RNA

The reaction components were assembled as follows: 5.0 μL 2× RNase H Reaction buffer, 1.0 μL RNA substrates (2.0 μM), 1.0 μL DNAzymes, 0.1 μL RNase H, 2.9 μL DEPC $H_2O$. The reaction mixes were incubated at 37 °C for 10 min. Finally, the products were placed on a 16% PAGE denaturing gel for electrophoresis and analysis.

#### 3.2.7. Statistical Analysis

Student's *t*-tests were used in GraphPad Prism 5.0 to analyze variances between groups (* $p < 0.05$, ** $p < 0.01$, *** $p < 0.001$ and **** $p < 0.0001$ represented significant results).

## 4. Conclusions

In summary, our research results provided insights into the catalytic activities and cellular antisense mechanisms of Dz8–17. In the presence of minimal divalent cations, it is catalytically active, and its gene-silencing activity is additionally enhanced by its RNase H compatibility in cells. In addition, we found that chemical modifications (such as LNA) can greatly enhance its activity in cells. Our investigations can also provide guidance for designing Dz8–17, such as in optimizing the arm lengths and placing modifications at the terminal positions to take advantage of the RNase H compatibility. Our studies demonstrated that under physiological conditions, the modified 8–17 DNAzyme can demonstrate high activities collectively regarding catalysis and antisense, making it a useful tool for exploring genetic functions and biomedical applications.

**Supplementary Materials:** The following supporting information can be downloaded at: https://www.mdpi.com/article/10.3390/molecules28010286/s1, Figure S1: EGFP mRNA sequence and the target sites of Dz04 and Dz08.; Figure S2: Agarose gel analysis of EGFP mRNA by in vitro transcription; Figure S3: The original real time PCR trace of a typical experiment in Figure 4d. Table S1: The sequences of short RNA substrates; Table S2: The sequences of Dz04, Dz4M, Dz4S and mismatched controls; Table S3: The sequences of Dz04 and Scr control with different modifications at both arms; Table S4: Dz04 sequences with various number of PS linkages on each arm.

**Author Contributions:** Conceptualization, Z.H. and Z.Z.; methodology, Z.Z.; software, Z.Z.; validation, Z.Z. and W.S.; investigation, Z.Z. and W.S.; resources, Z.H.; data curation, Z.H. and Z.Z.; writing—original draft preparation, Z.Z.; writing—review and editing, Z.H.; visualization, Z.Z.; supervision, Z.H.; project administration, Z.Z. and W.S.; funding acquisition, Z.H. All authors have read and agreed to the published version of the manuscript.

**Funding:** This research was funded by the National Natural Science Foundation of China, grant number 22077089.

**Institutional Review Board Statement:** Not applicable.

**Informed Consent Statement:** Not applicable.

**Data Availability Statement:** Not applicable.

**Conflicts of Interest:** The authors declare no conflict of interest.

## References

1. El-Sappah, A.H.; Yan, K.; Huang, Q.; Islam, M.M.; Li, Q.; Wang, Y.; Khan, M.S.; Zhao, X.; Mir, R.R.; Li, J.; et al. Comprehensive Mechanism of Gene Silencing and Its Role in Plant Growth and Development. *Front. Plant Sci.* **2021**, *12*, 705249. [CrossRef]
2. de Fougerolles, A.; Vornlocher, H.-P.; Maraganore, J.; Lieberman, J. Interfering with disease: A progress report on siRNA-based therapeutics. *Nat. Rev. Drug Discov.* **2007**, *6*, 443–453. [CrossRef]
3. Felippes, F.F.D.; Wang, J.-W.; Weigel, D. MIGS: miRNA-induced gene silencing. *Plant J.* **2012**, *70*, 541–547. [CrossRef]
4. Dowdy, S.F. Overcoming cellular barriers for RNA therapeutics. *Nat. Biotechnol.* **2017**, *35*, 222–229. [CrossRef]
5. Khvorova, A.; Watts, J.K. The chemical evolution of oligonucleotide therapies of clinical utility. *Nat. Biotechnol.* **2017**, *35*, 238–248. [CrossRef]
6. Shen, X.; Corey, D.R. Chemistry, mechanism and clinical status of antisense oligonucleotides and duplex RNAs. *Nucleic Acids Res.* **2018**, *46*, 1584–1600. [CrossRef]
7. Watts, J.K.; Corey, D.R. Silencing disease genes in the laboratory and the clinic. *J. Pathol.* **2012**, *226*, 365–379. [CrossRef]
8. Borggräfe, J.; Victor, J.; Rosenbach, H.; Viegas, A.; Gertzen, C.G.W.; Wuebben, C.; Kovacs, H.; Gopalswamy, M.; Riesner, D.; Steger, G.; et al. Time-resolved structural analysis of an RNA-cleaving DNA catalyst. *Nature* **2022**, *601*, 144–149. [CrossRef]
9. Rudeejaroonrung, K.; Hanpanich, O.; Saito, K.; Shimada, N.; Maruyama, A. Cationic copolymer enhances 8-17 DNAzyme and MNAzyme activities. *Biomater. Sci.* **2020**, *8*, 3812–3818. [CrossRef]
10. Liu, H.; Yu, X.; Chen, Y.; Zhang, J.; Wu, B.; Zheng, L.; Haruehanroengra, P.; Wang, R.; Li, S.; Lin, J.; et al. Crystal structure of an RNA-cleaving DNAzyme. *Nat. Commun.* **2017**, *8*, 2006. [CrossRef]

11. Reyes-Gutierrez, P.; Alvarez-Salas, L.M. Cleavage of HPV-16 E6/E7 mRNA mediated by modified 10-23 deoxyribozymes. *Oligonucleotides* **2009**, *19*, 233–242. [CrossRef]
12. Wang, Y.; Nguyen, K.; Spitale, R.C.; Chaput, J.C. A biologically stable DNAzyme that efficiently silences gene expression in cells. *Nat. Chem.* **2021**, *13*, 319–326. [CrossRef] [PubMed]
13. Mazumdar, D.; Nagraj, N.; Kim, H.-K.; Meng, X.; Brown, A.K.; Sun, Q.; Li, W.; Lu, Y. Activity, folding and Z-DNA formation of the 8–17 DNAzyme in the presence of monovalent ions. *J. Am. Chem. Soc.* **2009**, *131*, 5506–5515. [CrossRef]
14. Victor, J.; Steger, G.; Riesner, D. Inability of DNAzymes to cleave RNA in vivo is due to limited Mg(2+) concentration in cells. *Eur. Biophys. J.* **2018**, *47*, 333–343. [CrossRef]
15. Alama, A.; Barbieri, F.; Cagnoli, M.; Schettini, G. Antisense oligonucleotides as therapeutic agents. *Pharmacol. Res.* **1997**, *36*, 171–178. [CrossRef]
16. Wu, Y.; Yu, L.; McMahon, R.; Rossi, J.J.; Forman, S.J.; Snyder, D.S. Inhibition of bcr-abl oncogene expression by novel deoxyribozymes (DNAzymes). *Hum. Gene Ther.* **1999**, *10*, 2847–2857. [CrossRef]
17. Fokina, A.A.; Chelobanov, B.P.; Fujii, M.; Stetsenko, D.A. Delivery of therapeutic RNA-cleaving oligodeoxyribonucleotides (deoxyribozymes): From cell culture studies to clinical trials. *Expert Opin. Drug Deliv.* **2017**, *14*, 1077–1089. [CrossRef]
18. Smith, C.I.E.; Zain, R. Therapeutic Oligonucleotides: State of the Art. *Annu. Rev. Pharmacol. Toxicol.* **2019**, *59*, 605–630. [CrossRef]
19. Schubert, S.; Gül, D.C.; Grunert, H.-P.; Zeichhardt, H.; Erdmann, V.A.; Kurreck, J. RNA cleaving '10–23' DNAzymes with enhanced stability and activity. *Nucleic Acids Res.* **2003**, *31*, 5982–5992. [CrossRef]
20. Elmén, J.; Thonberg, H.; Ljungberg, K.; Frieden, M.; Westergaard, M.; Xu, Y.; Wahren, B.; Liang, Z.; Ørum, H.; Koch, T.; et al. Locked nucleic acid (LNA) mediated improvements in siRNA stability and functionality. *Nucleic Acids Res.* **2005**, *33*, 439–447. [CrossRef]
21. Shimo, T.; Tachibana, K.; Saito, K.; Yoshida, T.; Tomita, E.; Waki, R.; Yamamoto, T.; Doi, T.; Inoue, T.; Kawakami, J.; et al. Design and evaluation of locked nucleic acid-based splice-switching oligonucleotides in vitro. *Nucleic Acids Res.* **2014**, *42*, 8174–8187. [CrossRef]
22. Wu, S.; Xu, J.; Liu, J.; Yan, X.; Zhu, X.; Xiao, G.; Sun, L.; Tien, P. An efficient RNA-cleaving DNA enzyme can specifically target the 5′-untranslated region of severe acute respiratory syndrome associated coronavirus (SARS-CoV). *J. Gene Med.* **2007**, *9*, 1080–1086. [CrossRef]
23. Wang, F.; Saran, R.; Liu, J. Tandem DNAzymes for mRNA cleavage: Choice of enzyme, metal ions and the antisense effect. *Bioorg. Med. Chem. Lett.* **2015**, *25*, 1460–1463. [CrossRef]
24. Rosenbach, H.; Victor, J.; Etzkorn, M.; Steger, G.; Riesner, D.; Span, I. Molecular Features and Metal Ions That Influence 10–23 DNAzyme Activity. *Molecules* **2020**, *25*, 3100. [CrossRef]
25. Bondensgaard, K.; Petersen, M.; Singh, S.K.; Rajwanshi, V.K.; Kumar, R.; Wengel, J.; Jacobsen, J.P. Structural studies of LNA:RNA duplexes by NMR: Conformations and implications for RNase H activity. *Chemistry* **2000**, *6*, 2687–2695. [CrossRef] [PubMed]

*molecules*

MDPI

*Article*

# A Novel Aging-Related Prognostic lncRNA Signature Correlated with Immune Cell Infiltration and Response to Immunotherapy in Breast Cancer

**Zhixin Liu [1,2,†], Chongkang Ren [1,†], Jinyi Cai [1], Baohui Yin [3], Jingjie Yuan [1], Rongjuan Ding [1], Wenzhuo Ming [1], Yunxiao Sun [3,*] and Youjie Li [1,*]**

1 Department of Biochemistry and Molecular Biology, Binzhou Medical University, Yantai 264003, China
2 Department of Orthopedics, Qilu Hospital of Shandong University, Jinan 250012, China
3 Department of Pediatrics, Yantai Affiliated Hospital of Binzhou Medical University, Yantai 264100, China
* Correspondence: sunyunxiao1979@bzmc.edu.cn (Y.S.); liyoujie@bzmc.edu.cn (Y.L.)
† These authors contributed equally to this work.

**Abstract:** Breast cancer (BC) is among the most universal malignant tumors in women worldwide. Aging is a complex phenomenon, caused by a variety of factors, that plays a significant role in tumor development. Consequently, it is crucial to screen for prognostic aging-related long non-coding RNAs (lncRNAs) in BC. The BC samples from the breast-invasive carcinoma cohort were downloaded from The Cancer Genome Atlas (TCGA) database. The differential expression of aging-related lncRNAs (DEarlncRNAs) was screened by Pearson correlation analysis. Univariate Cox regression, LASSO–Cox analysis, and multivariate Cox analysis were performed to construct an aging-related lncRNA signature. The signature was validated in the GSE20685 dataset from the Gene Expression Omnibus (GEO) database. Subsequently, a nomogram was constructed to predict survival in BC patients. The accuracy of prediction performance was assessed through the time-dependent receiver operating characteristic (ROC) curves, Kaplan–Meier analysis, principal component analyses, decision curve analysis, calibration curve, and concordance index. Finally, differences in tumor mutational burden, tumor-infiltrating immune cells, and patients' response to chemotherapy and immunotherapy between the high- and low-risk score groups were explored. Analysis of the TCGA cohort revealed a six aging-related lncRNA signature consisting of MCF2L-AS1, USP30-AS1, OTUD6B-AS1, MAPT-AS1, PRR34-AS1, and DLGAP1-AS1. The time-dependent ROC curve proved the optimal predictability for prognosis in BC patients with areas under curves (AUCs) of 0.753, 0.772, and 0.722 in 1, 3, and 5 years, respectively. Patients in the low-risk group had better overall survival and significantly lower total tumor mutational burden. Meanwhile, the high-risk group had a lower proportion of tumor-killing immune cells. The low-risk group could benefit more from immunotherapy and some chemotherapeutics than the high-risk group. The aging-related lncRNA signature can provide new perspectives and methods for early BC diagnosis and therapeutic targets, especially tumor immunotherapy.

**Keywords:** breast cancer; signature; aging; cancer; nomogram; immunotherapy; tumor mutation burden

**Citation:** Liu, Z.; Ren, C.; Cai, J.; Yin, B.; Yuan, J.; Ding, R.; Ming, W.; Sun, Y.; Li, Y. A Novel Aging-Related Prognostic lncRNA Signature Correlated with Immune Cell Infiltration and Response to Immunotherapy in Breast Cancer. *Molecules* **2023**, *28*, 3283. https://doi.org/10.3390/molecules28083283

Academic Editors: Guohui Sun, Chongwen Wang and Luca Vannucci

Received: 20 January 2023
Revised: 15 March 2023
Accepted: 21 March 2023
Published: 7 April 2023



## 1. Introduction

Breast cancer (BC) is among the most universal malignancies in women worldwide, and its incidence is increasing yearly. It has exceeded lung cancer to become the world's largest cancer since 2020 [1]. BC is a highly heterogeneous disease, thereby indicating vast differences in gene expression profiles, clinicopathological characteristics, and biological behavior [2]. Tumor stage, histological grade, and molecular subtype are often used as prognostic factors for evaluating BC patients in clinical practice. However, the prognostic information of BC patients cannot be accurately predicted through these clinical characteristics, thereby resulting in inaccurate diagnosis of patient prognosis. Some low-risk patients may be treated unnecessarily or excessively, whereas those with high risk of recurrence or metastasis may be undertreated [3]. Accordingly, it is necessary to utilize novel prognostic biomarkers and a corresponding predictive model to risk-stratify the heterogeneous population with BC and to guide individualized treatment.

Aging is a complex phenomenon caused by many factors, and it is an important and inevitable biological process. It manifests as a gradual loss or degradation of functions at all levels of the human organism [4]. The link between aging and cancer has become stronger in recent years [5]. Aging is a comorbidity of BC, strongly indicating that aging-related transcriptomes contribute to BC progression [6]. Aging-related genes (ARGs) play critical roles in the generation and regulation of senescent cells. They also influence tumor cell progression. Alteration of ARGs may provide novel implications for tumor pathogenesis, diagnosis, and therapy [7]. More and more researchers have demonstrated that aging-related prognostic biomarkers can help obtain the potential diagnostic or prognostic value of malignancies, including colorectal, lung, and ovarian cancers [8–10].

Long non-coding RNA (lncRNA) are non-coding RNA with a length greater than 200 nucleotides. Studies have shown that lncRNAs play an important role in the dosage compensation effect, epigenetic regulation, cell cycle and cell differentiation regulation, etc. These lncRNAs have crucial effects in all kinds of biological processes, including cell multiplication, differentiation, invasion, apoptosis, and metastasis [11]. Aging-related lncRNAs (ARlncRNAs) participate in the genesis, metastasis, invasion, chemotherapy resistance, and prognosis of BC [12–15]. These studies focus on a single lncRNA in BC. A comprehensive signature that includes multiple genes is more predictive than a signature that includes only one gene [16]. A novel BC prognosis model based on aging-related lncRNA has not been reported. Therefore, it is crucial to screen for prognostic ARlncRNAs in BC.

We set up a prognostic model ground on six differentially expressed aging-related lncRNAs and validated their prognostic performance in the GSE20685 dataset. Afterward, we developed a novel nomogram that combines the risk scores with clinicopathological information to provide precise prognostic information for BC patients. Finally, we explored the immune landscape of the tumor micro-environment (TME), analyzed tumor mutational burden (TMB), and predicted the sensitivity to drug therapy with different risk scores (Supplementary Figure S1). Our findings provided new views to accurately predict the prognosis for BC patients and improve treatment programs.

## 2. Results

### 2.1. *Differential Expression of Age-Related lncRNA Identification*

First, 1135 ARlncRNAs were appraised on account of co-expression analysis of ARGs and lncRNAs expression levels in BC samples ($|\text{Pearson } R| > 0.4$, $p < 0.001$). We screened 287 significant DEarlncRNAs between 1022 BC patients and 112 normal patients for subsequent analysis ($|\log_2 FC| > 1$ and $p < 0.05$). We incorporated 198 up-regulated genes and 89 down-regulated genes (Figure 1A; Supplementary Figure S2).

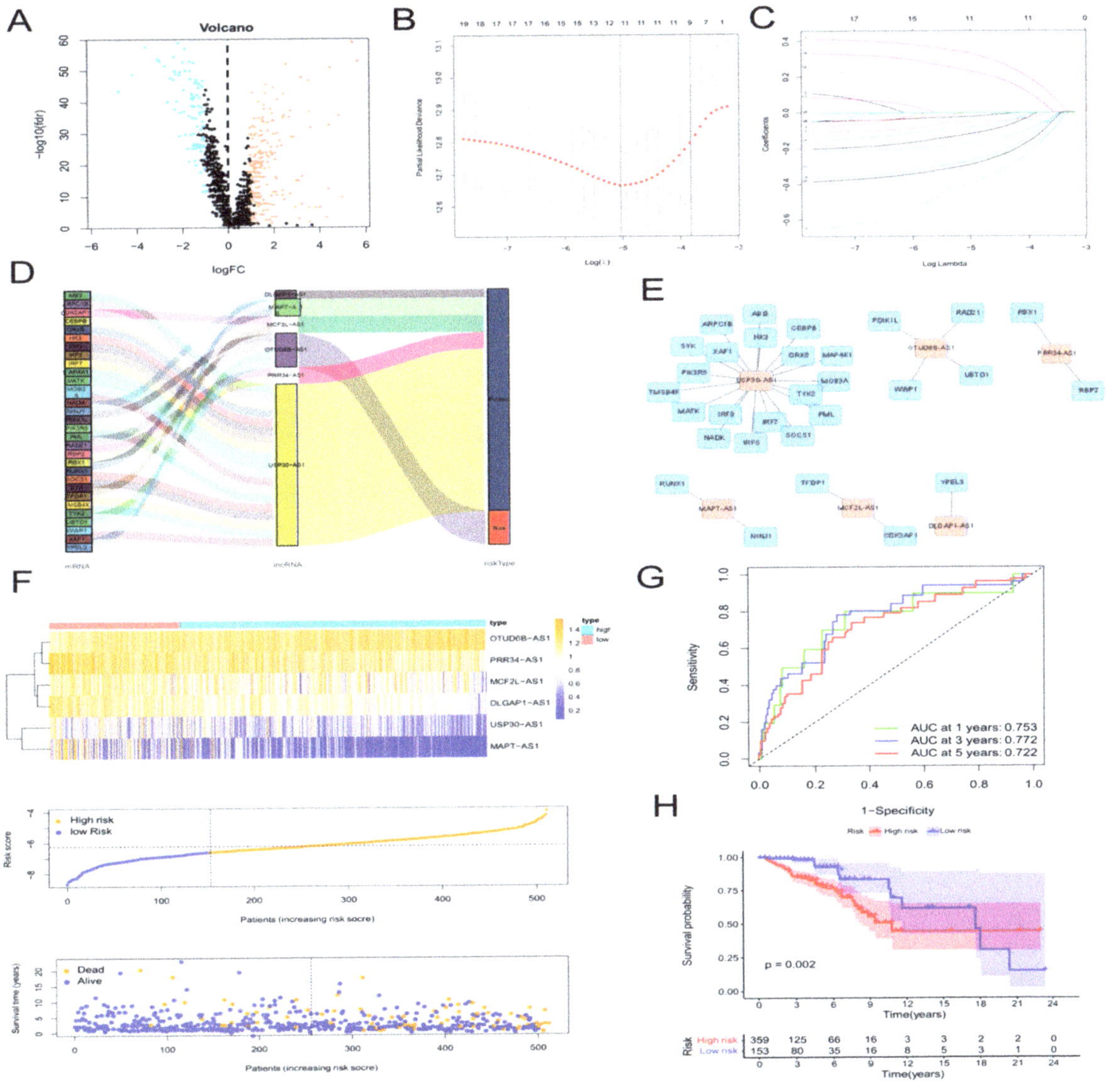

**Figure 1.** Construction of an aging-related lncRNA prognostic signature. (**A**) Volcano plot of differentially expressed lncRNAs in tumor samples and normal samples using the Pearson correlation analysis. Red dots represent up-regulated genes and blue dots represent down-regulated genes. (**B**,**C**) LASSO analysis to screen 20 prognostic aging-related lncRNAs by tenfold cross validation. (**D**) Sankey diagram of the relationship of 6 aging-related lncRNAs with related mRNAs and risk types. (**E**) Construction and visualization of co-expression network of aging-related lncRNAs and mRNAs using Cytoscape. Red rectangles represent prognostic lncRNAs, and blue rectangles represent aging-related mRNAs. (**F**) Expression patterns of 6 aging-related genes, relationship between risk scores and survival times, and distribution of risk scores in high- and low-risk-score groups. (**G**) Time-dependent receiver operating characteristic (ROC) curves and area under the curve (AUC) analyses in the TCGA training cohort. (**H**) Kaplan–Meier curves of overall survival of BC patients in high-risk and low-risk score groups by the log-rank test.

### 2.2. The Construction of Age-Related lncRNA Signature and Verification

First, 64 prognostic lncRNAs related with aging were appraised in the training set on account of the univariate Cox regression analysis. Second, we implemented the LASSO regression to reduce overfitting. The quantity of aging-related lncRNAs was reduced to 20 (tenfold cross validation (10-CV) gained prognostic lncRNAs by minimum lambda values)(Figure 1B,C). Subsequently, a risk signature composed of six lncRNAs was established by the multivariate COX analysis, including MCF2L-AS1, USP30-AS1, OTUD6B-AS1, MAPT-AS1, PRR34-AS1, and DLGAP1-AS1 (Supplementary Table S2). In the light of the risk score formula, we sequentially scored the risk of TCGA-BRCA patients. Risk Score = MCF2L-AS1 × (−0.6581) + USP30-AS1 × (−0.9132) + OTUD6B-AS1 × (1.4151) + MAPT-AS1 × (−0.8676) + PRR34-AS1 × (−1.0859) + DLGAP1-AS1 × (−1.5074). Based on the optimal X-tile truncation of risk score, the training set was split into low-risk and high-risk groups. In the training cohort, the optimal cutoff was identified as −6.6. Furthermore, between the two risk subgroups, the heatmap presented the expression pattern of the six lncRNAs related with aging (Figure 1F). The Sankey diagram indicated the correlation among risk types, genes related with aging, and prognostic lncRNAs related with aging (Figure 1D). The co-expression network between these six lncRNAs and genes related with aging in BC was visualized using Cytoscape 3.7.2 software (Figure 1E). Compared with the subgroup with low risk, the subgroup with high risk had observably lower OS, which was indicated by the Kaplan–Meier analysis. In the two different risk groups, BC patients with higher risk score showed lower overall survival (Figure 1H). Furthermore, the first-rank predictability of the prognosis risk score in BC patients with AUCs of 0.753, 0.772, and 0.722 in 1, 3, and 5 years, respectively, was proved by the ROC curve (Figure 1G). To prove the precision of the aging-related lncRNA feature, every patient in the test cohort was given a risk score on the same basis as the training queue formula. The testing cohort patients were then split into two different risk subgroups employing an identical cutoff value (Figure 2A). On the basis of the training cohort, risk scores can distinguish BC patients from two different risk subgroups, and the lower the risk score was, the better prognosis the BC patient had (Figure 2C). The satisfactory predictability of the BC patients' prognosis with AUCs of 0.662, 0.646, and 0.637 in 1, 3, and 5 years, respectively, was demonstrated by the risk score (Figure 2F).

We downloaded the GSE20685 dataset, but we excluded patients with a lack of clinical information. Finally, we had 327 BRCA patients (Supplementary Table S3). The TCGA-BRCA cohort was split into a low-risk subgroup and a high-risk subgroup on the basis of the same cutoff as the TCGA-training (Figure 2B). Compared with the subgroup with low risk, the OS of the subgroup with high risk was remarkably lower, which was indicated by the Kaplan–Meier test (Figure 2D). The AUCs at 1, 3, and 5 years were 0.719, 0.712, and 0.680, respectively (Figure 2G). The expression values of prognostic lncRNAs of BC patients by the external validation cohort was extracted, and the risk score of 327 patients in the GSE20685 was calculated on the basis of the risk score formula. Patients in the GSE2068 cohort were split into two different subgroups in the light of the truncation value of the training queue (Supplementary Figure S3). We validated the reliability of the aging-related lncRNA signature in predicting prognosis through the ROC curve (Figure 2E,H). The protein levels of these eight aging markers were used to express normal and BC tissues (Figure 3).

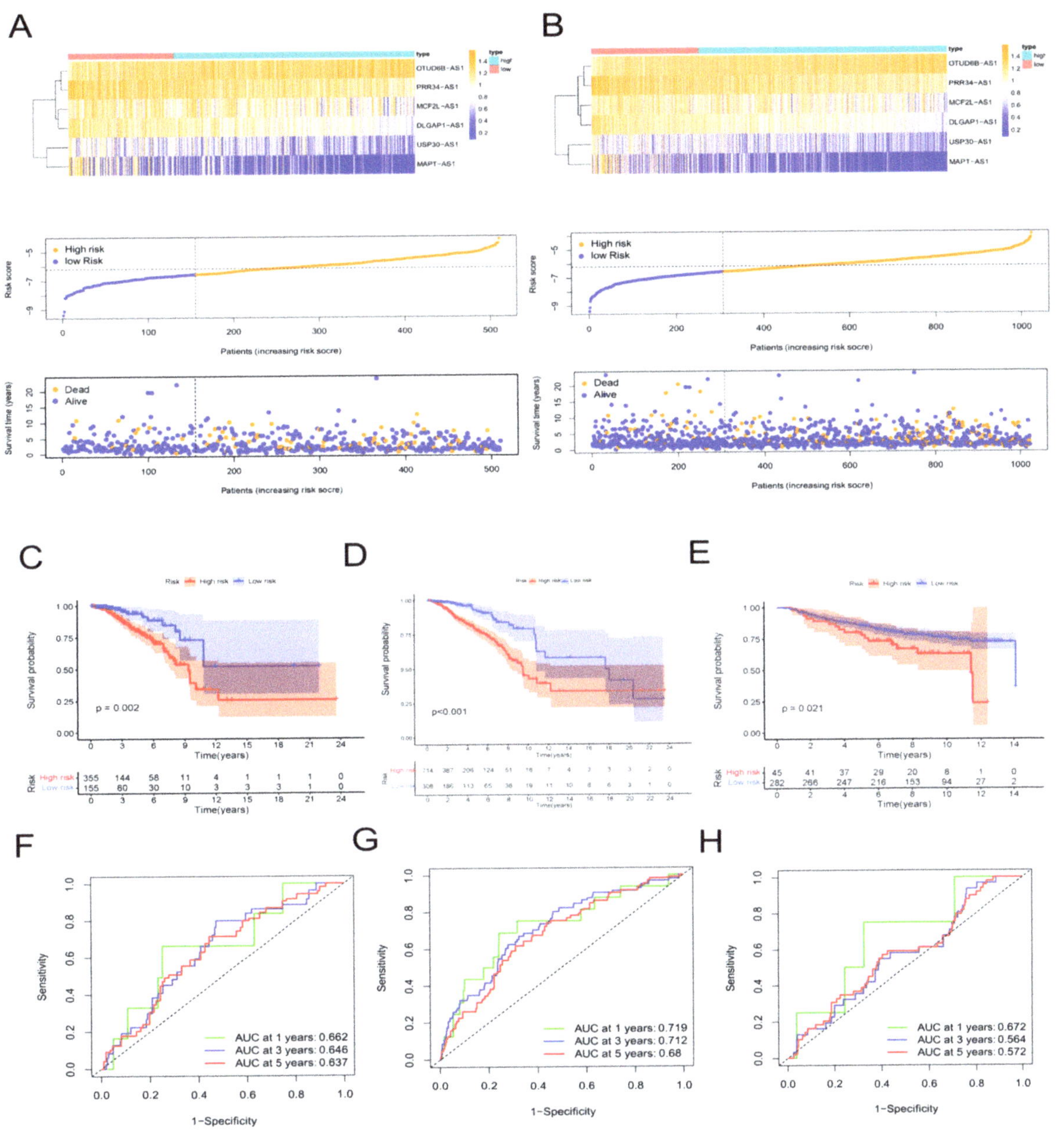


**Figure 2.** Validation of the aging−related lncRNA prognostic signature. Expression patterns of 6 aging−related genes, relationship between risk scores and survival times, and distribution of risk scores in the TCGA test cohort (**A**) and in the TCGA−BRCA cohort (**B**). Kaplan−Meier curves of overall survival of BC patients in the TCGA test cohort (**C**), in the TCGA−BRCA cohort (**D**), and in the GSE20685 cohort (**E**) by the log−rank test. Time−dependent ROC curves and AUC analyses in the TCGA test cohort (**F**), in the TCGA−BRCA cohort (**G**), and in the GSE20685 cohort (**H**).

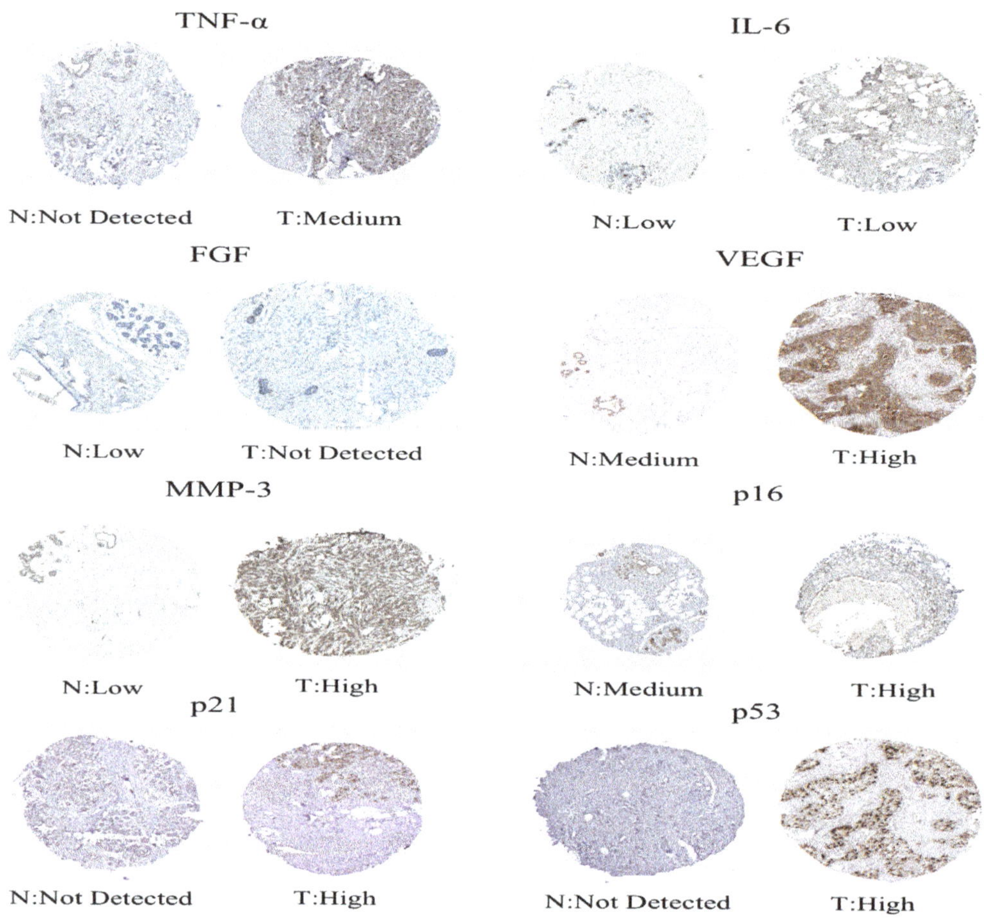


**Figure 3.** Immunohistochemical images obtained from the Human Protein Atlas.

### 2.3. Construction of a Nomogram of BC Patients

Then, we assessed whether a signature-based risk score was a separate prognostic factor for BC patients. According to the univariate Cox regression analysis, risk score was a prognostic indicator for BC patients (HR = 1.46, 95% CI = 1.30–1.64, $p < 0.001$) (Figure 4A). The risk score remained a separate prognostic marker for OS (HR = 1.47, 95% CI = 1.31–1.65, $p < 0.001$) in the multivariate Cox regression analysis, which adjusted for other clinical characteristics (Figure 4B). Simultaneously, the ROC curve—which is time-dependent—indicated that, compared with other clinical characteristics, the AUS of the risk score in 1, 3, and 5 years achieved 0.713, 0.707, and 0.673, respectively (Figure 4C–E). Compared with other independent clinical characteristic variables, the risk score was better. To further improve the clinical application of the aging-associated lncRNA signature, we formed a novel nomogram that combined the clinicopathological information with risk score to forecast OS at 1, 3, and 5 years in BC patients (Figure 4F). The ROC curve demonstrated excellent predictability, with AUCs at 1, 3, and 5 years reaching 0.863, 0.805, and 0.771, respectively (Figure 4H). The predictive performance and clinical value of the prognostic chart was evaluated by calibration curve and DCA analyses. The predicted possibility of 1-year, 3-year, and 5-year OS in the nomogram was in accord with the actual observation, which was indicated by the calibration curve (Figure 4G). This nomogram had compelling specificity and sensitivity in predicting clinical outcomes. The DCA curves confirmed our expectations. The nomogram had optimal predictive net benefit compared with traditional clinicopathological features (Figure 4J). Meanwhile, the C-index proved that the prognostic nomogram had better discrimination ability (Figure 4I). These abovementioned outcomes demonstrated that, for predicting the prognosis in BC patients clinically, the nomogram was appropriate.

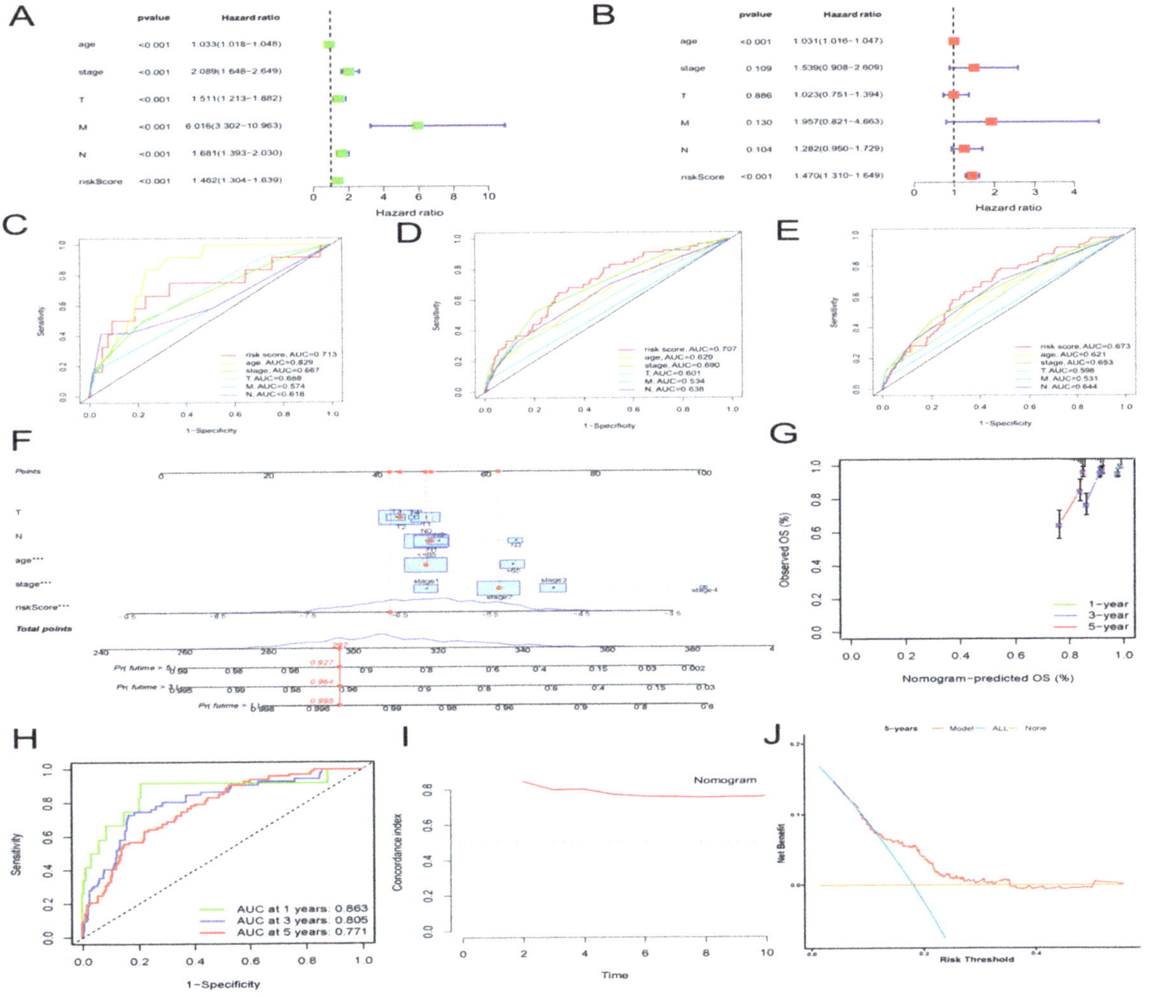


**Figure 4.** Construction and validation of a clinical prognostic nomogram for survival prediction. (**A**) Results of the univariate Cox regression analyses. (**B**) Results of the multivariate Cox regression analyses. (**C–E**) 1-, 3-, and 5-year ROC analyses of risk scores and clinical features. (**F**) The nomogram is based on the TCGA-BRCA cohort. (**G**) Calibration curves of the nomogram. (**H**) Time-dependent ROC curves and AUC analyses. (**I**) Concordance index (C-index) of the nomogram. (**J**) Decision curve analyses (DCA) comparing the predictive capacities.

### *2.4. Correlation of the Risk Score with Clinicopathological Features*

We evaluated the correlation between the risk score and clinicopathologic features in BC patients. The low-risk group highly expressed MCF2L-AS1, MAPT-AS1, USP30-AS1, PRR34-AS1, and DLGAP1-AS1, whereas the high-risk group highly expressed OTUD6B-AS1, as indicated in the heat map (Figure 5A). The risk score showed different patterns of distribution among different clinicopathological features of BC patients, involving TNM staging, survival status, age, and clinical stage (Figure 5B,C; Supplementary Figure S4). Risk scores were higher in BC patients with deeper invasion (stage T3-T4), advanced stage (stage III-IV), or who had died (Figure 5B,C). We also conducted survival analyses of aging features in different clinical feature subgroups. The BC patients had a higher probability of survival across age, M stage, T stage, and clinical stage in the low-risk group. Overall, the aging-associated model was well-correlated with clinical features and was able to predict survival in BC patients who had different clinical features (Figure 5D–J). We explored the independent prognostic value of the six molecular subtypes while considering the molecular heterogeneity of BC. The use of the Kaplan–Meier curve was assessing the connection between the expression of the six lncRNAs in the prognosis of BC patients and our risk model. In BC patients, only the high expression of OTUD6B-AS1 was linked with poor prognosis, whereas in BC patients, the high expression of the other five lncRNAs was linked with a good prognosis (Figure 5K–P). The risk score was inseparable from the clinicopathological characteristics of BC and can be used as an efficient auxiliary tool for forecasting the prognosis of BC patients, which these results demonstrated.

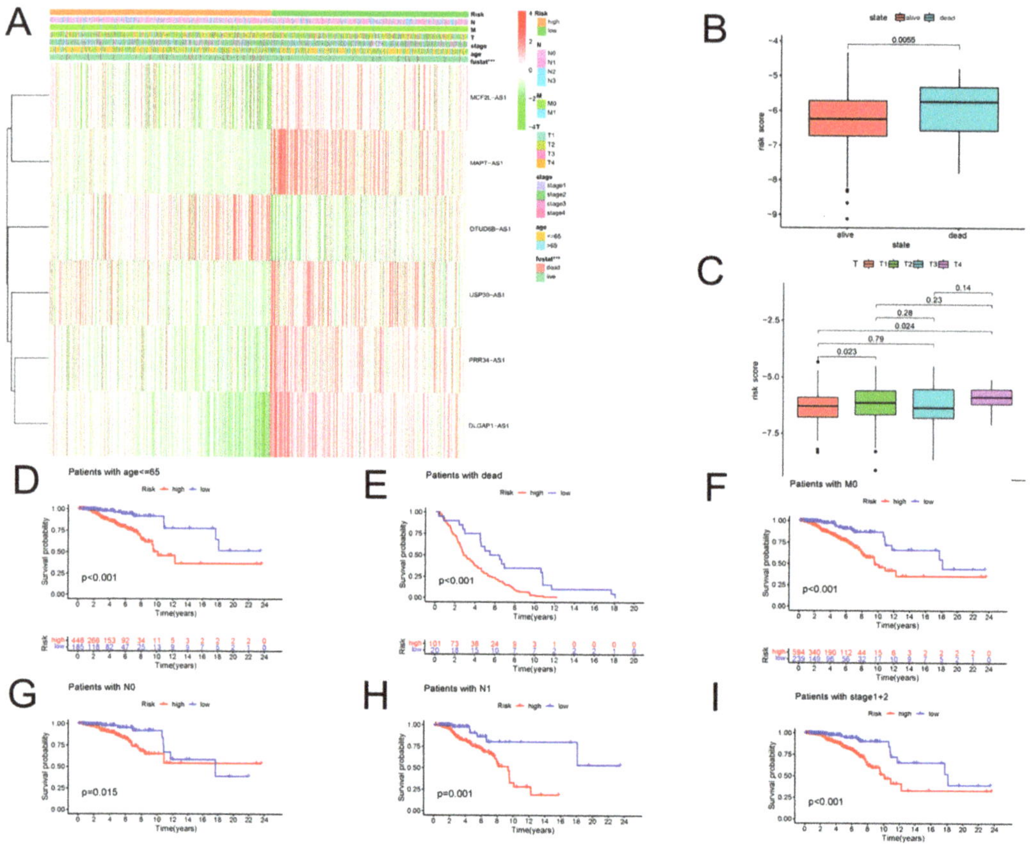


**Figure 5.** *Cont.*

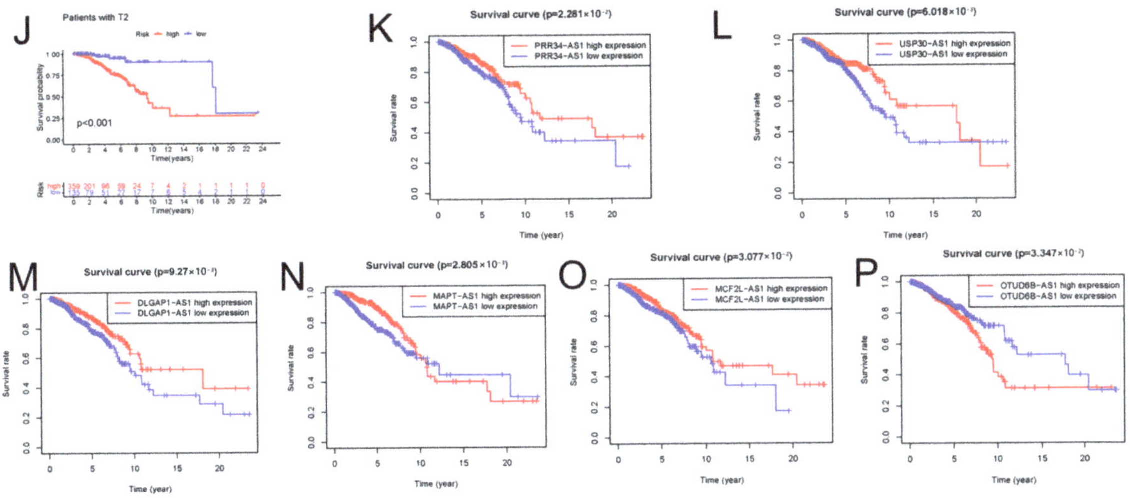


**Figure 5.** The relationship between risk scores and clinicopathological features. (**A**) The heatmap shows the distribution of six aging-related lncRNAs and clinicopathological features in high- and low-risk groups in TCGA-BRCA cohort. (**B**) Aging-related lncRNAs in the cohorts stratified by survival outcome. (**C**) Aging-related lncRNAs in the cohorts stratified by tumor size. (**D–J**) Correlation between different clinicopathological parameters and prognosis. (**K–P**) The expression of the six aging-related lncRNAs is significantly different in high- and low-risk groups using the log-rank test, *** $p < 0.001$.

### 2.5. *Differential Analysis of Signaling Pathway and Immune Functions in the Groups with High Risk and Low Risk*

We utilized GSEA to investigate the discrepancies in the KEGG and GO between the two groups. In the subgroup with high risk, the enriched signaling pathways were linked with tumor initiation and progression, such as PPAR signaling pathways, ECM receptor interactions, and TGF-β signaling pathways, which the KEGG analysis proved (Figure 6A). The high-risk subgroup was enriched with genetically altered cellular components, such as chromosomal region chromosomes and centromeric regions, as shown in the GO analysis (Figure 6C). In addition, the enriched signaling pathways in the group with low risk, such as Graft-versus-host disease, autoimmune thyroid disease, and primary immunodeficiency, were related to immunity according to the KEGG analysis (Figure 6B). The GO analysis indicated enriched biological processes of mRNA splice junction alterations, such as mRNA splice site selection and mRNA five splice site recognition (Figure 6D). Considering the variety of enriched immune-related pathways, we performed a detailed quantification of 16 immune cells and the corresponding 13 immune pathways and functions by the ssGSEA. We further searched the correlation between the different risk score and immune pathways. Five immune functions and six immune cell types were obviously linked with the aging-related risk score (Figure 6E,F). Moreover, the PCA revealed that, contrasted with all genes (Figure 7A) and all lncRNAs (Figure 7B), the BC patients were obviously divided into two different groups by the six lncRNAs in the risk model (Figure 7C).

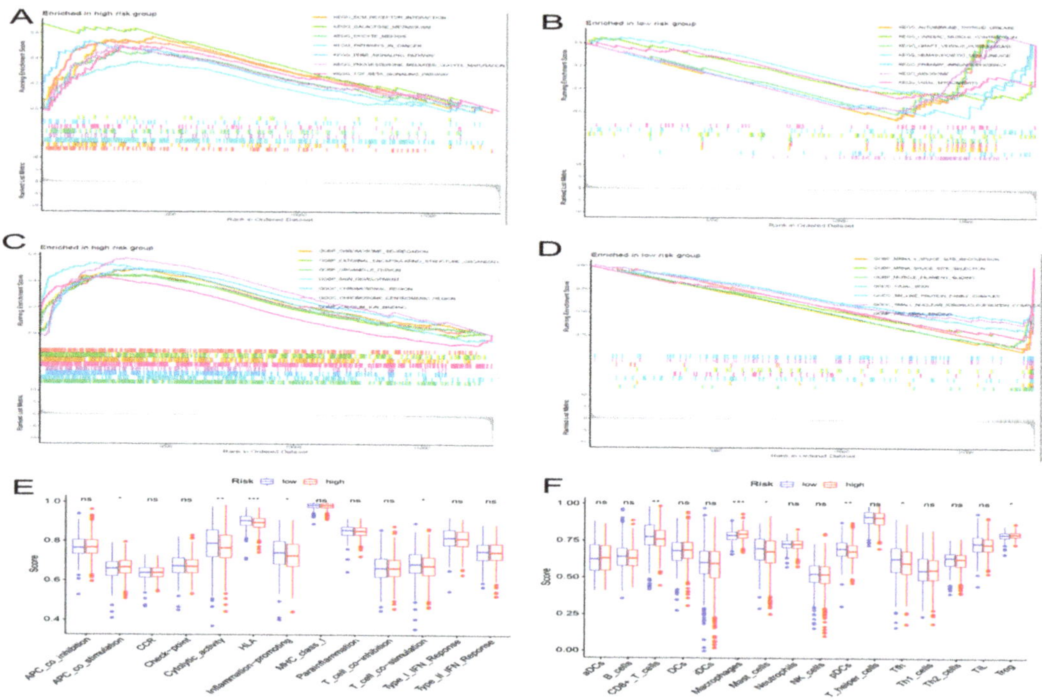


**Figure 6.** The gene set enrichment analysis (GSEA) and immune function between the high- and low-risk groups. The GSEA method was performed to identify and visualize the differential Kyoto Encyclopedia of Genes and Genomes (KEGG) (**A**,**B**) and Gene Ontology (GO) (**C**,**D**) enrichment analyses in the high- and low-risk groups. (**E**) Results of immune functions. (**F**) Results of infiltrating fractions of immune cells. * $p < 0.05$, ** $p < 0.01$, *** $p < 0.001$, ns (no significance).

## *2.6. Analysis of Immune Cell Infiltration*

We further explored whether the aging-related lncRNA signature was connected with tumor immunity. We employed the XCELL, QUANTISEQ, EPIC, TIMER, and MCP-COUNTER immune databases to calculate the differences between high- and low-risk subgroups in the proportion of tumor-infiltrating immune cells for the aging signature (Figure 7D). The relative proportions of the 22 tumor-infiltrating immune cells were further analyzed by the CIBERSORT in each sample. Interestingly, in the two different risk groups, the enrichment of the tumor-infiltrating immune cells showed different dimensions (Figure 8B; Supplementary Figure S5). Subsequently, the connection of the tumor-infiltrating immune cells and risk score was explored through Spearman's analysis [17]. Figure 8C,D indicates that resting NK cells, M2 macrophages, and M0 macrophages were positively related with risk score. Plasma cells, B memory cells, monocytes, activated NK cells, resting mast cells, regulatory T cells, and CD8+ T cells were all negatively associated with risk score. Meanwhile, eight immune cells were remarkably related with the risk score on the basis of the Pearson correlation analysis (Figure 8E–L). The outcomes showed that the six lncRNAs in our risk model can discriminate different features of immune cell infiltration in BC.

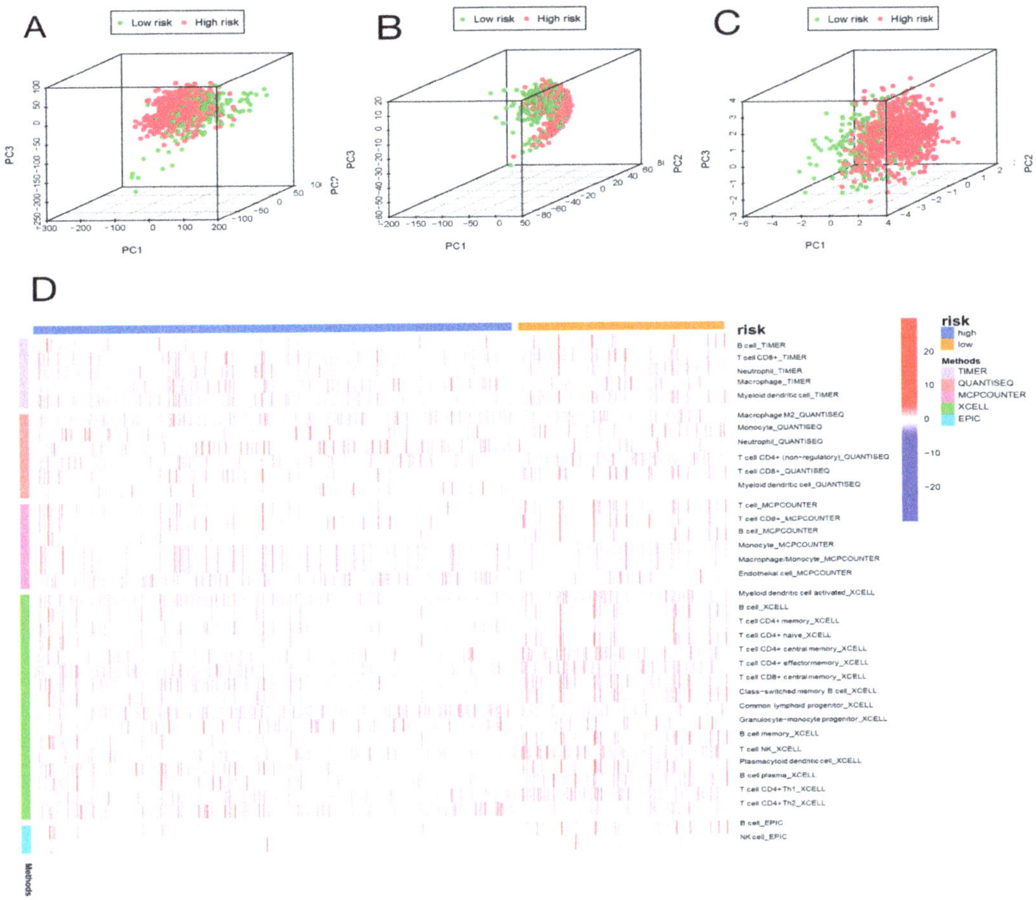


**Figure 7.** Principal component analysis (PCA) and correlation of risk scores with immune cell infiltration in different immune databases. PCA of low-risk and high-risk groups based on whole-genome (**A**), aging-related lncRNAs (**B**), and the risk signature including six aging-related lncRNAs (**C**). Patients with high risk scores are represented by red dots, patients with low risk scores are indicated by green dots. (**D**) The heatmap shows the correlation of risk scores with immune cell infiltration in different immune databases.

### *2.7. Genetic Alteration in Aging-Related Genes*

We examined the prevalence of somatic mutations and CNVs in entire genes of BC and discovered that the most frequent variant class was missense mutations, whereas single-nucleotide polymorphisms (SNPs) were the most common variant types. The highest SNV classification was C > T (Supplementary Figure S6). Then, the frequency and class of mutations in total genes were investigated in two different risk groups. In Figure 9A,B, genetic mutations were found in 247 of 308 (80.2%) BC samples in the low-risk subgroup and 714 of 714 (100%) BC samples in the high-risk subgroup. The most frequent variant class was missense mutations. Among high-risk subgroups, the mutation rate of PIK3CA was high (33.9%) and was second only to the mutation rate of TP53 (44.4%). The PIK3CA had the most alterations (34%) in the subgroup with low risk (Figure 9A,B). Additionally, when we figured the TMB for every BC patient, we discovered that, in the high-risk group, TMB was remarkably higher (Figure 9C). However, no association was observed between TMB and OS in BC patients (Figure 9D).

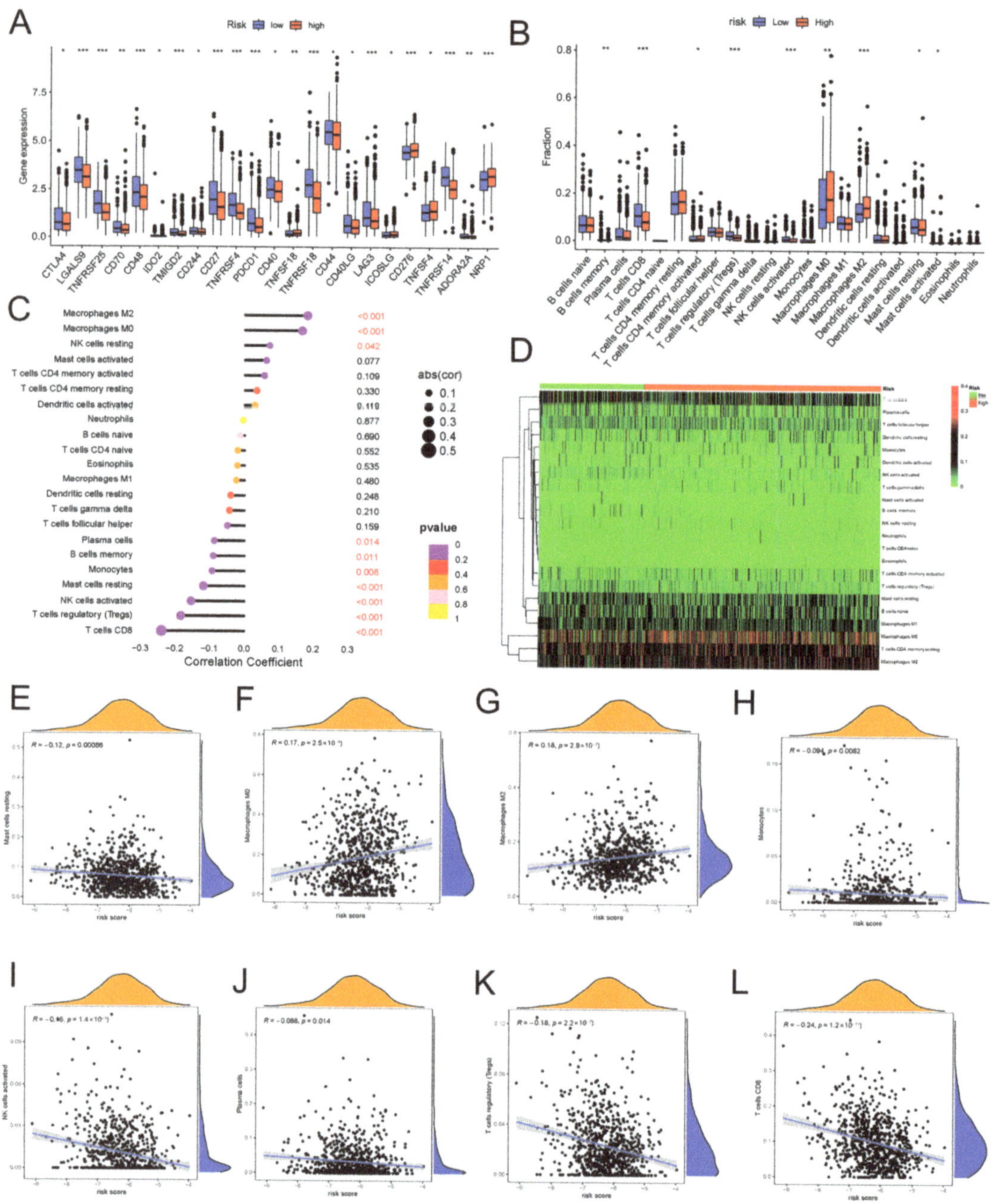


**Figure 8.** Expression of immune checkpoint molecules and the relationship between risk scores and immune cell infiltration in two different risk groups. (**A**) The distribution of immune checkpoints was significantly different between the high- and low-risk groups as determined by Wilcoxon test. (**B**) Differential analyses of the proportion of immune cell infiltration. Lollipop chart (**C**) and heatmap (**D**) showing correlation between risk scores and immune cell infiltration by Spearman rank correlation test. (**E**–**L**) Estimation of risk score coefficients for 8 types of immune cells using Spearman rank correlation test. * $p < 0.05$, ** $p < 0.01$, *** $p < 0.001$.

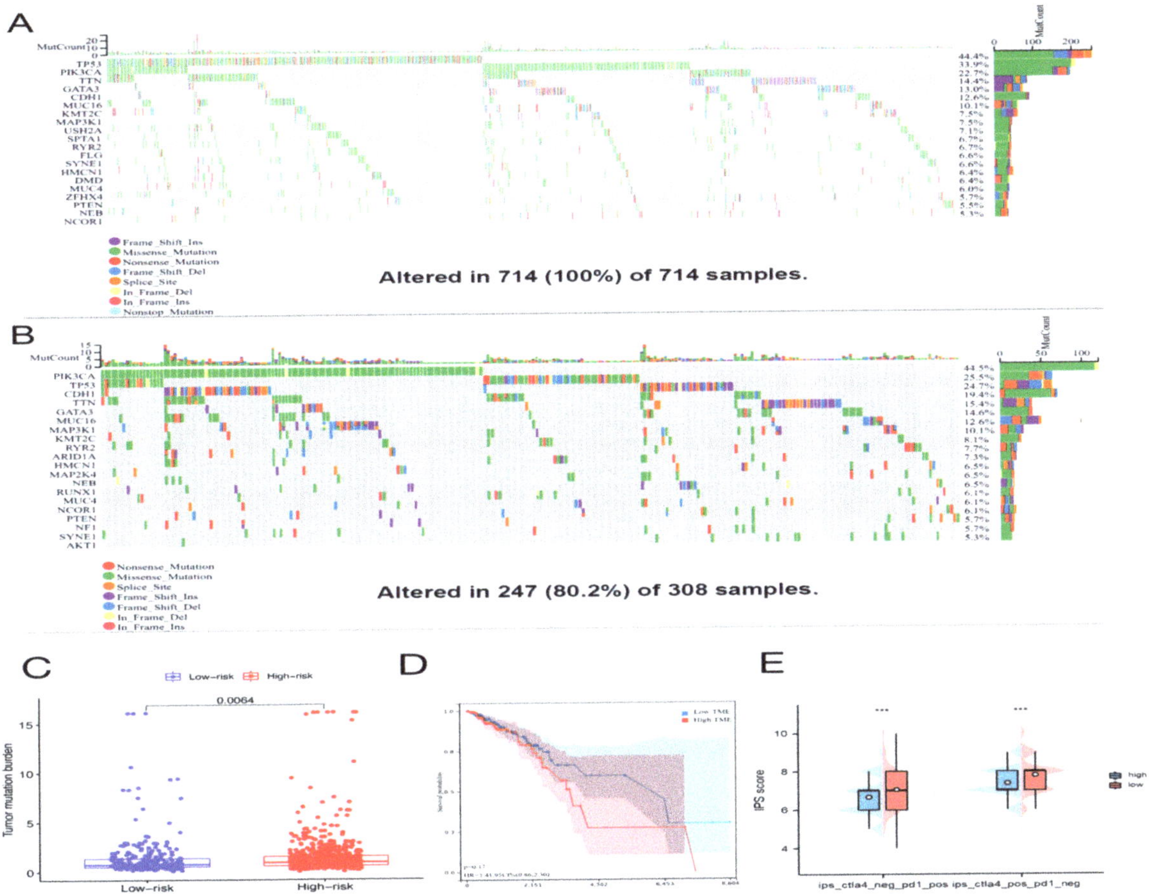


**Figure 9.** The mutation profile, tumor mutational burden (TMB), and immunophenoscore (IPS) in high-risk and low-risk groups. Mutation profile of BC patients in high-risk group (**A**) and in low-risk group (**B**). (**C**) The difference of TMB between two different risk groups as determined by the Wilcoxon test. (**D**) The relationship between TMB and overall survival of BC patients using the log-rank test. (**E**) Results of IPS in high-risk and low-risk groups using the Wilcoxon test, *** $p < 0.001$.

### *2.8. Immunotherapy Effect and Drug Sensitivity*

Immune checkpoint inhibitors (ICIs) are currently approved for the therapy of PD-L1+ metastatic triple-negative BC [18]. Between high-risk and low-risk groups, discrepancies in the expression of immune molecules related with checkpoint were investigated. In the low-risk group, the expression of all-important immune checkpoint-related molecules was up-regulated according to the result (Figure 8A). This provided a potential immunotherapy target for BC patient whose risk scores were different. Subsequently, the use of IPS was predicting the reaction of different BC patients to ICI. In order to gauge how BC patients in the two separate risk categories responded to anti-PD-1/PD-L1 and anti-CTLA-4 medication, two subtypes of IPS values—IPS-PD-1/PD-L1 pos and IPS-CTLA-4 pos—were utilized [19]. The immunophenoscore (IPS) of the TCGA-BRCA cohort was downloaded from the Cancer Immunome Atlas (TCIA) database. The form downloaded from the official website (https://tcia.at/home, accessed on 13 January 2023) already defines the positive and negative information for each patient. IPS scores in the low-risk group were higher than those in the high-risk group (Figure 9E). The outcomes showed that the immunotherapy effect was better in the low-risk group. Based on the potential role of lncRNAs in modulating drug sensitivity, we assessed the potential of the 6-ARlncRNAs signature as biomarkers for predicting drug response in BC patients (Figure 10). In TCGA-BRCA patients, we deduced the IC50 values for 138 medications. Patients in the high-risk group may react to Nutlin more strongly than Temirolimus, 3a, and other drugs. BC patients in the low-risk group may be more susceptible to drugs such as AZD6482 and Thapsigargin.

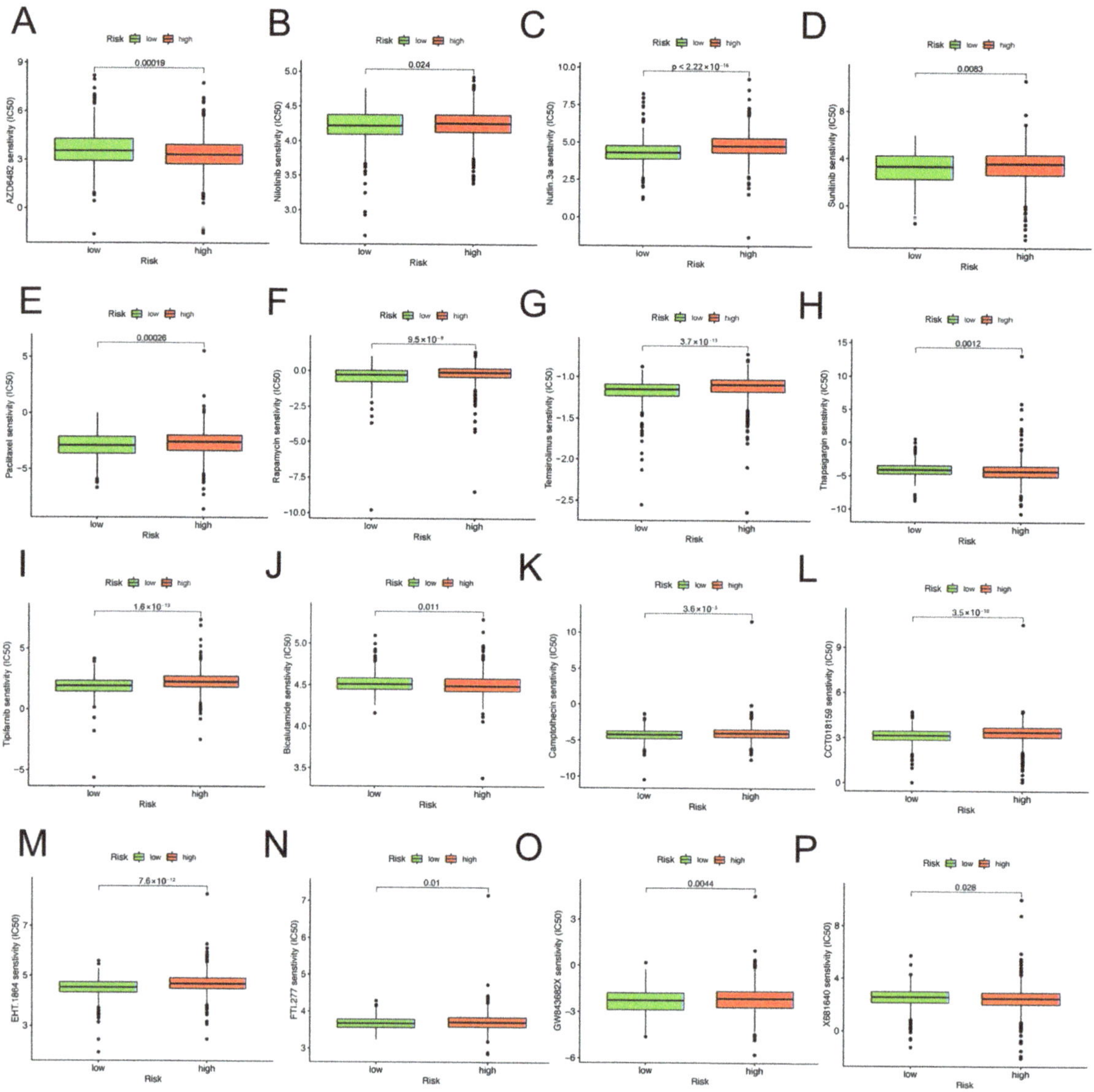


**Figure 10.** Drug sensitivity predictions as determined by the pRRophetic algorithm in R using the Wilcoxon test. (**A**–**P**) Different drug sensitivities associated with 6-ARlncRNAs signature.

## 3. Discussion

BC is the primary cause of cancer-related death in women worldwide, and the incidence rate is rising yearly. Despite improvements in diagnosis and therapy, invasive BC's high death rate remains a global concern [20]. Clinical consequences are highly variable in patients, which is likely due to BC's heterogeneity [21]. Therefore, it is urgent to search for and evaluate prognostic biomarkers for the early diagnosis of BC. The lncRNAs can promote tumor initiation and progression and can serve as biomarkers to predict the prognosis of cancer patients [22,23]. Aging is an important and unavoidable biological process; it leads to the progressive deterioration of the function of many tissues [24]. Studies have pointed out that ARGs can promote tumor initiation, progression, and metastasis. Cancer may be inhibited by managing senescence in tumor cells [25,26]. Aging is an independent risk factor for some cancers [24]. In addition, aging-related markers have prognostic potential for predicting cancer [27]. The above facts indicated that we urgently need to identify more aging-related molecular markers in BC patients.

A prognostic model of aging-related lncRNAs including six lncRNA molecules (MCF2L-AS1, USP30-AS1, OTUD6B-AS1, MAPT-AS1, PRR34-AS1 and DLGAP1-AS1) was constructed. Most of the aging-related lncRNAs included in this prognostic signature are closely linked to tumorigenesis, metastasis, and proliferation. MCF2L-AS1 promotes colorectal cancer development through the miR-105-5p/RAB 22A axis and regulates miR-873-5p levels to promote cancer stem-like features of non-small-cell lung carcinoma cells [28,29]. However, whether MCF2L-AS1 plays a role in BC has not been reported. USP30-AS1 was silenced to promote mitochondrial uncoupler-induced mitophagy as a negative regulator of mitochondrial homeostasis. Mitochondrial autophagy leads to decreased mitochondrial function, a hallmark of cancer [30]. For some malignant tumors, USP30-AS1 is a potential prognostic biomarker [31,32]. Multiple tumor types include the lncRNA ovarian tumor domain that contains 6B antisense RNA1 (OTUD6B-AS1). OTUD6B-AS1 has also been discovered to be a new tumor-associated lncRNA [33]. Several reports are consistent with our view that OTUD6B-AS1 can serve as a predictor of BC prognosis [34,35]. MAPT-AS1 can affect BC proliferation, migration, and drug sensitivity [36,37]. PRR34-AS1 was upregulated in hepatocellular carcinoma and pediatric medulloblastoma [38,39]. However, no reports explored the role of PRR34-AS1 in BC. According to a recent study, DLGAP1-AS1 may activate the Wnt signaling pathway to speed up the proliferation of glioblastoma and hepatocellular cancer [40,41]. However, the relationship between BC and DLGAP1-AS1 has not been pointed out. Using a ROC curve, we assessed the prognostic model composed of these six IncRNAs' capacity to forecast the prognosis of BC. This model did well in prediction (AUCs of 0.753, 0.772, and 0.722 for 1, 3, and 5 years, respectively). We built a novel nomogram associated with the risk score and clinical characteristics according to the lncRNA signature. The resulting verification demonstrated the compelling sensitivity and specificity of the nomogram in predicting clinical results. Correlation between clinical characteristics and the aging signature in BC patients was evaluated. The risk score correlated well with tumor clinical stage (T and M stages) and age, with the low-risk group showing better survival. This aging model indicated better predictive performance for early-stage BC patients. Therefore, this prognostic model enables better risk stratification for early BC patients.

The immune response in tumors can be triggered by aging. The tumor-infiltrating immune cell microenvironment contributes to tumor development [42]. For example, cytotoxic T, NK, and B cells disrupt tumor cells, whereas myeloid-derived suppressor cells (MDSCs), tumor-associated macrophages (TAMs), and regulatory T cells (Tregs) coordinate tumor growth and immune escape [43]. The relationship between aging and the immune cells that infiltrate tumors, however, was rarely covered in studies. The percentage of immune cells that infiltrated the tumor was determined using the CIBERSORT algorithm analysis to see whether an aging-related signature is connected to tumor immunity and immunotherapy. The prognostic model can recognize the different features of tumor-infiltrating immune cells well. Primitive B cells, regulatory T cells, and CD8+ T cells—tumor-killing immune cells—were less prevalent in the high-risk group than in the low-risk group, but M0 and M2 macrophages, which encourage tumor growth and progression, were more prevalent [44]. Five immune functions and six immune cell types were substantially linked with the aging-related risk score, according to the ssGSEA. After that, we looked into variations in immune checkpoint molecule expression between groups at high risk and those at low risk. The outcomes showed that the expression of all important immune checkpoints, including CTLA-4 and PD-1, was up-regulated in the low-risk group, indirectly suggesting that the low-risk group had pre-existing T cell activation. Therefore, BC patients who had a low risk score may benefit more from ICI treatment. Microsatellite instability, TMB, PD-L1 expression, and mismatch repair deficiency are used for patient selection before ICI therapy [45]. In this study, TMB was higher in high-risk BC patients, but no significant association was found between TMB and OS in BC. In addition, the IPS quantitatively predicts patient response to anti-PD-1/PD-L1 and anti-CTLA-4 therapies [46]. The IPS score of the low-risk group was higher than that of high-risk group ($p < 0.001$). The 6-ARlncRNAs signature can be used for patient selection before ICI treatment, and ICI treatment is more appropriate for BC patients with lower aging-related risk scores. To predict possible drug targets, analysis was conducted. The IC50 of Nutlin.3a, Temsirolimus, and others in the high-risk group are higher than those in the low-risk group, i.e., higher sensitivity, whereas the patients who are in the low-risk group are more sensitive to AZD6482, Thapsigargin, and others. Guidance was provided for the selection of therapeutic drugs.

In previous studies, some aging-related molecular markers that can predict invasive BC have been verified [27], but the utilization of aging-related lncRNAs in BRCA has not been presented. We utilized many samples in the TCGA and GEO databases to identify and validate the risk model, which can predict the survival, pathological characteristics, and treatment methods of BC patients well, thereby providing powerful guidance for clinical practice. The data utilized in our research came from several public databases. The 6-ARlncRNAs signature offers a fresh viewpoint on BC diagnosis, prognosis, and treatment, but further clinical trials are required to confirm the therapeutic relevance of these results.

## 4. Materials and Methods

### 4.1. Data and Clinical Information Acquisition and Collation

Transcriptome expression data and clinical data of BC patients were downloaded through the Gene Expression Omnibus (GEO) database (https://www.ncbi.nlm.nih.gov/geo/, accessed on 13 January 2023) (National Center for Biotechnology Information, NCBI) and The Cancer Genome Atlas (TCGA) database (https://portal.gdc.cancer.gov/repository, accessed on 13 January 2023) (National Cancer Institute, NCI; National Human Genome Research Institute, NHGRI). Patients with missing clinical information were removed. We then downloaded the gene transfer format (GTF) file through the Ensembl database (http://asia.ensembl.org, accessed on 13 January 2023) (Wellcome Sanger Institute). Transcriptome sequencing data of mRNAs and lncRNAs were annotated and differentiated [47]. We obtained a total of 279 ARGs through the Human Aging Genome Resource (HAGR, https://genomics.senescence.info/cells/, accessed on 13 January 2023) (senescence.info). The ARGs are shown in Supplementary Table S1.

### 4.2. Identification and Differential Analysis of Aging-Related lncRNAs

We sifted the differential expression of aging-related lncRNAs (DEarlncRNAs) by using Pearson correlation analysis [48]. Screening criteria were $|\text{Pearson } R| > 0.4$ and $p < 0.001$. Filtering of DEarlncRNAs was performed through the R package "limma" with filter criteria of $|\log_2\text{Fold Change}| > 1$ and FDR $< 0.05$ [49].

### 4.3. Construction and Validation of the Aging-Related lncRNA Prognostic Signature

We obtained 1022 BC samples and 112 paired normal samples from TCGA database. These BC patients were randomized 1:1 to the training set (n = 511) or the test set (n = 511). Meanwhile, there was no statistical difference in the clinical characteristics of BC patients between the training subgroup and the test subgroup ($p > 0.05$). The "forestplot," "survminer," and "survival" packages in R were employed to execute the univariate and the multivariate Cox regression analysis [50]. First, we screened DEarlncRNAs associated with overall survival (OS) in the TCGA breast cancer (BRCA) cohort by using the univariate Cox proportional hazard regression analysis. The LASSO–Cox regression analysis compressed differentially relevant regulators of prognosis, and we removed redundant genes by using the R package "glmnet" [51]. Afterward, the multivariate Cox analysis and stepwise variables were employed to further screen variables associated with OS in the TCGA-BRCA cohort, and we established a predictive signature [52]. The Akaike information criteria were the basis for the signature building. The TCGA-BRCA patients were scored according to the risk scoring formula. Risk score = coef gene (1) × exprgene (1) + coef gene (2) × exprgene (2) + . . . + coef gene (n) × exprgene (n) (8). We obtained coefficients of each gene by multivariate Cox regression analysis. On the basis of the best cut-off value confirmed by X-tile software, the TCGA-BRCA cohort was split into high- and low-risk subgroups. Afterward, we applied the "survival" R package and compared the diversity in OS in the two subgroups by the log-rank test. The R "pheatmap" package is used to visualize the allocation of corresponding risk values between the two groups [53]. The "timeROC" package was used to obtain receiver operating characteristic (ROC) curves to determine the accuracy of prediction for 1-, 3-, and 5-year OS of BC. The prediction accuracy was judged by the area under the curve (AUC) value [54].

The GSE20685 dataset was downloaded through the GEO database. A total of 327 patients with BRCA were included. The GSE20685 dataset was employed as an external validation set. We performed batch-to-batch correction of gene expression values between the GSE20685 cohort and the TCGA-BRCA cohort to eliminate batch differences with the ComBat function and the R "sva" package [55]. Afterward, the GSE20685 cohort was also split into the two different risk subgroups on the basis of risk cutoff values in the training cohort. The Kaplan–Meier survival curve was also drawn to assess differences of OS in the two different risk subgroups. The ROC curve assessed the ability to predict prognosis in the GSE20685 cohort. Afterwards, protein expression levels of aging-related markers (including p21, p16, p53, TNF-α, IL-6, FGF, VEGF, and MMP-3) were compared between BC and common samples from the Human Protein Atlas (HPA) database (https://www.proteinatlas.org/, accessed on 13 January 2023) (Knut and Alice Wallenberg Foundation) [56].

### *4.4. Development and Evaluation of a Nomogram in BC Patients*

We employed the R package "rms" to build a nomogram by combining an aging-related signature with clinical characteristics to predict survival in BC patients [57]. The total score was summed up by each point that corresponded to each variate in the nomogram scoring system to forecast 1-, 3-, and 5-year OS in patients with BC. Meanwhile, we plotted the calibration curve by using the "bootstrap" package. The decision curve analysis (DCA) was drawn by the R package "rmda" to assess the precision of the nomogram [58]. In addition, the consistency index (C-index) and the ROC curve over time were yielded.

### *4.5. Kyoto Encyclopedia of Genes and Genomes, Gene Ontology, Gene Set Enrichment Analysis and Single-Sample Gene Set Enrichment Analysis*

First, we carried out the Kyoto Encyclopedia of Genes and Genomes (KEGG) and the Gene Ontology (GO) functional enrichment analyses based on DEarlncRNAs with the packages "GOplot" and "KEGGplot" in R [59]. Second, employing GSEA-3.0 software (http://software.broadinstitute.org/gsea/index.jsp, accessed on 13 January 2023), we performed the Gene Set Enrichment Analysis (GSEA) to show the potential pathways and mechanisms of two different subgroups in the TCGA-BRCA cohort [60]. The GO and KEGG analyses were conducted and visualized through the packages "enrichplot" and "lusterProfiler" in the aging-related signature [61]. In addition, immune infiltration of 13 immune-related pathways and 16 cell types in two different risk subgroups was evaluated employing single-sample genome enrichment analysis (ssGSEA) using the package "GSVA" [62].

### 4.6. Immune Cell Infiltration Analysis

We assessed the scores of 22 immune cell types in the high- and low-risk subgroups in the TCGA-BRCA cohort by using the CIBERSORT algorithm, and the results were filtered ($p < 0.05$) [63]. Through the R package "pheatmap," differential tumor-infiltrating immune cells between the two subgroups were mapped. Afterward, the association between the proportion of tumor-infiltrating immune cells and the risk score was compared based on the Spearman rank correlation test in TCGA-BRCA cohort [64]. In addition, discrepancies in the proportion of immune cell infiltration in the groups were analyzed based on QUANTISEQ, XCELL, EPIC, TIMER, and MCPCOUNTER immunization databases. Moreover, the principal component analysis (PCA) was applied for grouping visualization of high-dimensional data, model identification, and effective dimensionality reduction. We used the R package "stats" for analysis, Z-Score on the expression spectrum, and further dimensionality reduction analysis using the PRCOMP function to obtain the dimensionality reduction matrix.

### 4.7. Immunotherapy and Drug Sensitivity Prediction

We evaluated the differential expression of immune-checkpoint-related molecules. Different expressions of immune checkpoints in the prognostic risk model can be applied to provide new treatment ideas and predict the clinical treatment effect of corresponding inhibitors. Afterward, the immunophenoscore (IPS) of the TCGA-BRCA cohort was downloaded from the Cancer Immunome Atlas (TCIA) database (Institute of Bioinformatics). Tumor immunogenicity was positively correlated with the IPS value. IPS predicts patient response to immune checkpoint inhibitor (ICI) therapy [46]. Data of IPS were subsequently extracted for analysis. We downloaded the single nucleotide variation dataset for TCGA-BRCA patients from the TCGA database. According to the six lncRNAs in our risk model, we used the R package "maftools" to explore copy number variations (CNVs), somatic mutation, and TMB in the two different risk subgroups [65]. We calculated the TMB of each BC patient to compare differences in the risk subgroups (mutations per million bases). To assess drug susceptibility, we downloaded an anticancer drug dataset from the Genomics of Cancer Drug Sensitivity (GDSC) database (Wellcome Sanger Institute, Genome Research Limited). We deduced the half maximal inhibitory concentration (IC50) values for 138 drugs using the pRRophetic algorithm [66]. Finally, the Wilcoxon test of these drugs was conducted to predict the level of response to the drug in patients at different risk levels.

### 4.8. Statistical Analysis

R is a language for statistical analysis and drawing, and it is an excellent tool for statistical calculation and statistical drawing. Its full name is The R Programming Language. We used R (R Programming Language version 4.1.2) with associated R package for all statistical analyses and plots (Institute for Statistics and Mathematics of WU). For comparison between the two groups, if the normal distribution is met, a $t$-test is used. If it is a non-normal distribution, the nonparametric test is used. The Pearson correlation analysis was employed to assess correlation. The log-rank test, Cox regression, and Kaplan–Meier curves were applied to evaluate prognostic value. The Wilcoxon rank sum test was applied to evaluate the relationship between immune checkpoints, TMB, immune cell infiltration, and IC50 values of chemotherapeutic drugs between these two groups. All statistical analyses were two-sided, and $p < 0.05$ was considered statistically significant if not specified.

## 5. Conclusions

We created a brand-new lncRNA predictive signature for aging. The training set was used to build the signature. It was well-validated on the validation set and has the potential to be a predictive biomarker for patients with invasive BC. The 6-ARlncRNAs signature can offer fresh perspectives and approaches for early diagnosis and therapeutic target identification of BC, including tumor immunotherapy. This research provided a new reference for further research on aging, tumor immunity, and chemotherapy drug selection.

**Supplementary Materials:** The following supporting information can be downloaded at: https://www.mdpi.com/article/10.3390/molecules28083283/s1, Figure S1. A flowchart describing the research process. Figure S2. Differentially expressed lncRNAs in tumor samples and normal samples. Figure S3. The distribution of risk score, survival status, and the 6 AG expression panel of breast cancer (BC) patients with different risk scores in the GSE20685 cohort. Figure S4. Aging-related lncRNAs in the cohorts stratified by age, stage, and TNM staging. Figure S5. The prevalence of somatic mutations and CNVs in entire genes of BC. Figure S6. The infiltration of immune cells in BC patients and the corresponding prognostic risk score and survival status. Supplementary Table S1. 279 genes related to the aging. Supplementary Table S2. Univariate and Multivariate cox regression analyses of the prognostic aging-related lncRNAs in BC patients. Supplementary Table S3. Clinical characteristics of BC patients in the two cohorts.

**Author Contributions:** Y.S. and Y.L. conceived and planned this study. Z.L. and C.R. performed the bioinformatics analysis of the study. J.C., J.Y. and B.Y. carried out the experiment based on B.C. tissues. R.D. and W.M. collected the samples and did statistical analysis. All authors provided vital feedback to shape the final version of the manuscript. All authors have read and agreed to the published version of the manuscript.

**Funding:** Our study was supported by The Support Plan for Youth Entrepreneurship and Technology of Colleges and Universities in Shandong (grant no. 2019KJK014), Innovation and Entrepreneurship Training Program for College Students (grant nos. X202210440474, S202110440145), The National Natural Science Foundation of China (grant no. 81800169), The Shandong Province Taishan Scholar Project (grant no. ts201712067).

**Institutional Review Board Statement:** Not applicable.

**Informed Consent Statement:** Not applicable.

**Data Availability Statement:** All data analyzed in the present study are publicly available in the Cancer Genome Atlas (TCGA; https://portal.gdc.cancer.gov/repository (accessed on 13 January 2023)), Gene Expression Omnibus (GEO; https://www.ncbi.nlm.nih.gov/geo/ (accessed on 13 January 2023)), and Human Protein Atlas (HPA; https://www.proteinatlas.org/ (accessed on 13 January 2023)).

**Acknowledgments:** The authors expressed their sincere thanks to all databases used in this study.

**Conflicts of Interest:** The authors declare no conflict of interest.

## References

1. Siegel, R.L.; Miller, K.D.; Jemal, A. Cancer statistics, 2020. *CA Cancer J. Clin.* **2020**, *70*, 7–30. [CrossRef] [PubMed]
2. Zhao, S.; Liu, X.Y.; Jin, X.; Ma, D.; Xiao, Y.; Shao, Z.M.; Jiang, Y.Z. Molecular portraits and trastuzumab responsivenessofestrogen receptor-positive, progesterone receptor-positive, and HER2-positivebreast cancer. *Theranostics* **2019**, *9*, 4935–4945. [CrossRef] [PubMed]
3. Wang, X.; Sun, C.; Huang, X.; Li, J.; Fu, Z.; Li, W.; Yin, Y. The advancing roles of exosomes in breast cancer. *Front. Cell Dev. Biol.* **2021**, *9*, 731062. [CrossRef] [PubMed]
4. Carroll, J.E.; Van Dyk, K.; Bower, J.E.; Scuric, Z.; Petersen, L.; Schiestl, R.; Irwin, M.; Ganz, P.A. Cognitive performance in survivors of breast cancer and markers of biological aging. *Cancer* **2018**, *125*, 298–306. [CrossRef] [PubMed]
5. Smetana, K.; Lacina, L.; Szabo, P.; Dvořánková, B.; Brož, P.; Šedo, A. Ageing as an Important Risk Factor for Cancer. *Anticancer. Res.* **2016**, *36*, 5009–5017. [CrossRef]
6. Gu, X.; Wang, B.; Zhu, H.; Zhou, Y.; Horning, A.M.; Huang, T.H.-M.; Chen, Y.; Houghton, P.; Lai, Z.; Michalek, J.E.; et al. Age-associated genes in human mammary gland drive human breast cancer progression. *Breast Cancer Res.* **2020**, *22*, 1–15. [CrossRef]
7. Ren, S.; Lin, P.; Wang, J.; Yu, H.; Lv, T.; Sun, L.; Du, G. Circular RNAs, Promising Molecular Biomarkers of Human Aging-Related Diseases via Functioning as an miRNA Sponge. *Mol. Ther.-Methods Clin. Dev.* **2020**, *18*, 215–229. [CrossRef]
8. Yue, T.; Chen, S.; Zhu, J.; Guo, S.; Huang, Z.; Wang, P.; Zuo, S.; Liu, Y. The aging-related risk signature in colorectal cancer. *Aging* **2021**, *13*, 7330–7349. [CrossRef]
9. Xu, Q.; Chen, Y. An Aging-Related Gene Signature-Based Model for Risk Stratification and Prognosis Prediction in Lung Adenocarcinoma. *Front. Cell Dev. Biol.* **2021**, *9*, 685379. [CrossRef]

10. Liu, L.; Zhao, J.; Du, X.; Zhao, Y.; Zou, C.; Zhou, H.; Li, W.; Yan, X. Construction and validation of a novel aging-related gene signature and prognostic nomogram for predicting the overall survival in ovarian cancer. *Cancer Med.* **2021**, *10*, 9097–9114. [CrossRef]
11. Ransohoff, J.D.; Wei, Y.; Khavari, P.A. The functions and unique features of long intergenic non-coding RNA. *Nat. Rev. Mol. Cell Biol.* **2018**, *19*, 143–157. [CrossRef]
12. Du, C.; Zhang, J.-L.; Wang, Y.; Zhang, Y.-Y.; Zhang, J.-H.; Zhang, L.-F.; Li, J.-R. The Long Non-coding RNA LINC01705 Regulates the Development of Breast Cancer by Sponging miR-186-5p to Mediate TPR Expression as a Competitive Endogenous RNA. *Front. Genet.* **2020**, *11*, 779. [CrossRef]
13. Wan, Q.; Tang, M.; Sun, S.-L.; Hu, J.; Sun, Z.-J.; Fang, Y.-T.; He, T.-C.; Zhang, Y. SNHG3 promotes migration, invasion, and epithelial-mesenchymal transition of breast cancer cells through the miR-186-5p/ZEB1 axis. *Am. J. Transl. Res.* **2021**, *13*, 585–600.
14. Huang, Y.; Xie, B.; Cao, M.; Lu, H.; Wu, X.; Hao, Q.; Zhou, X. LncRNA RNA Component of Mitochondrial RNA-Processing Endoribonuclease Promotes AKT-Dependent Breast Cancer Growth and Migration by Trapping MicroRNA-206. *Front. Cell Dev. Biol.* **2021**, *9*, 2591. [CrossRef]
15. D'Amico, S.; Krasnowska, E.; Manni, I.; Toietta, G.; Baldari, S.; Piaggio, G.; Ranalli, M.; Gambacurta, A.; Vernieri, C.; Di Giacinto, F.; et al. DHA Affects Microtubule Dynamics Through Reduction of Phospho-TCTP Levels and Enhances the Antiproliferative Effect of T-DM1 in Trastuzumab-Resistant HER2-Positive Breast Cancer Cell Lines. *Cells* **2020**, *9*, 1260. [CrossRef]
16. Srivastava, S.; Gopal-Srivastava, R. Biomarkers in cancer screening: A public health perspective. *J Nutr.* **2002**, *132*, 2471S–2475S. [CrossRef]
17. Xu, Z.; Peng, B.; Liang, Q.; Chen, X.; Cai, Y.; Zeng, S.; Gao, K.; Wang, X.; Yi, Q.; Gong, Z.; et al. Construction of a Ferroptosis-Related Nine-lncRNA Signature for Predicting Prognosis and Immune Response in Hepatocellular Carcinoma. *Front Immunol.* **2021**, *12*, 719175. [CrossRef]
18. Franzoi, M.A.; Romano, E.; Piccart, M. Immunotherapy for early breast cancer, Too soon, too superficial, or just right? *Ann. Oncol.* **2021**, *32*, 323–336. [CrossRef]
19. Yi, M.; Li, A.; Zhou, L.; Chu, Q.; Luo, S.; Wu, K. Immune signature-based risk stratification and prediction of immune checkpoint inhibitor's efficacy for lung adenocarcinoma. *Cancer Immunol. Immunother.* **2021**, *70*, 1705–1719. [CrossRef]
20. Siegel, R.L.; Miller, K.D.; Fuchs, H.E.; Jemal, A. Cancer Statistics, 2021. *CA Cancer J Clin.* **2021**, *71*, 7–33. [CrossRef]
21. Sachs, N.; de Ligt, J.; Kopper, O.; Gogola, E.; Bounova, G.; Weeber, F.; Balgobind, A.V.; Wind, K.; Gracanin, A.; Begthel, H.; et al. A Living Biobank of Breast Cancer Organoids Captures Disease Heterogeneity. *Cell* **2018**, *172*, 373–386.e10. [CrossRef] [PubMed]
22. Ye, R.; Tang, R.; Gan, S.; Li, R.; Cheng, Y.; Guo, L.; Zeng, C.; Sun, Y. New insights into long non-coding RNAs in non-small cell lung cancer. *Biomed Pharmacother.* **2020**, *131*, 110775. [CrossRef] [PubMed]
23. Wang, M.-Q.; Zhu, W.-J.; Gao, P. New insights into long non-coding RNAs in breast cancer, Biological functions and therapeutic prospects. *Exp. Mol. Pathol.* **2021**, *120*, 104640. [CrossRef] [PubMed]
24. Calcinotto, A.; Kohli, J.; Zagato, E.; Pellegrini, L.; Demaria, M.; Alimonti, A. Cellular Senescence, Aging, Cancer, and Injury. *Physiol. Rev.* **2019**, *99*, 1047–1078. [CrossRef] [PubMed]
25. Mosteiro, L.; Pantoja, C.; Alcazar, N.; Marión, R.M.; Chondronasiou, D.; Rovira, M.; Fernandez-Marcos, P.J.; Muñoz-Martin, M.; Blanco-Aparicio, C.; Pastor, J.; et al. Tissue damage and senescence provide critical signals for cellular reprogramming in vivo. *Science* **2016**, *354*, aaf4445. [CrossRef]
26. Lee, S.; Schmitt, C.A. The dynamic nature of senescence in cancer. *Nat. Cell Biol.* **2019**, *21*, 94–101. [CrossRef]
27. Lv, W.; Zhao, C.; Tan, Y.; Hu, W.; Yu, H.; Zeng, N.; Zhang, Q.; Wu, Y. Identification of an Aging-Related Gene Signature in Predicting Prognosis and Indicating Tumor Immune Microenvironment in Breast Cancer. *Front. Oncol.* **2021**, *11*, 5320. [CrossRef]
28. Kong, W.; Li, H.; Xie, L.; Cui, G.; Gu, W.; Zhang, H.; Ma, W.; Zhou, Y. LncRNA MCF2L-AS1 aggravates the malignant development of colorectal cancer via targeting miR-105-5p/RAB22A axis. *BMC Cancer* **2021**, *21*, 1069. [CrossRef]
29. Li, S.; Lin, L. Long noncoding RNA MCF2L-AS1 promotes the cancer stem cell-like traits in non-small cell lung cancer cells through regulating miR-873-5p level. *Environ Toxicol.* **2021**, *36*, 1457–1465. [CrossRef]
30. Wang, N.; Li, J.; Xin, Q.; Xu, N. USP30-AS1 contributes to mitochondrial quality control in glioblastoma cells. *Biochem. Biophys. Res. Commun.* **2021**, *581*, 31–37. [CrossRef]
31. Sun, Z.; Jing, C.; Xiao, C.; Li, T. An autophagy-related long non-coding RNA prognostic signature accurately predicts survival outcomes in bladder urothelial carcinoma patients. *Aging* **2020**, *12*, 15624–15637. [CrossRef]
32. Gao, M.; Wang, X.; Han, D.; Lu, E.; Zhang, J.; Zhang, C.; Wang, L.; Yang, Q.; Jiang, Q.; Wu, J.; et al. A Six-lncRNA Signature for Immunophenotype Prediction of Glioblastoma Multiforme. *Front Genet.* **2021**, *11*, 604655. [CrossRef]
33. Kong, S.; Xue, H.; Li, Y.; Li, P.; Ma, F.; Liu, M.; Li, W. The long noncoding RNA OTUD6B-AS1 enhances cell proliferation and the invasion of hepatocellular carcinoma cells through modulating GSKIP/Wnt/β-catenin signalling via the sequestration of miR-664b-3p. *Exp. Cell Res.* **2020**, *395*, 112180. [CrossRef]
34. Lv, W.; Wang, Y.; Zhao, C.; Tan, Y.; Xiong, M.; Yi, Y.; He, X.; Ren, Y.; Wu, Y.; Zhang, Q. Identification and Validation of m6A-Related lncRNA Signature as Potential Predictive Biomarkers in Breast Cancer. *Front. Oncol.* **2021**, *11*, 745719. [CrossRef]
35. Ma, W.; Zhao, F.; Yu, X.; Guan, S.; Suo, H.; Tao, Z.; Qiu, Y.; Wu, Y.; Cao, Y.; Jin, F. Immune-related lncRNAs as predictors of survival in breast cancer: A prognostic signature. *J. Transl. Med.* **2020**, *18*, 442. [CrossRef]

36. Pan, Y.; Pan, Y.; Cheng, Y.; Yang, F.; Yao, Z.; Wang, O. Knockdown of LncRNA MAPT-AS1 inhibites proliferation and migration and sensitizes cancer cells to paclitaxel by regulating MAPT expression in ER-negative breast cancers. *Cell Biosci.* **2018**, *8*, 7. [CrossRef]
37. Wang, D.; Li, J.; Cai, F.; Xu, Z.; Li, L.; Zhu, H.; Liu, W.; Xu, Q.; Cao, J.; Sun, J.; et al. Overexpression of MAPT-AS1 is associated with better patient survival in breast cancer. *Biochem. Cell Biol.* **2019**, *97*, 158–164. [CrossRef]
38. Yang, X.; Song, D.; Zhang, J.; Feng, H.; Guo, J. PRR34-AS1 sponges miR-498 to facilitate TOMM20 and ITGA6 mediated tumor progression in HCC. *Exp. Mol. Pathol.* **2021**, *120*, 104620. [CrossRef]
39. Kesherwani, V.; Shukla, M.; Coulter, D.W.; Sharp, J.G.; Joshi, S.S.; Chaturvedi, N.K. Long non-coding RNA profiling of pediatric Medulloblastoma. *BMC Med. Genom.* **2020**, *13*, 87. [CrossRef]
40. Lin, Y.; Jian, Z.; Jin, H.; Wei, X.; Zou, X.; Guan, R.; Huang, J. Long non-coding RNA DLGAP1-AS1 facilitates tumorigenesis and epithelial-mesenchymal transition in hepatocellular carcinoma via the feedback loop of miR-26a/b-5p/IL-6/JAK2/STAT3 and Wnt/β-catenin pathway. *Cell Death Dis.* **2020**, *11*, 34. [CrossRef]
41. Wang, Z.; Han, Y.; Li, Q.; Wang, B.; Ma, J. LncRNA DLGAP1-AS1 accelerates glioblastoma cell proliferation through targeting miR-515-5p/ROCK1/NFE2L1 axis and activating Wnt signaling pathway. *Brain Behav.* **2021**, *11*, e2321. [CrossRef] [PubMed]
42. Zhao, L.; Zhang, Y.; Gao, Y.; Geng, P.; Lu, Y.; Liu, X.; Yao, R.; Hou, P.; Liu, D.; Lu, J.; et al. JMJD3 promotes SAHF formation in senescent WI38 cells by triggering an interplay between demethylation and phosphorylation of RB protein. *Cell Death Differ.* **2015**, *22*, 1630–1640. [CrossRef] [PubMed]
43. Taube, J.M.; Galon, J.; Sholl, L.M.; Rodig, S.J.; Cottrell, T.R.; A Giraldo, N.; Baras, A.S.; Patel, S.S.; A Anders, R.; Rimm, D.L.; et al. Implications of the tumor immune microenvironment for staging and therapeutics. *Mod. Pathol.* **2017**, *31*, 214–234. [CrossRef]
44. Sebestyen, Z.; Prinz, I.; Déchanet-Merville, J.; Silva-Santos, B.; Kuball, J. Translating gammadelta (γδ) T cells and their receptors into cancer cell therapies. *Nat. Rev. Drug Discov.* **2020**, *19*, 169–184. [CrossRef] [PubMed]
45. Yi, M.; Jiao, D.; Xu, H.; Liu, Q.; Zhao, W.; Han, X.; Wu, K. Biomarkers for predicting efficacy of PD-1/PD-L1 inhibitors. *Mol. Cancer* **2018**, *17*, 129. [CrossRef]
46. Charoentong, P.; Finotello, F.; Angelova, M.; Mayer, C.; Efremova, M.; Rieder, D.; Hackl, H.; Trajanoski, Z. Pan-cancer Immunogenomic Analyses Reveal Genotype-Immunophenotype Relationships and Predictors of Response to Checkpoint Blockade. *Cell Rep.* **2017**, *18*, 248–262. [CrossRef]
47. Yates, A.D.; Achuthan, P.; Akanni, W.; Allen, J.; Allen, J.; Alvarez-Jarreta, J.; Amode, M.R.; Armean, I.M.; Azov, A.G.; Bennett, R.; et al. Ensembl 2020. *Nucleic Acids Res.* **2020**, *48*, D682–D688. [CrossRef]
48. Schober, P.; Boer, C.; Schwarte, L.A. Correlation Coefficients, Appropriate Use and Interpretation. *Anesth. Analg.* **2018**, *126*, 1763–1768. [CrossRef]
49. Bian, Z.; Fan, R.; Xie, L. A Novel Cuproptosis-Related Prognostic Gene Signature and Validation of Differential Expression in Clear Cell Renal Cell Carcinoma. *Genes* **2022**, *13*, 851. [CrossRef]
50. Van De Vijver, M.J.; He, Y.D.; Van't Veer, L.J.; Dai, H.; Hart, A.A.; Voskuil, D.W.; Schreiber, G.J.; Peterse, J.L.; Roberts, C.; Marton, M.J.; et al. A gene-expression signature as a predictor of survival in breast cancer. *N. Engl. J. Med.* **2002**, *347*, 1999–2009. [CrossRef]
51. Liu, G.-M.; Zeng, H.-D.; Zhang, C.-Y.; Xu, J.-W. Identification of a six-gene signature predicting overall survival for hepatocellular carcinoma. *Cancer Cell Int.* **2019**, *19*, 138. [CrossRef]
52. Du, J.-X.; Chen, C.; Luo, Y.-H.; Cai, J.-L.; Cai, C.-Z.; Xu, J.; Ni, X.-J.; Zhu, W. Establishment and validation of a novel autophagy-related gene signature for patients with breast cancer. *Gene* **2020**, *762*, 144974. [CrossRef]
53. Drevets, W.C.; Wittenberg, G.M.; Bullmore, E.T.; Manji, H.K. Immune targets for therapeutic development in depression: Towards precision medicine. *Nat. Rev. Drug Discov.* **2022**, *21*, 224–244. [CrossRef]
54. Lu, T.; Xu, R.; Li, Q.; Zhao, J.-Y.; Peng, B.; Zhang, H.; Guo, J.-D.; Zhang, S.-Q.; Li, H.-W.; Wang, J.; et al. Systematic profiling of ferroptosis gene signatures predicts prognostic factors in esophageal squamous cell carcinoma. *Mol. Ther.-Oncolytics* **2021**, *21*, 134–143. [CrossRef]
55. Chen, S.; Yang, D.; Lei, C.; Li, Y.; Sun, X.; Chen, M.; Wu, X.; Zheng, Y. Identification of crucial genes in abdominal aortic aneurysm by WGCNA. *PeerJ* **2019**, *7*, e7873. [CrossRef]
56. Ogrodnik, M. Cellular aging beyond cellular senescence, Markers of senescence prior to cell cycle arrest in vitro and in vivo. *Aging Cell* **2021**, *20*, e13338. [CrossRef]
57. Iasonos, A.; Schrag, D.; Raj, G.V.; Panageas, K.S. How to Build and Interpret a Nomogram for Cancer Prognosis. *J. Clin. Oncol.* **2008**, *26*, 1364–1370. [CrossRef]
58. Van Calster, B.; Wynants, L.; Verbeek, J.F.; Verbakel, J.Y.; Christodoulou, E.; Vickers, A.J.; Roobol, M.J.; Steyerberg, E.W. Reporting and Interpreting Decision Curve Analysis, A Guide for Investigators. *Eur. Urol.* **2018**, *74*, 796–804. [CrossRef]
59. Chen, Z.-A.; Tian, H.; Yao, D.-M.; Zhang, Y.; Feng, Z.-J.; Yang, C.-J. Identification of a Ferroptosis-Related Signature Model Including mRNAs and lncRNAs for Predicting Prognosis and Immune Activity in Hepatocellular Carcinoma. *Front. Oncol.* **2021**, *11*, 738477. [CrossRef]
60. Wang, Y.; Chen, L.; Ju, L.; Qian, K.; Liu, X.; Wang, X.; Xiao, Y. Novel Biomarkers Associated With Progression and Prognosis of Bladder Cancer Identified by Co-expression Analysis. *Front. Oncol.* **2019**, *9*, 1030. [CrossRef]
61. Chen, X.; Lan, H.; He, D.; Xu, R.; Zhang, Y.; Cheng, Y.; Chen, H.; Xiao, S.; Cao, K. Multi-Omics Profiling Identifies Risk Hypoxia-Related Signatures for Ovarian Cancer Prognosis. *Front. Immunol.* **2021**, *12*, 645839. [CrossRef] [PubMed]

62. Fu, Y.; Bao, Q.; Liu, Z.; He, G.; Wen, J.; Liu, Q.; Xu, Y.; Jin, Z.; Zhang, W. Development and Validation of a Hypoxia-Associated Prognostic Signature Related to Osteosarcoma Metastasis and Immune Infiltration. *Front. Cell Dev. Biol.* **2021**, *9*, 633607. [CrossRef] [PubMed]
63. Newman, A.M.; Liu, C.L.; Green, M.R.; Gentles, A.J.; Feng, W.; Xu, Y.; Hoang, C.D.; Diehn, M.; Alizadeh, A.A. Robust enumeration of cell subsets from tissue expression profiles. *Nat. Methods* **2015**, *12*, 453–457. [CrossRef] [PubMed]
64. Aslam, M. Clinical laboratory medicine measurements correlation analysis under uncertainty. *Ann. Clin. Biochem. Int. J. Biochem. Lab. Med.* **2021**, *58*, 377–383. [CrossRef]
65. Mayakonda, A.; Lin, D.-C.; Assenov, Y.; Plass, C.; Koeffler, H.P. Maftools: Efficient and comprehensive analysis of somatic variants in cancer. *Genome Res.* **2018**, *28*, 1747–1756. [CrossRef]
66. Geeleher, P.; Cox, N.J.; Huang, R.S. Clinical drug response can be predicted using baseline gene expression levels and in vitro drug sensitivity in cell lines. *Genome Biol.* **2014**, *15*, R47. [CrossRef]

molecules

MDPI

Review

# Toll-like Receptor 4 Inflammatory Perspective on Doxorubicin-Induced Cardiotoxicity

**Natticha Sumneang [1,2,*], Pongpan Tanajak [3] and Thura Tun Oo [4]**

1 Department of Medical Science, School of Medicine, Walailak University, Nakhon Si Thammarat 80160, Thailand
2 Research Center in Tropical Pathobiology, Walailak University, Nakhon Si Thammarat 80160, Thailand
3 Department of Physical Therapy, Rehabilitation Center, Apinop Wetchakam Hospital, Kaeng-Khoi District, Saraburi 18110, Thailand; ptanajak2531@gmail.com
4 Department of Biomedical Sciences, University of Illinois at Chicago, College of Medicine Rockford, Rockford, IL 61107, USA; drthura@uic.edu
* Correspondence: natticha.su@wu.ac.th; Tel.: +66-75-477-487

**Abstract:** Doxorubicin (Dox) is one of the most frequently used chemotherapeutic drugs in a variety of cancers, but Dox-induced cardiotoxicity diminishes its therapeutic efficacy. The underlying mechanisms of Dox-induced cardiotoxicity are still not fully understood. More significantly, there are no established therapeutic guidelines for Dox-induced cardiotoxicity. To date, Dox-induced cardiac inflammation is widely considered as one of the underlying mechanisms involved in Dox-induced cardiotoxicity. The Toll-like receptor 4 (TLR4) signaling pathway plays a key role in Dox-induced cardiac inflammation, and growing evidence reports that TLR4-induced cardiac inflammation is strongly linked to Dox-induced cardiotoxicity. In this review, we outline and address all the available evidence demonstrating the involvement of the TLR4 signaling pathway in different models of Dox-induced cardiotoxicity. This review also discusses the effect of the TLR4 signaling pathway on Dox-induced cardiotoxicity. Understanding the role of the TLR4 signaling pathway in Dox-induced cardiac inflammation might be beneficial for developing a potential therapeutic strategy for Dox-induced cardiotoxicity.

**Keywords:** doxorubicin; Toll-like receptor 4; cardiotoxicity; heart

**Citation:** Sumneang, N.; Tanajak, P.; Oo, T.T. Toll-like Receptor 4 Inflammatory Perspective on Doxorubicin-Induced Cardiotoxicity. *Molecules* **2023**, *28*, 4294. https://doi.org/10.3390/molecules28114294

Academic Editors: Guohui Sun and Chongwen Wang

Received: 29 April 2023
Revised: 18 May 2023
Accepted: 22 May 2023
Published: 24 May 2023



## 1. Introduction

Cancer is a major public health problem worldwide. Cancer is a complicated disease caused by internal factors (e.g., inherited mutations, hormones, and immune conditions) and environmental factors (e.g., diet, radiation, and infectious organisms) [1,2]. According to a report from the World Health Organization, cancer is one of the leading causes of death worldwide [3]. In the United States of America, there were approximately 1.98 million new cases of diagnosed cancer in 2022 [4]. Furthermore, the number of cancer cases around the world is expected to reach around 26 million by 2030, with 17 million deaths per year [5].

Currently, doxorubicin (Dox) is one of the most used chemotherapeutic drugs for various types of cancer, such as breast, lung, ovarian, and thyroid cancers [6,7]. Although it is able to combat tumor cells, it is also harmful to normal cells, including cardiomyocytes [7–9]. In addition, several studies have reported the potential mechanisms of Dox-induced cardiotoxicity mediated by mitochondrial dysfunction, DNA damage, oxidative stress, and apoptosis [2,7,9]. A growing body of research suggests that aseptic inflammation might also play a plausible role in Dox-induced cardiotoxicity due to the activation of the innate immune system after Dox treatment [10–12]. A number of studies showed that Dox treatment upregulates the expressions of pro-inflammatory cytokines in cardiac tissue, such as tumor nuclear factor (TNF)-α, interleukin (IL)-1β, and IL-6, via activation of the nuclear factor-κB (NF-κB), triggering the progression of cardiovascular diseases and other adverse cardiac events [13–15]. Dox-induced cardiotoxicity mainly limits the cumulative dose of Dox in clinical settings. Cardiotoxicity is a potentially lethal condition and also a well-known adverse effect of Dox; however, the underlying molecular mechanism of Dox-induced cardiotoxicity, particularly related to inflammation, is not fully understood. Although the underlying mechanisms of Dox-induced cardiotoxicity remain complex, the role of cardiac inflammation in Dox-induced cardiotoxicity has become a focus for researchers in recent years.

Toll-like receptor 4 (TLR4), an important member of the TLR family, is part of the innate immune system that responds to the endogenous and exogenous signals and triggers pathophysiological functions in organs, including the heart [16–19]. The precise molecular mechanisms of TLR4 signaling have been elucidated. Upon binding to a specific ligand with the help of myeloid differentiation factor 2 (MD2), the TLR4 signaling pathway is activated, followed by recruiting the intracellular adaptor molecules and releasing the inflammatory mediators [15,20,21]. Interestingly, there is increasing evidence that Dox-induced TLR4 signaling pathway activation is implicated in cardiac adverse effects, which manifest as left ventricle (LV) impairment [20,22,23]. In support of this evidence, a previous study demonstrated that Dox-induced cardiac adverse effects were completely alleviated in TLR4 knockout mice [22]. Therefore, TLR4 signaling is thought to be involved in the mechanism contributing to Dox-induced cardiotoxicity, and inhibition of TLR4 is considered to be one the potential interventions against Dox-induced cardiomyopathy [22,23].

The aim of this review is to comprehensively summarize and discuss in vitro, in vivo, and clinical reports on the plausible mechanism of TLR4 on Dox-induced cardiotoxicity, including inflammatory mediators, oxidative stress, and apoptosis in cardiomyocytes, as well as cardiac remodeling and cardiac function. To support the upcoming clinical trials that will help to either prevent or lessen Dox-induced cardiotoxicity, we also aim to thoroughly compile the currently available evidence of TLR4 inhibition.

## 2. The Mechanism of TLR4 on Dox-Induced Cardiotoxicity

Growing evidence has shown that Dox activates the innate immune system, which is one of the expected components of the response against tumor cells [2,9]. However, Dox-induced innate immune activation provokes the release of inflammatory cytokines in a number of tissues, including the heart, intestine, brain, and liver, which results in inflammation in non-targeted organs [9,10,24,25]. To date, several studies have demonstrated that Dox-induced cardiac inflammation is strongly linked to Dox-induced cardiotoxicity [10,14,26,27]. Dox induces NF-κB activity to promote cardiac inflammation by upregulating TNF-α, IL-1β, and IL-6 expression [15,26].

TLR4 is responsible for the innate immune response; thus, the activation of TLR4 has become one of the most attractive targets in recent years [28,29]. Moreover, the expression of TLR4 is also found in the cardiomyocytes in addition to the immune cells [21]. This receptor responds to exogenous and endogenous ligands: pathogen-associated molecular patterns (PAMPs) and damage-associated molecular patterns (DAMPs) [28]. TLR4 recognizes the bacterial LPS as a major PAMP with the help of its co-receptor MD2 [21]. In addition, the TLR4 signaling pathway can also be activated by various endogenous DAMPs, namely alarmin protein, including high-mobility group protein box 1 (HMGB1) and the heat shock protein family (Hsps) [28,30]. Upon ligand binding, TLR4 dimerization occurs, followed by activation of the myeloid differentiation primary response protein 88 (MyD88); the downstream signaling pathways propagate NF-κB phosphorylation via reducing the inhibitory κB kinase (IKK) response, leading to upregulation of inflammatory cytokines and cardiac inflammation [21]. In addition to NF-κB activity, components of mitogen-activated protein kinases (MAPKs), including extracellular signal-regulated kinase 1 and 2 (ERK1/2), c-Jun N-terminal kinase (JNK), and p38, are also activated as downstream signal transducers of MyD88 to contribute to the regulation of pro-inflammatory responses [18,21].

To date, a growing body of evidence demonstrates that Dox induces the release of PAMP and DAMP, resulting in TLR4-mediated cardiac inflammation that contributes significantly to cardiotoxicity [10,15,28,31]. Based on evidence from in vitro and in vivo studies, we review and discuss how Dox causes cardiac TLR4 upregulation and inflammation in this review. A simplified overview of the mechanistic activation of TLR4 in Dox-induced cardiotoxicity is shown in Figure 1. Therefore, inhibition of TLR4 would have a therapeutic benefit against Dox-induced cardiotoxicity.

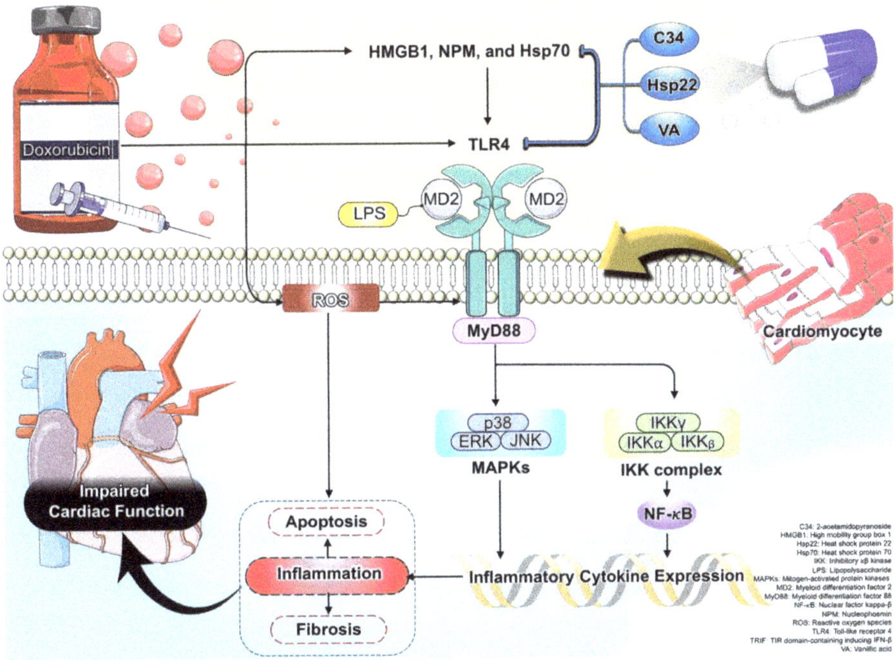


**Figure 1.** A schematic presentation of oxidative stress and inflammation-related mechanisms via TLR4 in Dox-induced cardiotoxicity. Activation of TLR4 was exhibited by HMGB1, NPM, and Hsp70, leading to increased cardiac inflammation, apoptosis, and fibrosis, and impaired cardiac function in Dox treatment. Its subsequent detrimental effects were effectively attenuated by treatment with C34, Hsp22, and VA.

## 3. The Effects of Dox on TLR4 Expression in Cardiomyocytes: Reports from In Vitro Studies

To date, the relationship between the role of TLR4 and Dox administration is still scarce; however, increasing evidence from in vitro studies suggests that Dox-mediated cardiotoxicity is related to the upregulation of cardiomyocyte TLR4 expressions [32–34]. Li et al. reported that Dox administration upregulated the expression of TLR4 in human cardiomyocyte cell lines [32]. In addition, Feng et al. demonstrated that Dox treatment significantly enhanced the TLR4 signaling pathway in H9c2 cardiomyocytes, as evidenced by increased TLR4 expression and its downstream signaling pathways, including MyD88, NF-κB, IL-1 β, IL-6, and TNF-α, along with decreased IkBα expression [27]. According to these two in vitro results, we speculate that Dox administration triggered cardiac inflammation via the TLR4-MyD88 signaling pathway. However, further research is necessary to fully understand the underlying molecular signaling of Dox-induced TLR4 signaling pathway activation in cardiomyocytes.

Traditionally, TLR4 ligands, PAMPs, and DAMPs bind to MD2 to activate the TLR4 signaling pathway in cardiomyocytes, resulting in cardiac inflammation [18,35]. The DNA-binding nuclear protein HMGB1, which regulates gene transcription and nucleosome stability, has a variety of biological functions in clinicopathological conditions [36]. It is considered a cytokine involved in the activation of innate immunity after being actively released from cells or passively released upon cell injury [36]. Interestingly, TLR4 functions as the major HMGB1 receptor [37]. Disulfide HMGB1 activates the TLR4 complex, binding to MD2, which triggers dimerization of TLR4 that can stimulate downstream signal transduction molecules (e.g., NF-κB) to produce pro-inflammatory cytokines [37]. A previous study demonstrated that the release of HMGB1 from the cells was increased following Dox treatment [38]. Moreover, the oxidative stress and DAMPs were considered to be major stimulators of HMGB1 release, activating inflammation through the TLR4 signaling pathway [38]. Therefore, Dox is closely related to altered HMGB1 levels, leading to induced cardiotoxicity.

In addition to HMGB1, there is also nucleophosmin (NPM), which behaves similarly to an alarmin protein released in response to cellular excess due to severe injury [39]. Similar to HMGB1, NPM can also bind to the TLR4 signaling cascade, leading it to exerting pro-inflammatory cytokines [39]. Interestingly, a previous study showed that Dox induced nucleolar stress, subsequently disrupting and releasing the NPM from the nucleolus. Then, the extracellular NPM induced inflammation via TLR4 signaling pathway activation [39]. To date, there is only one study that has shown that NPM can bind to the TLR4 signaling cascade, promoting pro-inflammatory function following the Dox treatment of human cardiac mesenchymal progenitor cells (hCmPCs) [39]. Thus, NPM is a novel ligand of TLR4 that activates inflammation in Dox-treated cardiotoxicity [39].

Inhibition of TLR4 via genetic deletion suppressed Dox-induced cardiotoxicity [27,31,32]. This evidence was provided by previous studies, showing that genetic ablation of TLR4 not only reduced inflammation but also decreased apoptosis in cardiomyocytes after exposure to Dox [27,31,32]. Although molecular signaling through TLR4 has not been demonstrated in the genetic deletion of TLR4 on apoptosis in cardiomyocytes after Dox exposure, the aforementioned discussions suggest that pro-inflammatory cytokines [27], mediated by TLR4, were suppressed, resulting in a decrease in cardiomyocyte apoptosis, as evidenced by the decrease in apoptotic proteins (Bax and cleaved caspase-3) and the increase in anti-apoptotic proteins (Bcl-2) [31,32]. These in vitro studies demonstrate the plausible involvement of TLR4 in Dox-induced cardiotoxicity via the implication in cardiomyocyte apoptosis.

C34, a potent and selective antagonist of TLR4, reduced NPM secretion in Dox-treated hCmPcs, further confirming that TLR4 regulated the secretion of NPM via autoregulatory feedback between TLR4 and NPM following Dox treatment [39]. Therefore, these findings revealed that inhibition of TLR4 might be a novel therapeutic strategy in reducing Dox-induced cardiotoxicity. Reports regarding the effects of Dox on TLR4 expression in cardiomyocytes in in vitro studies are summarized in Table 1.

**Table 1.** The effects of Dox on TLR4 expression in cardiomyocytes: reports from in vitro studies.

| Model | Protocol (Drug, Dose, Duration) | Major Findings | | Interpretation | Ref. |
|---|---|---|---|---|---|
| | | Inflammatory Markers | Apoptosis | | |
| AC16 cells | • Dox (5 μM, 24 h)<br>• si-TLR4 | • Dox ↑ TLR4<br>• Dox + si-TLR4 ↓ TLR4 | • Dox ↑ Bax ↑ Caspase-3 ↓ Bcl-2<br>• Dox + si-TLR4 ↓ Bax ↓ Caspase-3 ↑ Bcl-2 | Dox treatment increased the expressions of TLR4 and apoptotic proteins in AC16 cells. | [32] |

**Table 1.** *Cont.*

| Model | Protocol (Drug, Dose, Duration) | Major Findings | | Interpretation | Ref. |
|---|---|---|---|---|---|
| | | Inflammatory Markers | Apoptosis | | |
| H9c2 | • Dox (5 μM, 24 h)<br>• si-TLR4 | • Dox<br>↑ TLR4<br>↑ MyD88<br>↑ IL-1<br>↑ IL-6<br>↑ NF-kB<br>↑ TNF-α<br>↓ IkBα<br>• Dox + si-TLR4<br>↓ TLR4<br>↓ MyD88<br>↓ IL-1<br>↓ IL-6<br>↓ NF-kB<br>↓ TNF-α<br>↑ IkBα | • Dox<br>n/a<br>• Dox + si-TLR4<br>n/a | Dox administration induced TLR4 signaling pathway activation in H9c2. | [27] |
| Neonatal cardiomyocytes | • Dox (0.5 μM, 6 h)<br>• $TLR4^{-/-}$ | • Dox<br>↑ HMGB1<br>• Dox + $TLR4^{-/-}$<br>n/a | • Dox<br>↑ Caspase-3<br>• Dox + $TLR4^{-/-}$<br>↓ Caspase-3 | Dox treatment increased the release of HMGB1 and caspase-3 expression in neonatal cardiomyocytes. | [31] |
| hCmPCs cells | • Dox (1 μM, 8 h)<br>• C34 (100 μM, 30 min) | • Dox<br>↑ NPM<br>• Dox + C34<br>↓ NPM | • Dox<br>n/a<br>• Dox + C34<br>n/a | Dox treatment increased the level of NPM in hCmPCs cells. | [39] |

n/a: Data are not available; AC16: human cardiomyocyte cell line; Bax: Bcl-2-associated X; Bcl-2: B-cell lymphoma-2; C34: 2-acetamidopyranoside; Dox: doxorubicin; hCmPCs: human cardiac mesenchymal progenitor cells; HMGB1: high-mobility group box 1; IkBα: inhibitory κB (IκB) kinase; IL-1: interleukin-1; IL-6: interleukin-6; MyD88: myeloid differentiation factor 88; NF-kB: nuclear factor kappa-light-chain-enhancer of activated B cells; NPM: nucleophosmin; TLR4: Toll-like receptor 4; TNF-α: tumor necrosis factor-alpha; ↑: Increase; ↓: Decrease.

## 4. The TLR4 Expression in Dox-Induced in Cardiomyocytes: Reports from In Vivo Studies

Several studies have demonstrated that Dox treatment leads to impaired cardiac function in rodents, as assessed via echocardiography and invasive hemodynamic assessment [14,23,40]. Cellular and molecular studies have also been carried out using the cardiomyocytes isolated from these Dox-treated animals, and the results are largely consistent with the findings from the in vitro models discussed in the previous section. Previous in vivo studies have shown that Dox enhanced the expression of TLR4 in the cardiac tissue of rodents [14,23,40]. Furthermore, the levels of TLR4 ligands, such as HMGB1 and Hsp70, were also significantly increased in the cardiac tissue of Dox-treated mice [40]. This result further confirms that the TLR4 signaling pathway was activated following Dox treatment, leading to the triggering of NF-κB activity, which leads to the generation of pro-inflammatory cytokines in cardiac tissue, as evidenced by increases in TNF-α, IL-6, IL-13, monocyte chemotactic protein (MCP-1), and transforming growth factor (TGF)-β1 [14,40]. Consistently, the expression of cardiac macrophage markers, including CD45 and CD68, was also elevated in Dox-treated mice [14]. Taken together, elevated cardiac macrophages expressing TLR4 can lead to overwhelming pro-inflammatory cytokines in Dox-treated animals [14,40]. In addition to cardiac inflammation, cardiac remodeling was also observed in Dox-treated mice [40], which was attributed to the fact that TGF-β1 was increased in the hearts of these mice [40]. TGF-β1 is involved in the cardiac remodeling process, since it is a multifunctional cytokine and a growth factor that plays multiple roles in inflammation and fibrosis [41]. Therefore, an increase in cardiac TGF-β1 expression promoted collagen accumulation in the heart of Dox-treated mice, as indicated by increasing α-smooth muscle actin (α-SMA) [40].

Dox induces ROS through redox reactions due to its quinone component, with the resulting production of the superoxide anion ($O_2^-$), hydrogen peroxide ($H_2O_2$), and hydroxyl radical (·OH) [8]. This ROS can lead to lipid peroxidation, as indicated by an increase in malondialdehyde (MDA) that was observed in the hearts of Dox-treated rats [14]. In addition, this ROS can trigger apoptosis in cardiac tissue in Dox-treated mice, as evidenced by an increase in Bax, cytochrome *c*, and TUNEL$^+$ cells, as well as a decrease in Bcl-2 in cardiac tissue [14]. According to in vitro studies [31,32], an in vivo study has also shown that increasing TLR4 activation was implicated in cardiac apoptosis in Dox-treated mice [14]. This could be due to the oxidative stress associated with the increase in TLR4 expression that further promotes inflammation [42], which, in turn, contributes to apoptosis. Therefore, oxidative stress not only directly induces cardiac apoptosis per se but also promotes inflammation via increasing TLR4 expression, leading to cardiac apoptosis in Dox-treated mice.

Therefore, in vivo studies have suggested that enhanced activation of the TLR4 signaling pathway is involved in Dox-induced cardiotoxicity, which includes cardiac inflammation, remodeling, oxidative stress, and apoptosis via activation of TLR4, resulting in impaired cardiac function [14,23,40]. The in vivo evidence pertinent to TLR4 expression in Dox-induced cardiotoxicity is summarized in Table 2.

**Table 2.** TLR4 expression in Dox-induced cardiotoxicity: reports from in vivo studies.

| Model | Protocol (Dose, Route, Duration) | Major Findings | | | | | Interpretation | Ref. |
|---|---|---|---|---|---|---|---|---|
| | | Cardiac Function | Inflammatory Markers | Cardiac Remodeling/Fibrosis | Oxidative Stress | Apoptosis | | |
| C57BL/6J mice | Dox (3.4 mg/kg/wk, i.p., 8 wk) | ↓ %LVEF<br>↓ %LVFS | ↑ TLR4<br>↑ HMGB1<br>↑ Hsp70<br>↑ MCP-1<br>↑ IL-13<br>↑ TGF-β1 | ↑ Fibrosis<br>↑ α-SMA | n/a | n/a | Dox induced cardiac inflammation via increasing TLR4 signaling pathway, leading to cardiac dysfunction in mice. | [40] |
| C57BL/6J mice | Dox (15 mg/kg, i.p., single dose) | ↓ %LVEF | ↑ TLR4<br>↑ TNF-α<br>↑ IL-6<br>↑ NF-kB<br>↑ CD68<br>↑ CD45 | n/a | n/a | ↑ Bax<br>↓ Bcl-2<br>↑ Cyt *c*<br>↑ TUNEL$^+$ | Dox induced cardiac inflammation and apoptosis via increasing TLR4/NF-kB signaling pathway, leading to impaired cardiac function in mice. | [14] |
| Wistar rats | Dox (2.5 mg/kg/ 3 doses/wk, i.p., 2 wk) | n/a | ↑ TLR4 | n/a | ↑ MDA | n/a | Dox induced cardiac inflammation and oxidative stress via increasing TLR4 and MDA in rats. | [23] |

n/a: Data are not available; Bax: Bcl-2-associated X; Bcl-2: B-cell lymphoma-2; Cyt *c*: cytochrome c; Dox: doxorubicin; HMGB1: high-mobility group box 1; Hsp70: heat shock protein 70; i.p.: intraperitoneal injection; IL-13: interleukin 13; IL-6: interleukin-6; LVEF: left ventricular ejection fraction; LVFS: left ventricular fractional shortening; MCP-1: monocyte chemotactic protein 1; MDA: malondialdehyde; NF-kB: nuclear factor kappa-light-chain-enhancer of activated B cells; TGF-β1: tumor growth factor β1; TLR4: Toll-like receptor 4; TNF-α: tumor necrosis factor-alpha; TUNEL: terminal deoxynucleotidyl transferase mediated dUTP nick end labeling; α-SMA: α-smooth muscle actin; ↑: Increase; ↓: Decrease.

## 5. The Potential Role of TLR4 Inhibition in Dox-Induced Cardiotoxicity: Reports from In Vivo Studies

The evidence from in vivo studies on the effects of TLR4 inhibition in Dox-induced cardiotoxicity is summarized in Table 3. Since several studies, including both in vitro and in vivo studies, have revealed that activation of the TLR4 signaling pathway is partly involved in Dox-induced cardiotoxicity, targeting the TLR4 pathway has become a possible therapeutic strategy [14,22,23]. In vitro reports have shown that either silencing or knockout TLR4 attenuated Dox-induced cardiotoxicity [27,31,32,39], and in vivo study reports have also demonstrated that genetic ablation of TLR4 improved cardiac function in Dox-treated mice, as indicated by increased LV end-systolic pressure–volume relation (LVESPVR), cardiac output (CO), and stroke volume (SV) [22,31]. In addition, the molecular levels, including inflammatory markers, fibrosis accumulation, oxidative stress, and apoptotic markers, in cardiac tissue of Dox treatment were alleviated in TLR4-deficient mice [22,31] (Table 3).

In addition to genetic ablation of TLR4 models in cardiomyocytes, the blockade of TLR4 by using the pharmacological compound vanillic acid (VA), a TLR4 antibody, and adeno-associated virus–heat shock protein 22 (AAV-Hsp22) injection has also exerted protective effects in Dox-induced cardiotoxicity [14,23]. A previous study has shown that VA, a natural compound of the phenolic acid family, can suppress the TLR4-induced inflammatory response in Dox-treated mice [23]. This study was the first report of the helpful effect of VA via inhibiting TLR4 [23]; however, future studies need to validate which structure of VA blocks TLR4 signaling. In addition, a reduction in TLR4 by VA resulted in decreased cardiac MDA levels and further improved cardiac function in Dox-treated mice [23]. In addition to intervention with a pharmacological agent, Hsp22, a novel TLR4 ligand, has been shown to exert anti-apoptotic and anti-inflammatory effects in other diseases [43,44]. According to this study, adeno-associated virus Hsp22 alleviated cardiac inflammation and apoptosis by blocking TLR4 activation, leading to an improvement in cardiac function in Dox-induced cardiotoxicity in mice [14]. It is proposed that the TLR4-dependent signaling pathway renders Dox-mediated adverse cardiac effects, and the findings implied that inhibitions of TLR4 might attenuate pathophysiological key mechanisms of Dox-induced cardiotoxicity.

As previously discussed, Dox-induced cardiotoxicity can be due to several complex mechanisms. The TLR4-independent signaling pathway, involving oxidative stress, apoptosis, and so on, might play a significant role in Dox-induced cardiotoxicity [8,45]; however, only inhibition of TLR4 could attenuate the adverse effects on the heart after Dox administration [14,22,23,31,32,39]. Therefore, both TLR4-dependent and -independent signaling pathways could be considered as the fundamental mechanisms linked to the Dox-mediated cardiotoxicity, and targeting TLR4 is sufficient to reduce Dox-mediated cardiotoxicity.

Although several studies have shown that targeting TLR4 exerted cardioprotective effects in Dox-treated mice [14,22,23], only one study demonstrated completely contradictory findings [40]. Immunomodulation of TLR4 signaling with a TLR4-neutralizing antibody exacerbated cardiac fibrosis and impaired cardiac function by promoting higher inflammatory gene expression and, subsequently, an increase in inflammatory cytokines [40]. This might be possible due to the isotope of the TLR4 antibody used in that study [40]. Ma et al. injected TLR4 immunoglobulin G (IgG) antibody in Dox-treated mice [40]. Since IgG enhances pro-inflammatory response and stimulates immune cells, this might be the reason why TLR4 antibody administration did not attenuate Dox-induced cardiotoxicity. Furthermore, immunomodulation with TLR4 antibody administration resulted in a disruption of downregulating p38, which eventually suppressed autophagy and led to further cardiac dysfunction in Dox-treated mice [40]. Knockout of the TLR4 gene, a technique used by Raid et al. and Yao et al., showed greater efficacy on Dox-induced cardiotoxicity compared to TLR4 immunomodulation with TLR4 antibodies in Dox-treated mice [22,31]. Moreover, the administration of TLR4 inhibitors with Hsp22 and VA showed a more positive effect than the use of TLR4 antibodies [14,23,40]. Therefore, depletion of TLR4 by either genetic modification or pharmacological agents might be more effective than TLR4-neutralizing antibodies in Dox administration.

**Table 3.** The effects of TLR4 inhibition in Dox-induced cardiotoxicity: reports from in vivo studies.

| Model | Protocol (Dose, Duration) | Intervention of TLR4 Inhibition (Agent, Dose, Route, Duration) | Major Findings | | | | | Interpretation | Ref. |
|---|---|---|---|---|---|---|---|---|---|
| | | | Cardiac Function | Inflammatory Markers | Cardiac Remodeling/Fibrosis | Oxidative Stress | Apoptosis | | |
| C57BL/6J mice | Dox (20 mg/kg, i.p., single dose) -Sacrifice on day 5 | TLR4$^{-/-}$ | ↑ LVESPVR | n/a | n/a | n/a | ↓ TUNEL$^+$ | TLR4 deficiency attenuated Dox-induced cardiac apoptosis and dysfunction in mice. | [31] |
| C57BL/10ScSn mice | Dox (20 mg/kg, i.p., single dose) -Sacrifice on day 5 | TLR4$^{-/-}$ | ↑ SV<br>↑ CO | ↓ TNF-α<br>↓ CD3+<br>↓ CD11b+<br>↓ CD8a+ | n/a | ↓ Lipid peroxidation<br>↓ Nitrotyrosine | ↓ Bax<br>↑ Bcl-2<br>↓ TUNEL$^+$ | TLR4 deficiency rescued Dox-induced cardiotoxicity in mice. | [22] |
| C57BL/6J mice | Dox (3.4 mg/kg/wk, i.p., 8 wk) -Sacrifice on day 103 | TLR4ab (first dose was 200 µg/mg and following doses were 100 µg/mg, tv.i., on day 64, 68, 72, 79, 86, 96, and 100 after Dox injection) | ↓ %LVEF<br>↓ %LVFS | ↔TLR4<br>↔ HMGB1<br>↔ Hsp70<br>↑ MCP-1<br>↑ IL-13<br>↑ TGF-β1 | ↑ Fibrosis<br>↑ α-SMA | n/a | n/a | Immunomodulation of TLR4 exacerbated cardiac dysfunction in Dox-treated mic by increasing inflammation and fibrosis. | [40] |
| C57BL/6J mice | Dox (15 mg/kg, i.p., single dose) -Sacrifice on day 21 | AAV-Hsp22 ($5 \times 10^{10}$ viral genome particles, tv.i., single dose, before Dox injection for 4 wk) | ↑ %LVEF | ↓ TLR4<br>↓ TNF-α<br>↓ IL-6<br>↓ NF-kB<br>↓ CD68<br>↓ CD45 | n/a | n/a | ↓ Bax<br>↑ Bcl-2<br>↓ Cyt *c*<br>↓ TUNEL$^+$ | Hsp22 protected the heart against Dox-induced cardiotoxicity via inhibited TLR4/NF-kB signaling pathway in mice. | [14] |
| Wistar rats | Dox (2.5 mg/kg/ 3 doses/wk, i.p., 2 wk) -Sacrifice on day 28 | VA (10, 20, and 40 mg/kg/d, p.o., 4 wk, before Dox injection for 14 days) | n/a | ↓ TLR4 | n/a | ↓ MDA | n/a | VA protected the heart against Dox-induced cardiotoxicity via inhibited TLR4 signaling pathway in rats. | [23] |

n/a: Data are not available; AAV-Hsp22: adeno-associated virus–heat shock protein 22; Bax: Bcl-2-associated X; Bcl-2: B-cell lymphoma-2; CO: cardiac output; Cyt *c*: cytochrome c; Dox: doxorubicin; HMGB1: high-mobility group box 1; Hsp70: heat shock protein 70; i.p.: intraperitoneal injection; IL-13: interleukin 13; IL-6: interleukin-6; LVEF: left ventricular ejection fraction; LVESPVR: left ventricular end-systolic pressure-volume relation; LVFS: left ventricular fractional shortening; MCP-1: monocyte chemotactic protein 1; MDA: malondialdehyde; NF-kB: nuclear factor kappa-light-chain-enhancer of activated B cells; p.o.: per orally; SV: stroke volume; TGF-β1: tumor growth factor β1; TLR4: Toll-like receptor 4; TNF-α: tumor necrosis factor-alpha; TUNEL: terminal deoxynucleotidyl transferase mediated dUTP nick end labeling; tv.i.: tail vein injection; VA: vanillic acid; α-SMA: α-smooth muscle actin; ↑: Increase; ↓: Decrease; ↔: No change.

## 6. The Effects of Dox on Systemic TLR4 Expression: Reports from Clinical Studies

Clinical studies have shown the important role of systemic TLR4 on cardiac function in patients with hematological malignancy who received a Dox regimen [46,47]. After 6 months of Dox administration, blood sampling was collected to determine TLR4 expression, together with measuring the non-invasive cardiac function in Dox-treated patients [46,47]. The results of these two studies were consistently showed that these patients had impaired cardiac function after 6 months of Dox treatment, whereas TLR4 expressions were significantly increased in peripheral blood sampling [46,47]. In addition, a negative correlation between left ventricular ejection fraction and TLR4 expression was observed after 6 months of Dox treatment [47]. These findings were explained by the fact that Dox disrupts the intestinal mucosa integrity, leading to leakage of endotoxin, LPS, into the systemic circulation [48–50]. Therefore, the systemic inflammation was stimulated via the interaction of LPS and TLR4 [51–53]. This suggested that a decline in cardiac function is associated, in part, with systemic TLR4 expression. Therefore, Dox-induced cardiotoxicity is not only caused by cardiac TLR4 inflammation and cardiac dysfunction but also linked with systemic TLR4 elevation. To date, there is no available clinical evidence regarding the role of TLR4 in Dox-induced cardiotoxicity. In addition, multi-omics integration [54,55] to reveal the mechanism of TLR4 on Dox-induced cardiotoxicity is needed to better understand health and disease, and in certain cases, as part of medical care in the future. A summary of clinical studies is shown in Table 4. In addition, a schematic presentation of oxidative stress and inflammation-related mechanisms via TLR4 in Dox-induced cardiotoxicity is summarized and shown in Figure 1.

**Table 4.** The effects of Dox on systemic TLR4 expression: reports from clinical studies.

| Model | Methods | Major Findings | Interpretation | Ref. |
|---|---|---|---|---|
| Patients with hematological malignancy who received treatment with doxorubicin (n = 25);<br>- (Doxorubicin at 100–250 mg/m$^2$, 6 months)<br>- LVEF > 50% | - The blood was collected to determine TLR4 gene expression.<br>- Echocardiography was used to determine cardiac function. | 16 patients (64%) developed left ventricular diastolic dysfunction, associated with high gene expression of TLR4 after 6 months of Dox treatment. | The TLR4 expression may play as a marker for risk of doxorubicin-induced cardiotoxicity. | [46] |
| Patients with hematological malignancy who received treatment with doxorubicin (n = 25);<br>- (Doxorubicin at 100–250 mg/m$^2$, 6 wk)<br>- LVEF > 50% | - The blood was collected to determine TLR4 gene expression.<br>- Echocardiography was used to determine cardiac function. | There is a strong negative linear relationship between TLR4 expression and LVEF in patients after 6 weeks of Dox treatment. | Elevation of TLR4 levels were implicated in Dox-induced left ventricular dysfunction. | [47] |

Dox: doxorubicin; LVEF: left ventricular ejection fraction; TLR4: Toll-like receptor 4.

## 7. Conclusions

There is an accumulation of evidence from in vitro and in vivo studies and clinical reports demonstrating that the intensive arm of the TLR4 in Dox-induced cardiotoxicity, especially cardiac inflammation, leads to cardiac remodeling and impaired cardiac function. Therefore, the mechanism of understanding the role of the TLR4 signaling pathway in Dox-induced cardiac inflammation might be beneficial for developing a potential therapeutic strategy for Dox-induced cardiotoxicity in the near future.

**Author Contributions:** Conceptualization, N.S.; Data curation, N.S.; Formal analysis, N.S; Methodology, N.S.; Supervision, N.S.; Writing—original draft, N.S.; Writing—review and editing, N.S.; Writing—review and editing; Validation; P.T. and T.T.O. All authors have read and agreed to the published version of the manuscript.

**Funding:** This research received no external funding.

**Institutional Review Board Statement:** Not applicable.

**Informed Consent Statement:** Not applicable.

**Data Availability Statement:** Not applicable.

**Conflicts of Interest:** The authors declare no conflict of interest.

## Abbreviations

| | |
|---|---|
| OH | hydroxyl radical |
| AAV-Hsp22 | adeno-associated virus-heat shock protein 22 |
| AC16 | human cardiomyocyte cell line |
| Ad-Hsp22 | adenovirus-heat-shock protein 22 |
| Bax | Bcl-2-associated X |
| Bcl-2 | B-cell lymphoma-2 |
| C34 | 2-acetamidopyranoside |
| CO | cardiac output |
| Cyt *c* | cytochrome c |
| DAMP | damage-associated molecular pattern |
| DIC | Dox-induced cardiotoxicity |
| DNA | deoxyribonucleic acid |
| Dox | doxorubicin |
| ERK | extracellular signal-regulated kinase |
| $H_2O_2$ | hydrogen peroxide |
| hCmPCs | human cardiac mesenchymal progenitor cells |
| HMGB1 | high mobility group box 1 |
| Hsp | heat shock protein |
| Hsp70 | heat shock protein 70 |
| i.p. | intraperitoneal injection |
| IKK | inhibitory κB kinase |
| IL-1β | interleukin-1 beta |
| IL-6 | interleukin-6 |
| JNK | c-Jun N-terminal kinase |
| LPS | lipopolysaccharide |
| LVESPVR | left ventricular end-systolic pressure-volume relation |
| MAPK | mitogen-activated protein kinase |
| MCP-1 | monocyte chemotactic protein 1 |
| MD2 | myeloid differentiation factor 2 |
| MDA | malondialdehyde |
| MyD88 | myeloid differentiation factor 88 |
| NPM | nucleophosmin |
| $O_2^-$ | superoxide anion |
| PAMP | pathogen-associated molecular pattern |
| ROS | reactive oxygen species |
| SV | stroke volume |
| TGF-β1 | tumor growth factor β1 |
| TLR4 | Toll-like receptor 4 |
| TNF-α | tumor necrosis factor-alpha |
| TUNEL | Terminal deoxynucleotidyl transferase dUTP nick end labeling |
| VA | vanillic acid |
| α-SMA | α-smooth muscle actin |

## References


1. Anand, P.; Kunnumakkara, A.B.; Sundaram, C.; Harikumar, K.B.; Tharakan, S.T.; Lai, O.S.; Sung, B.; Aggarwal, B.B. Cancer is a preventable disease that requires major lifestyle changes. *Pharm. Res.* **2008**, *25*, 2097–2116. [CrossRef]
2. Behranvand, N.; Nasri, F.; Zolfaghari Emameh, R.; Khani, P.; Hosseini, A.; Garssen, J.; Falak, R. Chemotherapy: A double-edged sword in cancer treatment. *Cancer Immunol. Immunother.* **2022**, *71*, 507–526. [CrossRef]
3. Ferlay, J.; Colombet, M.; Soerjomataram, I.; Parkin, D.M.; Piñeros, M.; Znaor, A.; Bray, F. Cancer statistics for the year 2020: An overview. *Int. J. Cancer* **2021**, *149*, 778–789. [CrossRef]
4. Siegel, R.L.; Miller, K.D.; Fuchs, H.E.; Jemal, A. Cancer statistics, 2022. *CA Cancer J. Clin.* **2022**, *72*, 7–33. [CrossRef]

5. Thun, M.J.; DeLancey, J.O.; Center, M.M.; Jemal, A.; Ward, E.M. The global burden of cancer: Priorities for prevention. *Carcinogenesis* **2010**, *31*, 100–110. [CrossRef]
6. Sohail, M.; Sun, Z.; Li, Y.; Gu, X.; Xu, H. Research progress in strategies to improve the efficacy and safety of doxorubicin for cancer chemotherapy. *Expert Rev. Anticancer Ther.* **2021**, *21*, 1385–1398. [CrossRef] [PubMed]
7. Christidi, E.; Brunham, L.R. Regulated cell death pathways in doxorubicin-induced cardiotoxicity. *Cell Death Dis.* **2021**, *12*, 339. [CrossRef]
8. Benjanuwattra, J.; Siri-Angkul, N.; Chattipakorn, S.C.; Chattipakorn, N. Doxorubicin and its proarrhythmic effects: A comprehensive review of the evidence from experimental and clinical studies. *Pharmacol. Res.* **2020**, *151*, 104542. [CrossRef]
9. Tacar, O.; Sriamornsak, P.; Dass, C.R. Doxorubicin: An update on anticancer molecular action, toxicity and novel drug delivery systems. *J. Pharm. Pharmacol.* **2013**, *65*, 157–170. [CrossRef] [PubMed]
10. Wang, L.; Chen, Q.; Qi, H.; Wang, C.; Wang, C.; Zhang, J.; Dong, L. Doxorubicin-Induced Systemic Inflammation Is Driven by Upregulation of Toll-Like Receptor TLR4 and Endotoxin Leakage. *Cancer Res.* **2016**, *76*, 6631–6642. [CrossRef] [PubMed]
11. Yarmohammadi, F.; Karbasforooshan, H.; Hayes, A.W.; Karimi, G. Inflammation suppression in doxorubicin-induced cardiotoxicity: Natural compounds as therapeutic options. *Naunyn-Schmiedebergs Arch. Pharmacol.* **2021**, *394*, 2003–2011. [CrossRef] [PubMed]
12. Hadi, N.; Yousif, N.G.; Al-amran, F.G.; Huntei, N.K.; Mohammad, B.I.; Ali, S.J. Vitamin E and telmisartan attenuates doxorubicin induced cardiac injury in rat through down regulation of inflammatory response. *BMC Cardiovasc. Disord.* **2012**, *12*, 63. [CrossRef] [PubMed]
13. Singla, D.K.; Johnson, T.A.; Tavakoli Dargani, Z. Exosome Treatment Enhances Anti-Inflammatory M2 Macrophages and Reduces Inflammation-Induced Pyroptosis in Doxorubicin-Induced Cardiomyopathy. *Cells* **2019**, *8*, 1224. [CrossRef] [PubMed]
14. Lan, Y.; Wang, Y.; Huang, K.; Zeng, Q. Heat Shock Protein 22 Attenuates Doxorubicin-Induced Cardiotoxicity via Regulating Inflammation and Apoptosis. *Front. Pharmacol.* **2020**, *11*, 257. [CrossRef]
15. Amin, M.N.; Siddiqui, S.A.; Ibrahim, M.; Hakim, M.L.; Ahammed, M.S.; Kabir, A.; Sultana, F. Inflammatory cytokines in the pathogenesis of cardiovascular disease and cancer. *SAGE Open Med.* **2020**, *8*, 2050312120965752. [CrossRef]
16. Vijay, K. Toll-like receptors in immunity and inflammatory diseases: Past, present, and future. *Int. Immunopharmacol.* **2018**, *59*, 391–412. [CrossRef]
17. Spirig, R.; Tsui, J.; Shaw, S. The Emerging Role of TLR and Innate Immunity in Cardiovascular Disease. *Cardiol. Res. Pract.* **2012**, *2012*, 181394. [CrossRef]
18. Sumneang, N.; Oo, T.T.; Singhanat, K.; Maneechote, C.; Arunsak, B.; Nawara, W.; Pratchayasakul, W.; Benjanuwattra, J.; Apaijai, N.; Liang, G.; et al. Inhibition of myeloid differentiation factor 2 attenuates cardiometabolic impairments via reducing cardiac mitochondrial dysfunction, inflammation, apoptosis and ferroptosis in prediabetic rats. *Biochim. Biophys. Acta (BBA)—Mol. Basis Dis.* **2022**, *1868*, 166301. [CrossRef]
19. Oo, T.T.; Sumneang, N.; Ongnok, B.; Arunsak, B.; Chunchai, T.; Kerdphoo, S.; Apaijai, N.; Pratchayasakul, W.; Liang, G.; Chattipakorn, N.; et al. L6H21 protects against cognitive impairment and brain pathologies via toll-like receptor 4–myeloid differentiation factor 2 signalling in prediabetic rats. *Br. J. Pharmacol.* **2022**, *179*, 1220–1236. [CrossRef]
20. Alanazi, A.M.; Fadda, L.; Alhusaini, A.; Ahmad, R.; Hasan, I.H.; Mahmoud, A.M. Liposomal Resveratrol and/or Carvedilol Attenuate Doxorubicin-Induced Cardiotoxicity by Modulating Inflammation, Oxidative Stress and S100A1 in Rats. *Antioxidants* **2020**, *9*, 159. [CrossRef]
21. Sumneang, N.; Apaijai, N.; Chattipakorn, S.C.; Chattipakorn, N. Myeloid differentiation factor 2 in the heart: Bench to bedside evidence for potential clinical benefits? *Pharmacol. Res.* **2021**, *163*, 105239. [CrossRef]
22. Riad, A.; Bien, S.; Gratz, M.; Escher, F.; Westermann, D.; Heimesaat, M.M.; Bereswill, S.; Krieg, T.; Felix, S.B.; Schultheiss, H.P.; et al. Toll-like receptor-4 deficiency attenuates doxorubicin-induced cardiomyopathy in mice. *Eur. J. Heart Fail.* **2008**, *10*, 233–243. [CrossRef]
23. Baniahmad, B.; Safaeian, L.; Vaseghi, G.; Rabbani, M.; Mohammadi, B. Cardioprotective effect of vanillic acid against doxorubicin-induced cardiotoxicity in rat. *Res. Pharm. Sci.* **2020**, *15*, 87–96. [CrossRef] [PubMed]
24. Ahmed, O.M.; Galaly, S.R.; Mostafa, M.M.A.; Eed, E.M.; Ali, T.M.; Fahmy, A.M.; Zaky, M.Y. Thyme Oil and Thymol Counter Doxorubicin-Induced Hepatotoxicity via Modulation of Inflammation, Apoptosis, and Oxidative Stress. *Oxid. Med. Cell. Longev.* **2022**, *2022*, 6702773. [CrossRef] [PubMed]
25. Oo, T.T.; Pratchayasakul, W.; Chattipakorn, N.; Chattipakorn, S.C. Emerging roles of toll-like receptor 4 in chemotherapy-induced neurotoxicity. *Neurotoxicology* **2022**, *93*, 112–127. [CrossRef] [PubMed]
26. Rawat, P.S.; Jaiswal, A.; Khurana, A.; Bhatti, J.S.; Navik, U. Doxorubicin-induced cardiotoxicity: An update on the molecular mechanism and novel therapeutic strategies for effective management. *Biomed. Pharmacother.* **2021**, *139*, 111708. [CrossRef]
27. Feng, P.; Yang, Y.; Liu, N.; Wang, S. Baicalin regulates TLR4/IκBα/NFκB signaling pathway to alleviate inflammation in Doxorubicin related cardiotoxicity. *Biochem. Biophys. Res. Commun.* **2022**, *637*, 1–8. [CrossRef]
28. Kashani, B.; Zandi, Z.; Pourbagheri-Sigaroodi, A.; Bashash, D.; Ghaffari, S.H. The role of toll-like receptor 4 (TLR4) in cancer progression: A possible therapeutic target? *J. Cell. Physiol.* **2021**, *236*, 4121–4137. [CrossRef]
29. Vaure, C.; Liu, Y. A comparative review of toll-like receptor 4 expression and functionality in different animal species. *Front. Immunol.* **2014**, *5*, 316. [CrossRef]
30. Yang, D.; Han, Z.; Oppenheim, J.J. Alarmins and immunity. *Immunol. Rev.* **2017**, *280*, 41–56. [CrossRef]

31. Yao, Y.; Xu, X.; Zhang, G.; Zhang, Y.; Qian, W.; Rui, T. Role of HMGB1 in doxorubicin-induced myocardial apoptosis and its regulation pathway. *Basic Res. Cardiol.* **2012**, *107*, 267. [CrossRef] [PubMed]
32. Li, B.; Cai, X.; Wang, Y.; Zhu, H.; Zhang, P.; Jiang, P.; Yang, X.; Sun, J.; Hong, L.; Shao, L. Circ-SKA3 Enhances Doxorubicin Toxicity in AC16 Cells Through miR-1303/TLR4 Axis. *Int. Heart J.* **2021**, *62*, 1112–1123. [CrossRef]
33. Xu, L.; Wang, C.; Zou, Z.; Wu, Z. Ozone Attenuated H9c2 Cell Injury Induced by Doxorubicin. *J. Cardiovasc. Pharmacol.* **2021**, *78*, e86–e93. [CrossRef] [PubMed]
34. Tavakoli Dargani, Z.; Singla, D.K. Embryonic stem cell-derived exosomes inhibit doxorubicin-induced TLR4-NLRP3-mediated cell death-pyroptosis. *Am. J. Physiol. Heart Circ. Physiol.* **2019**, *317*, H460–H471. [CrossRef] [PubMed]
35. He, M.; Bianchi, M.E.; Coleman, T.R.; Tracey, K.J.; Al-Abed, Y. Exploring the biological functional mechanism of the HMGB1/TLR4/MD-2 complex by surface plasmon resonance. *Mol. Med.* **2018**, *24*, 21. [CrossRef]
36. Tripathi, A.; Shrinet, K.; Kumar, A. HMGB1 protein as a novel target for cancer. *Toxicol. Rep.* **2019**, *6*, 253–261. [CrossRef] [PubMed]
37. Yang, H.; Wang, H.; Andersson, U. Targeting Inflammation Driven by HMGB1. *Front. Immunol.* **2020**, *11*, 484. [CrossRef] [PubMed]
38. Syukri, A.; Budu; Hatta, M.; Amir, M.; Rohman, M.S.; Mappangara, I.; Kaelan, C.; Wahyuni, S.; Bukhari, A.; Junita, A.R.; et al. Doxorubicin induced immune abnormalities and inflammatory responses via HMGB1, HIF1-α and VEGF pathway in progressive of cardiovascular damage. *Ann. Med. Surg.* **2022**, *76*, 103501. [CrossRef]
39. Beji, S.; D'Agostino, M.; Gambini, E.; Sileno, S.; Scopece, A.; Vinci, M.C.; Milano, G.; Melillo, G.; Napolitano, M.; Pompilio, G.; et al. Doxorubicin induces an alarmin-like TLR4-dependent autocrine/paracrine action of Nucleophosmin in human cardiac mesenchymal progenitor cells. *BMC Biol.* **2021**, *19*, 124. [CrossRef] [PubMed]
40. Ma, Y.; Zhang, X.; Bao, H.; Mi, S.; Cai, W.; Yan, H.; Wang, Q.; Wang, Z.; Yan, J.; Fan, G.-C.; et al. Toll-like receptor (TLR) 2 and TLR4 differentially regulate doxorubicin induced cardiomyopathy in mice. *PLoS ONE* **2012**, *7*, e40763. [CrossRef]
41. Saadat, S.; Noureddini, M.; Mahjoubin-Tehran, M.; Nazemi, S.; Shojaie, L.; Aschner, M.; Maleki, B.; Abbasi-kolli, M.; Rajabi Moghadam, H.; Alani, B.; et al. Pivotal Role of TGF-β/Smad Signaling in Cardiac Fibrosis: Non-coding RNAs as Effectual Players. *Front. Cardiovasc. Med.* **2021**, *7*, 588347. [CrossRef]
42. Tawadros, P.S.; Powers, K.A.; Ailenberg, M.; Birch, S.E.; Marshall, J.C.; Szaszi, K.; Kapus, A.; Rotstein, O.D. Oxidative Stress Increases Surface Toll-Like Receptor 4 Expression in Murine Macrophages Via Ceramide Generation. *Shock* **2015**, *44*, 157–165. [CrossRef]
43. Hedhli, N.; Wang, L.; Wang, Q.; Rashed, E.; Tian, Y.; Sui, X.; Madura, K.; Depre, C. Proteasome activation during cardiac hypertrophy by the chaperone H11 Kinase/Hsp22. *Cardiovasc. Res.* **2008**, *77*, 497–505. [CrossRef] [PubMed]
44. Lizano, P.; Rashed, E.; Kang, H.; Dai, H.; Sui, X.; Yan, L.; Qiu, H.; Depre, C. The valosin-containing protein promotes cardiac survival through the inducible isoform of nitric oxide synthase. *Cardiovasc. Res.* **2013**, *99*, 685–693. [CrossRef] [PubMed]
45. Shi, S.; Chen, Y.; Luo, Z.; Nie, G.; Dai, Y. Role of oxidative stress and inflammation-related signaling pathways in doxorubicin-induced cardiomyopathy. *Cell Commun. Signal.* **2023**, *21*, 61. [CrossRef] [PubMed]
46. Pop-Moldovan, A.L.; Trofenciuc, N.M.; Dărăbanţiu, D.A.; Precup, C.; Branea, H.; Christodorescu, R.; Puşchiţă, M. Customized laboratory TLR4 and TLR2 detection method from peripheral human blood for early detection of doxorubicin-induced cardiotoxicity. *Cancer Gene Ther.* **2017**, *24*, 203–207. [CrossRef]
47. Trofenciuc, N.M.; Bordejevic, A.D.; Tomescu, M.C.; Petrescu, L.; Crisan, S.; Geavlete, O.; Mischie, A.; Onel, A.F.M.; Sasu, A.; Pop-Moldovan, A.L. Toll-like receptor 4 (TLR4) expression is correlated with T2* iron deposition in response to doxorubicin treatment: Cardiotoxicity risk assessment. *Sci. Rep.* **2020**, *10*, 17013. [CrossRef] [PubMed]
48. Sun, Z.; Wang, X.; Wallen, R.; Deng, X.; Du, X.; Hallberg, E.; Andersson, R. The influence of apoptosis on intestinal barrier integrity in rats. *Scand. J. Gastroenterol.* **1998**, *33*, 415–422. [CrossRef]
49. Cray, P.; Sheahan, B.J.; Cortes, J.E.; Dekaney, C.M. Doxorubicin increases permeability of murine small intestinal epithelium and cultured T84 monolayers. *Sci. Rep.* **2020**, *10*, 21486. [CrossRef]
50. Huang, J.; Wei, S.; Jiang, C.; Xiao, Z.; Liu, J.; Peng, W.; Zhang, B.; Li, W. Involvement of Abnormal Gut Microbiota Composition and Function in Doxorubicin-Induced Cardiotoxicity. *Front. Cell. Infect. Microbiol.* **2022**, *12*, 808837. [CrossRef]
51. Lu, Y.-C.; Yeh, W.-C.; Ohashi, P.S. LPS/TLR4 signal transduction pathway. *Cytokine* **2008**, *42*, 145–151. [CrossRef] [PubMed]
52. Kuzmich, N.N.; Sivak, K.V.; Chubarev, V.N.; Porozov, Y.B.; Savateeva-Lyubimova, T.N.; Peri, F. TLR4 Signaling Pathway Modulators as Potential Therapeutics in Inflammation and Sepsis. *Vaccines* **2017**, *5*, 34. [CrossRef]
53. Shimazu, R.; Akashi, S.; Ogata, H.; Nagai, Y.; Fukudome, K.; Miyake, K.; Kimoto, M. MD-2, a molecule that confers lipopolysaccharide responsiveness on Toll-like receptor 4. *J. Exp. Med.* **1999**, *189*, 1777–1782. [CrossRef] [PubMed]
54. Song, Q.; Zhu, X.; Jin, L.; Chen, M.; Zhang, W.; Su, J. SMGR: A joint statistical method for integrative analysis of single-cell multi-omics data. *NAR Genom. Bioinform.* **2022**, *4*, lqac056. [CrossRef] [PubMed]
55. Tang, Z.; Zhang, T.; Yang, B.; Su, J.; Song, Q. spaCI: Deciphering spatial cellular communications through adaptive graph model. *Brief Bioinform.* **2023**, *24*, bbac563. [CrossRef]

molecules

MDPI

*Article*

# Graphene Oxide-Sensitized Surface Plasmon Resonance Biosensor of Porcine Reproductive and Respiratory Syndrome Virus

**Xuemei Liu [1,2,†], Chao Xu [2,†], Chunyu Fu [1], Dongfang Xia [2], Fuchao Wang [2], Hongzong Yin [2,*] and Jun Peng [1,*]**

[1] Shandong Provincial Key Laboratory of Animal Biotechnology and Disease Control and Prevention, College of Veterinary Medicine, Shandong Agricultural University, Taian 271018, China; qian1214876244@163.com (X.L.); lhr980222@163.com (C.F.)
[2] College of Chemistry and Material Science, Shandong Agricultural University, Taian 271018, China; xuc@sdau.edu.cn (C.X.); xiadf@ihep.ac.cn (D.X.); 15610303603@163.com (F.W.)
* Correspondence: hzyin@sdau.edu.cn (H.Y.); jpeng@sdau.edu.cn (J.P.)
† These authors contributed equally to this work.

**Abstract:** Biosensor analysis based on the surface plasmon resonance (SPR) phenomenon enables label-free, highly sensitive analyte detection without prior sample purification or processing. However, potential applications of SPR biosensors in virus detection in biological samples remain to be explored. Owing to its excellent biocompatibility and abundance of hydroxyl and carboxyl functional groups, graphene oxide (GO) has been widely used as a biosensor of proteins and metal ions in living cells. The present work explored the effect of GO modification on the sensitivity of an SPR biosensor and used a GO-modified sensor to detect porcine reproductive and respiratory syndrome virus in cell culture, as shown. The GO modification markedly enhanced the sensitivity of the Fourier transform SPR sensor and enabled linear detection of porcine reproductive and respiratory syndrome virus (PRRSV) with a multiplicity of infection in the range 0.2–1.7 ($R^2$ = 0.998). Such a GO-modified sensor provides a promising alternative for virus detection.

**Keywords:** GO; PRRSV; FT-SPR; sensitization

**Citation:** Liu, X.; Xu, C.; Fu, C.; Xia, D.; Wang, F.; Yin, H.; Peng, J. Graphene Oxide-Sensitized Surface Plasmon Resonance Biosensor of Porcine Reproductive and Respiratory Syndrome Virus. *Molecules* **2022**, *27*, 3942. https://doi.org/10.3390/molecules27123942

Academic Editors: Guohui Sun and Chongwen Wang

Received: 24 May 2022
Accepted: 13 June 2022
Published: 20 June 2022



## 1. Introduction

Surface plasmon resonance (SPR) detects molecular adsorption events by measuring changes in the resonance of surface plasmon waves [1]. SPR biosensors have been developed to distinguish different target analytes and have become a conventional analytical tool for real-time, label-free analysis of biomolecular interactions [2]. Detection of protein adsorption depends on a change of the SPR signal in response to real-time interactions with materials on the sensor chip [3]. In the past two decades, SPR has been applied to many areas, such as medical diagnostics [4], food safety [5], biotechnology [6], and protein–protein interactions [5]. However, its low sensitivity to small molecules seriously hinders its application to these targets. Therefore, many new methods have been developed to enhance the SPR signal [7], and notable progress has recently been made using nanomaterials.

The large surface area and unique chemical properties of graphene oxide (GO) can be exploited to develop sensitive SPR biosensor interfaces with a strong capacity for target immobilization [8,9]. GO has been used for the analysis and detection of numerous targets, such as proteins, viable cells, and glucose [10]. SPR biosensor chips have been used to study a wide range of antibody–antigen interactions and improve the sensitivity of biosensors using GO linking layers. GO is most commonly coated on the biosensor surface using spray or immersion methods [11]. The spray method yields a higher sensitivity than the immersion method but is a time consuming and complex process. Thus, we sought to identify conditions that improve the sensitivity of SPR biosensors using simple methods and non-specialist instrumentation.

Porcine reproductive and respiratory syndrome virus (PRRSV) is a forward single-stranded RNA virus [12], which belongs to the family *Arteriviridae* [13]. PRRSV has caused huge economic losses in the global pig industry [14], and its high mortality, strong transmission, multiple transmission routes, and strong mutagenicity [15] have had a significant impact on the pig industry in China. Traditional PRRSV detection techniques include quantitative reverse transcription polymerase chain reaction (qRT-PCR), enzyme-linked immunosorbent assay (ELISA), immunohistochemistry, antibody detection, and so on [16–18]. These methods are time-consuming and virus sample damage treatment. Hence, developing a real-time dynamic monitoring and detection method and realizing nondestructive tests of viruses is of great significance for studying the interaction between viruses and biomolecules or antivirus drugs.

In the present work, as shown in Scheme 1, a GO−based SPR biosensor was constructed. The sensitivity enhancement of SPR biosensors was explored using GO. After identifying GO modifications resulting in optimum sensitization, the sensor was used to detect PRRSV in mock samples.

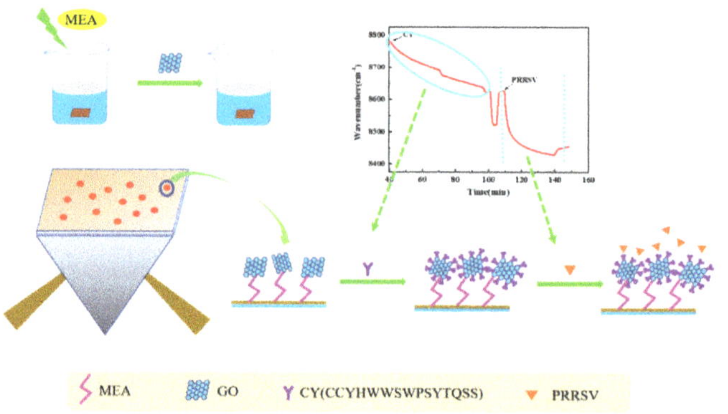


**Scheme 1.** Procedure for the preparation and application of GO-based SPR biosensors.

## 2. Material and Methods

### 2.1. Construction of a GO-Based SPR Biosensor

The GO-based SPR biosensor was constructed as shown in Scheme 1. Briefly, the naked gold SPR chip was obtained from Thermo Fisher Scientific, Waltham, MA, USA. The obtained SPR chip (Thermo Fisher Scientific) was immersed in a freshly prepared piranha solution (30% $H_2O_2$:$H_2SO_4$, 3:7 $v/v$) for 10 min. During this period, the chip was shaken periodically to remove surface air bubbles. The chip was then rinsed thoroughly with ultrapure water, dried under $N_2$ flow, then immersed in a 100 mM cysteamine hydrochloride solution (Macklin Biotech Ltd., Shanghai, China) with parafilm and stored in the dark for 12 h. Then the cysteamine hydrochloride was modified on the naked gold SPR chip by the strong interaction between sulfhydryl and gold, and the amino functional group of cysteamine hydrochloride was exposed to the surface of the chip. Finally, it was soaked in GO (XFNANO, Nanjing, China) to form a gold–cysteamine GO sensor.

In order to investigate the influence of modification conditions on SPR biosensor properties, the GO concentration, mode, and time of addition were varied to detect papain, respectively. After identifying the optimum conditions, the GO-modified sensor was probed using SEM, and the functional groups on the GO were characterized using FTIR spectroscopy.

### 2.2. Investigation of Sensitization Effect

As a mercapto protease enzyme, its papain sulfhydryl shows a strong affinity to the surface of the bare gold. In order to investigate the sensitization effect of GO and explore the possible mechanism of this sensitization, the papain was employed to investigate the sensitization effect of GO. The experimental process is briefly described as follows: Firstly, using the optimal conditions for fabrication, the GO-modified sensor was mounted in the SPR instrument and equilibrated with phosphate-buffered saline for 15 min. After the bioconjunction regents, 1-ethyl-3-(3-dimethylaminopropyl) carbodiimide (EDC) and Nhydroxysuccinimide hydrochloride (NHS) were employed to activate the surface carboxyl group of GO, and the activated GO were modified on the SPR chip by the interaction between the activated carboxyl group and the amino functional group of CY polypeptide (Sequence: $NH_2$−CCYHWWSWPSYTQSS−COOH, Wuhan Xinghao Pharmaceutical Co., Ltd., Wuhan, China, abbreviated as CY). To stabilize the signal, the unreacted carboxyl groups in GO were banned using ethanolamine hydrochloride. The SPR signal was then measured in response to papain. To determine the GO-induced signal enhancement, the procedure was repeated using without GO modification.

### 2.3. *Evaluates MOI Values of PRRSV*

MARC-145 cells were inoculated into 96−well plates and cultured to 80%. Serial tenfold dilutions of PRRSV solution were made with DMEM. The diluted virus was inoculated into 96-well plates. Inoculate one longitudinal row per dilution, 100 μL per well. Normal cells were used as control. The results were observed and recorded day by day for 5 to 7 days. Results were calculated according to the Reed–Muench method. MOI values were calculated as PUFs = 0.7 × $TCID_{50}$, MOI = (PUF/mL) × $V_{PRRSV}/N_{MACK\text{-}145}$, where $V_{PRRSV}$ is the Volume of PRRSV, and $N_{MACK\text{-}145}$ is the number of cells.

### 2.4. *Proliferation and Detection of PRRSV*

African green monkey kidney cells (MARC-145) were cultured in Dulbecco's modified Eagle medium containing 10% fetal bovine serum and infected with PRRSV (JXA1 strain; GenBank accession No. EF112445.1). When the cell lesion rate exceeded 80%, the cells were lysed using three freeze–thaw cycles, cell debris was removed by centrifugation, and the supernatant was collected and stored at −80 °C until further use.

The amount of virus required to cause lesions in 50% of the cells was calculated by the $TCID_{50}$ (half of tissue culture infection dose) assay. This was used to calculate the viral titer (MOI).

The GO-modified sensor was mounted in the SPR instrument, and solutions containing the carbodiimide/*N*-hydroxysuccinimide activator, CY polypeptide (Sequence: $NH_2$−CCYHWWSWPSYTQSS−COOH, Wuhan Xinghao Pharmaceutical Co., Ltd., abbreviated as CY), ethanolamine quencher, and PRRSV (in medium containing 10% serum and 1% double antibiotics) were added in turn. As a control experiment, the procedure was repeated using a PRRSV-free medium.

To determine the GO-induced signal enhancement, the measurements were repeated using an SPR sensor without GO modification.

### 2.5. *Specificity of PRRSV Detection*

To demonstrate that PRRSV specifically binds to CY peptides, the SPR signal in response to porcine circovirus was also measured under the same conditions. Briefly, GO-modified sensors prepared under optimal conditions were utilized. After the CY polypeptides were modified on the GO-modified SPR chip with the bioconjunction regents, EDC and NHS, and the unreacted carboxyl groups in GO were banned using ethanolamine hydrochloride. The SPR signal was then measured in response to PCV.

## 3. Results and Discussion

### 3.1. *Characterization of GO*

Figure 1a shows the morphology of the GO precursor material, which had a sheet structure (Figure 1b,c) and a wrinkled surface (Figure 1d), consistent with prior reports [19].

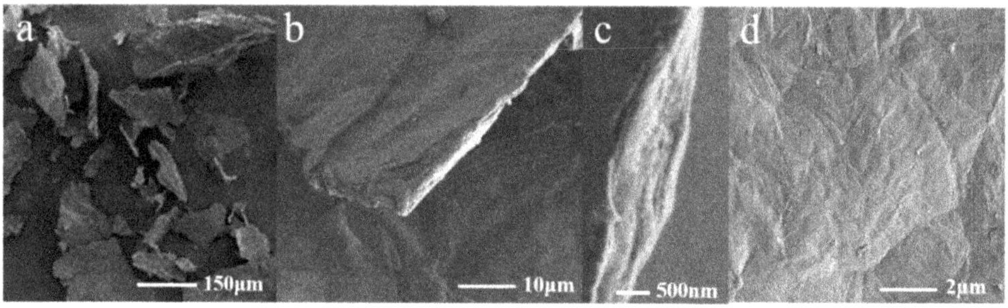


**Figure 1.** SEM images of the graphene oxide (GO) used to sensitize the surface plasmon resonance (SPR) biosensor. (**a**) Overall morphology of GO; (**b**,**c**) GO showing sheet-like structure; (**d**) GO surface with folds.

GO is a multifunctional material that contains numerous functional groups, such as epoxy (–O), hydroxyl (–OH), carboxyl (–COOH), and ether moieties. The FTIR spectrum of GO (Figure 2) contains peaks characteristic of O–H stretching vibrations (3430 $cm^{-1}$), C–OH bending vibrations (1630 $cm^{-1}$), saturated C–H bending vibrations (1400 $cm^{-1}$), and asymmetric C–O–C stretching vibrations (1110 $cm^{-1}$), showing it is rich in oxygen-containing functional groups.

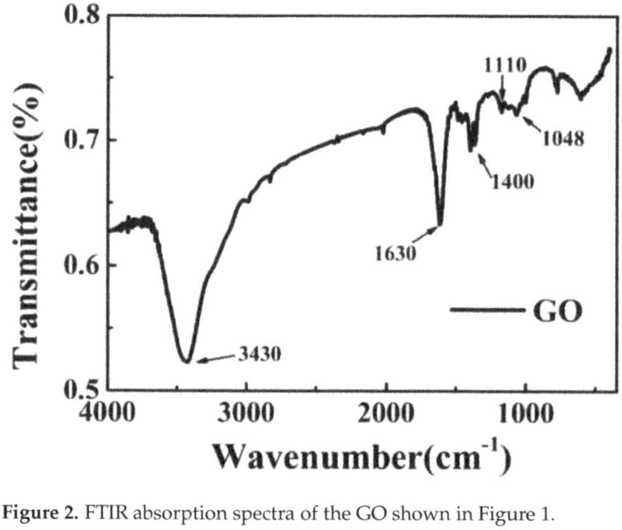


**Figure 2.** FTIR absorption spectra of the GO shown in Figure 1.

*3.2. Optimal Conditions for GO-Modified Sensor*

In this experiment, papain (500 mg/mL) was detected to explore the optimum conditions for SPR sensitization. By varying the modification modes, dynamic flow (flowing GO solution through the surface of the gold flake), immersion (immersing the gold flake in GO solution), modification time, and GO concentration, it was concluded that optimal sensitivity enhancement is obtained by immersion of the gold flakes (Figure 3a) for 5 h (Figure 3b) in 2.0 g/L GO (Figure 3c).

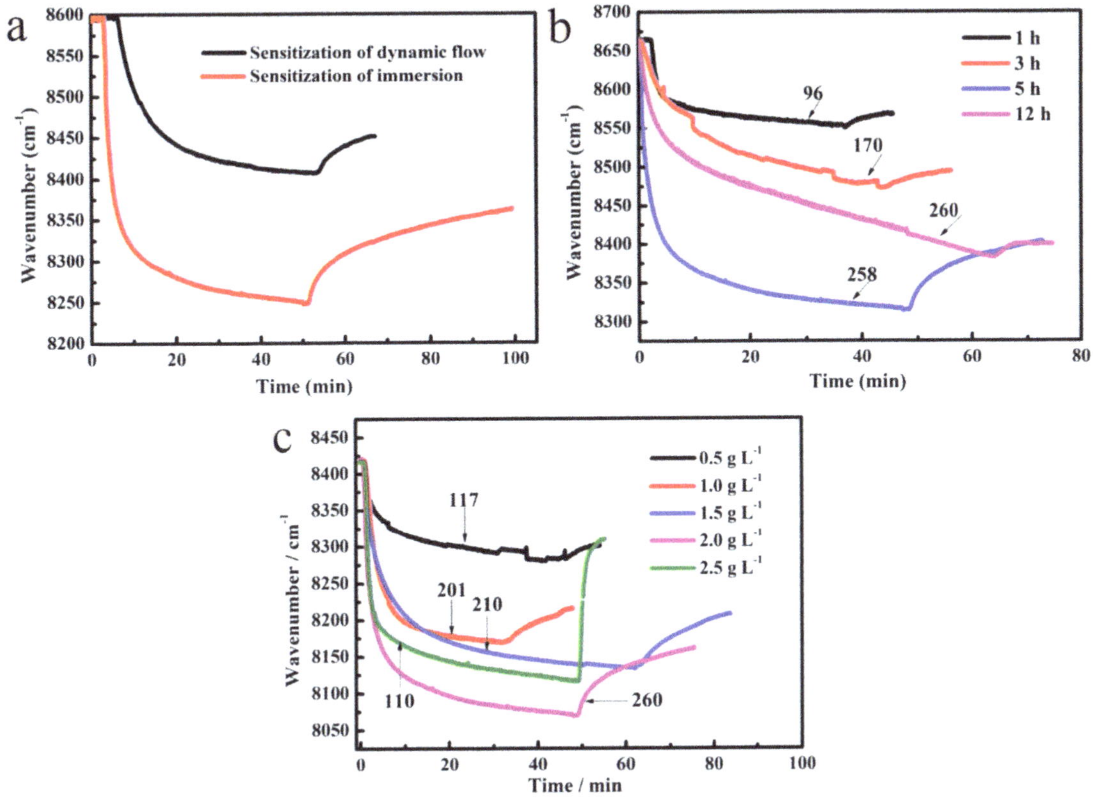


**Figure 3.** Exploration of optimal conditions for fabricating a GO-sensitized SPR biosensor. (**a**) SPR signal response for different GO modification methods, c(GO) = 2 mg/mL; (**b**) SPR signal response for various immersion time of gold flakes in GO aqueous solution (2 mg/mL); (**c**) SPR signal response to different concentrations of GO aqueous solution (0.5 to 2.5 mg/mL) after gold flakes immersed in GO aqueous solution for 5 h.

### *3.3. SEM Characterization of GO–Modified Sensor*

SEM images of the optimally modified GO–modified sensor (Figure 4a,b) show that the surface of the gold flake was successfully modified with GO. To determine whether the modification was reproducible, the preparation of the GO–modified sensor was repeated under the optimal conditions and used to detect a fixed concentration of papain. Figure 4c shows that there was no marked change in Fourier-transform SPR (FT-SPR) frequency, which indicated that the surface content of GO was reproducible from one preparation to the next.

### *3.4. GO-Sensitized SPR Detection of Papain*

In the early work, as a mercapto protease enzyme, the papain has been reported to investigate the sensitization effect of different size noble metal nanoclusters [20]. Hence, GO-modified sensors under optimal conditions were used to detect different papain concentrations. Figure 5a compares the FT–SPR signal in response to 500 mg/L papain, using sensors with and without GO modification. The FT–SPR frequency change in response to 100–500 mg/L papain was also measured using sensors with and without GO modification. The SPR sensitivity was enhanced approximately 2.9-fold by the GO-modified sensor (Figure 5b), and a strong linear relationship ($R^2$ = 0.994) between papain concentration and the FT–SPR frequency change was obtained (Figure 5c). The sensitizing effect of GO on

SPR is that GO is modified to gold flake, which provides more binding sites and improves the sensitivity of SPR to substance detection. Considering that the surface of the bare gold of the chip has more binding sites than the GO−modified chip, we hypothesize that it is mainly due to the surface charge or the sensitizing effect of GO.

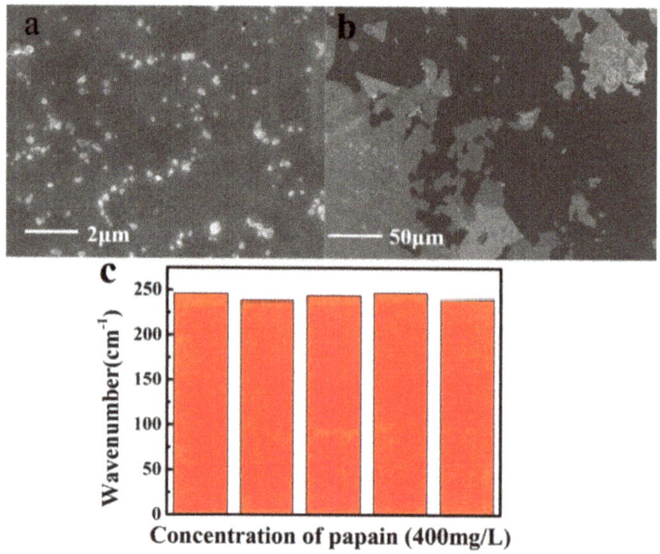


**Figure 4.** Characterization of GO−modified sensor (immersion for 5 h in 2.0 g/L$^{-1}$ GO). (**a**,**b**) SEM images. (**c**) Reproducibility of the FT−SPR frequency change using a fixed papain concentration and GO-modified sensor under optimal conditions.

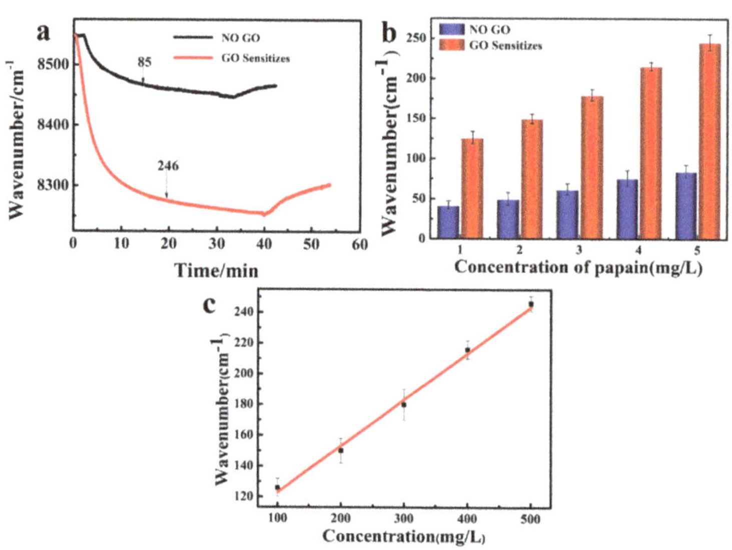


**Figure 5.** (**a**) Time dependence of FT−SPR signal in response to 500 mg/L papain, detected using sensors with and without GO modification. (**b**) Dependence of FT−SPR frequency change on papain concentration, measured using sensors with and without GO modification. (**c**) The linear relationship between papain concentration and FT−SPR frequency change of the GO-modified sensor ($R^2$ = 0.994).

### 3.5. Application of the GO-Modified SPR Biosensor to PRRSV Detection

SPR biosensors with and without GO modification were cross-linked with CY functional peptides targeting PRRSV, and the FT−SPR signals in response to PRRSV were compared. As shown in Figure 6a, the addition of a PRRSV− free medium did not cause a signal change, indicating there were no matrix effects. Additionally, in the presence of PRRSV, the change in FT−SPR frequency of the GO-modified sensor was 2.8-fold larger than that of the unmodified sensor, showing that PRRSV and CY peptides can interact. As shown in Figure 6b, there is an obvious relationship between the signal intensity of the SPR sensor with the concentration of the PRRSV virus ($R^2$ = 0.998) and the virus in the range of 0.4–1.7 MOI could be quantitatively detected.

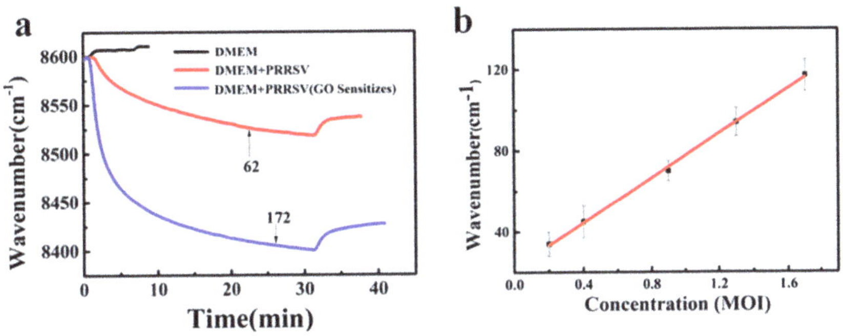


**Figure 6.** Specificity of porcine reproductive and respiratory syndrome virus (PRRSV) detection. (**a**) Time dependence of FT−SPR signal in response to PRRSV (2.6 MOI). (**b**) The linear relationship between the PRRSV concentration and the FT-SPR frequency change of the GO−modified sensor ($R^2$ = 0.998). DMEM—Dulbecco's modified Eagle medium; MOI—multiplicity of infection.

### 3.6. Specificity of PRRSV Detection

To demonstrate that PRRSV specifically binds to CY peptides, porcine circovirus (2 MOI) was applied to the SPR biosensor under the same conditions used for PRRSV detection. As shown in Figure 7, the FT−SPR signal was essentially unchanged in response to porcine circovirus (PCV), indicating it had no binding interaction with the CY peptide.

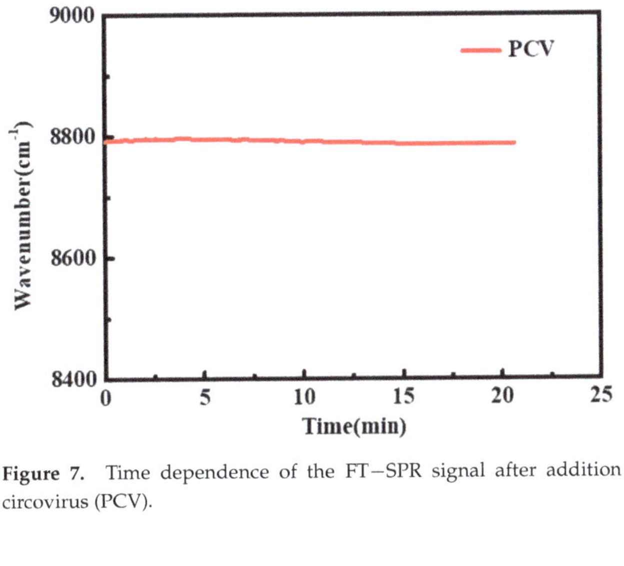


**Figure 7.** Time dependence of the FT−SPR signal after addition (arrow) of 2 MOI porcine circovirus (PCV).

## 4. Conclusions

In summary, a GO−modified SPR sensor was developed for fast, real-time detection of PRRSV. The GO−modified SPR chip has the advantages of simple operation, good plasmonic properties, real−time monitoring, and high sensitivity compared with traditional methods. By exploring different modification methods, optimal enhancement of the SPR sensitivity was obtained by immersion of gold flake in 2.0 g/L GO for 5 h. The GO−modified sensor under these conditions was confirmed using SEM and shown to enhance papain detection sensitivity 2.86−fold. The results of the TCID50 experiment were $10^{-5.21}$/0.1 mL. Compared with the traditional virus detection method, the prepared SPR sensor could realize the nondestructive and real-time online detection of viruses.

**Author Contributions:** Conceptualization, X.L., C.X. and H.Y.; methodology, X.L, C.X., H.Y. and J.P.; validation, C.F.; formal analysis, X.L., C.X. and C.F.; investigation, X.L., C.X. and H.Y.; data curation, X.L., C.X., C.F., F.W. and D.X.; writing original draft preparation, X.L. and C.X.; supervision, H.Y. and J.P.; project administration, H.Y. and J.P.; funding acquisition, J.P. All authors have read and agreed to the published version of the manuscript.

**Funding:** This research was funded by Agricultural improved varieties project in Shandong province of China (2021LZGC001) and Great scientific and technological innovation projects in Shandong province of China (2020CXGC010801).

**Institutional Review Board Statement:** Not applicable.

**Informed Consent Statement:** Not applicable.

**Data Availability Statement:** Data is contained within the review article.

**Conflicts of Interest:** The authors declare no conflict of interest.

## References

1. Bako, J.; Kelemen, M.; Hegedus, C. Analysis of benzoyl-peroxide and formaldehyde as dental allergens by FT-SPR method. *IOP Conf. Ser. Mater. Sci. Eng.* **2013**, *47*, 012001. [CrossRef]
2. Simon, L.; Bognár, Z.; Gyurcsányi, R.E. Finding the Optimal Surface Density of Aptamer Monolayers by SPR Imaging Detection-based Aptamer Microarrays. *Electroanalysis* **2020**, *32*, 851–858. [CrossRef]
3. Mauriz, E.; García-Fernández, M.C.; Lechuga, L.M. Towards the design of universal immunosurfaces for SPR-based assays: A review. *TrAC Trends Anal. Chem.* **2016**, *79*, 191–198. [CrossRef]
4. Teotia, P.K.; Kaler, R.S. 1-D grating based SPR biosensor for the detection of lung cancer biomarkers using Vroman effect. *Opt. Commun.* **2018**, *406*, 188–191. [CrossRef]
5. Ashley, J.; Piekarska, M.; Segers, C.; Trinh, L.; Rodgers, T.; Willey, R.; Tothill, I.E. An SPR based sensor for allergens detection. *Biosens. Bioelectron.* **2017**, *88*, 109–113. [CrossRef] [PubMed]
6. Rini, M.; Abdi, M.M.; Abdullah, L.C.; Sadrolhosseini, A.R.; Mat Yunus, W.M.; Moksin, M.M.; Tahir, P.M. Surface Plasmon Resonance Sensing Detection of Mercury and Lead Ions Based on Conducting Polymer Composite. *PLoS ONE* **2011**, *9*, e24578. [CrossRef]
7. Mudgal, N.; Yupapin, P.; Ali, J.; Singh, G. $BaTiO_3$-Graphene-Affinity Layer–Based Surface Plasmon Resonance (SPR) Biosensor for Pseudomonas Bacterial Detection. *Plasmonics* **2020**, *15*, 1221–1229. [CrossRef]
8. Chen, H.; Müller, M.B.; Gilmore, K.J.; Wallace, G.G.; Li, D. Mechanically Strong, Electrically Conductive, and Biocompatible Graphene Paper. *Adv. Mater.* **2008**, *20*, 3557–3561. [CrossRef]
9. Compton, O.C.; Nguyen, S.T. Graphene oxide, highly reduced graphene oxide, and graphene: Versatile building blocks for carbon-based materials. *Small* **2010**, *6*, 711–723. [CrossRef]
10. Struchkov, N.S.; Alexandrov, E.V.; Romashkin, A.V.; Silakov, G.O.; Rabchinskii, M.K. Uniform graphene oxide films fabrication via spray-coating for sensing application. *Fullerenes, Nanotubes and Carbon Nanostructures* **2019**, *28*, 214–220. [CrossRef]
11. Stebunov, Y.V.; Arsenin, A.V.; Volkov, V.S. SPR analysis of antibody-antigen interactions using graphene oxide linking layers. *Mater. Today Proc.* **2018**, *5*, 17442–17446. [CrossRef]
12. Snijder, E.J.; Meulenberg, J.J. The molecular biology of arteriviruses. *J. Gen. Virol.* **1998**, *79*, 961–979. [CrossRef]
13. Dokland, T. The structural biology of PRRSV. *Virus Res.* **2010**, *154*, 86–97. [CrossRef]
14. Albina, E. Epidemiology of porcine reproductive and respiratory syndrome (PRRS): An overview. *Vet. Microbiol.* **1997**, *5*, 309–316. [CrossRef]
15. Du, T.; Nan, Y.; Xiao, S.; Zhao, Q.; Zhou, E.M. Antiviral Strategies against PRRSV Infection. *Trends Microbiol.* **2017**, *25*, 968–979. [CrossRef] [PubMed]

16. Bai, Y.; Zhou, Y.; Liu, H.; Fang, L.; Liang, J.; Xiao, S. Glutathione-Stabilized Fluorescent Gold Nanoclusters Vary in Their Influences on the Proliferation of Pseudorabies Virus and Porcine Reproductive and Respiratory Syndrome Virus. *ACS Appl. Nano Mater.* **2018**, *1*, 969–976. [CrossRef]
17. Wang, Y.; Guo, J.; Qiao, S.; Li, Q.; Yang, J.; Jin, Q.; Zhang, G. GP5 Protein-based ELISA for the Detection of PRRSV Antibodies. *Pol. J. Vet. Sci.* **2016**, *19*, 495–501. [CrossRef]
18. Qiu, W.; Meng, K.; Liu, Y.; Zhang, Y.; Wang, Z.; Chen, Z.; Yang, J.; Sun, W.; Guo, L.; Ren, S.; et al. Simultaneous detection of classical PRRSV, highly pathogenic PRRSV and NADC30-like PRRSV by TaqMan probe real-time PCR. *J. Virol. Methods* **2020**, *282*, 113774. [CrossRef] [PubMed]
19. Meyer, J.C.; Geim, A.K.; Katsnelson, M.I.; Novoselov, K.S.; Booth, T.J.; Roth, S. The structure of suspended graphene sheets. *Nature* **2007**, *446*, 60–63. [CrossRef] [PubMed]
20. Liu, X.M.; Zhang, Y.; Wang, S.C.; Liu, C.; Wang, T.Y.; Qiu, Z.F.; Wang, X.L.; Waterhouse, G.I.N.; Xu, C.; Yin, H.Z. Performance com parison of surface plasmon resonance biosensors based on ultrasmall noble metal nanoparticles templated using bovine serum albumin. *Microchem. J.* **2020**, *155*, 104737. [CrossRef]

MDPI
St. Alban-Anlage 66
4052 Basel
Switzerland
www.mdpi.com

*Molecules* Editorial Office
E-mail: molecules@mdpi.com
www.mdpi.com/journal/molecules

www.ingramcontent.com/pod-product-compliance
Lightning Source LLC
Chambersburg PA
CBHW061619210926
51850CB00009B/2374
* 9 7 8 3 0 3 6 5 9 6 6 8 6 *